Lecture Notes in Artificial Intelligence   2592

Subseries of Lecture Notes in Computer Science
Edited by J. G. Carbonell and J. Siekmann

Lecture Notes in Computer Science

Edited by G. Goos, J. Hartmanis, and J. van Leeuwen

W0230414

Springer
Berlin
Heidelberg
New York
Barcelona
Hong Kong
London
Milan
Paris
Tokyo

Ryszard Kowalczyk   Jörg P. Müller
Huaglory Tianfield   Rainer Unland (Eds.)

Agent Technologies, Infrastructures, Tools, and Applications for E-Services

NODe 2002 Agent-Related Workshops
Erfurt, Germany, October 7-10, 2002
Revised Papers

 Springer

Series Editors

Jaime G. Carbonell, Carnegie Mellon University, Pittsburgh, PA, USA
Jörg Siekmann, University of Saarland, Saarbrücken, Germany

Volume Editors

Ryszard Kowalczyk
CSIRO Mathematical and Information Sciences
723 Swanston Street, Carlton, Victoria 3053, Australia
E-mail: ryszard.kowalczyk@cmis.csiro.au

Jörg P. Müller
Siemens AG
CT IC 6, Munich, Germany
E-mail: joerg.mueller@mchp.siemens.de

Huaglory Tianfield
Glasgow Caledonian University, Department of Computing
City Campus, 70 Cowcaddens Road, Glasgow G4 0BA, Scotland, UK
E-mail: lcsr@gcal.ac.uk

Rainer Unland
University of Essen, Institute for Computer Science
Schützenbahn 70, 45117 Essen, Germany
E-mail: UnlandR@informatik.uni-essen.de

Cataloging-in-Publication Data applied for

A catalog record for this book is available from the Library of Congress.

Bibliographic information published by Die Deutsche Bibliothek.
Die Deutsche Bibliothek lists this publication in the Deutsche Nationalbibliografie;
detailed bibliographic data is available in the Internet at <http://dnb.ddb.de>.

CR Subject Classification (1998): I.2.11, I.2, D.2, K.4.4, C.2.4, H.4

ISSN 0302-9743
ISBN 3-540-00742-3 Springer-Verlag Berlin Heidelberg New York

Springer-Verlag Berlin Heidelberg New York
a member of BertelsmannSpringer Science+Business Media GmbH

http://www.springer.de

© Springer-Verlag Berlin Heidelberg 2003
Printed in Germany

Typesetting: Camera-ready by author, data conversion by PTP-Berlin, Stefan Sossna e.K.
Printed on acid-free paper   SPIN: 10872491   06/3142   5 4 3 2 1 0

Preface

Net.ObjectDays (NODe) has established itself as one of the most significant events on Objects, Components, Architectures, Services and Applications for a Networked World in Europe and in the world. As in previous years, it took place in the Messekongresszentrum (Fair and Convention Center) in Erfurt, Thuringia, Germany, this time during 7–10 October 2002. Founded only three years ago as the official successor conference to JavaDays, STJA (Smalltalk and Java in Industry and Education) and JIT (Java Information Days), NODe has grown into a major international conference that attracts participants from industry, research and users in equal measure since it puts strong emphasis on the active exchange of concepts and technologies between these three communities.

Over the past few years, the NODe conference has developed a remarkable track record: a new paradigm (*Generative Programming*) was born at NODe (citation James Coplien), nearly all of the most prominent researchers and contributors in the object-oriented field (and beyond) have given keynotes at NODe, new topics have been integrated (like Agent Technology and Web-Services) and, now, for the first time, postconference proceedings are being published by Springer-Verlag. Altogether three volumes will be available. This volume is compiled from the best papers of the agent-related workshops (*Agent Technology and Software Engineering AgeS*, and *Agent Technologies for E-Services ATES 2002*) and the *3rd International Symposium on Multi-Agent Systems, Large Complex Systems, and E-Businesses (MALCEB 2002)*. Two additional volumes will be published, one containing the best contributions of the main conference and another one with the best contributions to the workshops relating to the Web, Databases and Web-Services that were cohosted with NODe 2002: M. Aksit, M. Mezini, R. Unland (editors), *Objects, Components, Architectures, Services, and Applications for a Networked World (LNCS 2591)*; and A. Chaudhri, M. Jeckle, E. Rahm, R. Unland (editors), *Web, Web-Services, and Database Systems (LNCS 2593)*.

This volume contains abstracts of the keynote speeches as well as 23 peer-reviewed, original papers that were chosen from the papers accepted for the workshops and the symposium. Hence, the papers in this volume are a subset of the papers presented at the conference, which in turn were selected by the respective programme committees from the submitted papers based on their scientific quality, the novelty of the ideas, the quality of the writing, and the practical relevance. This double selection process not only guaranteed high-quality papers but also allowed the authors to improve their original contributions using comments and suggestions they received during reviewing and at the conference. Furthermore, authors were allowed to extend their papers to fully fledged versions. We hope that you will find the results as convincing as we do, and that these proceedings give you many new inspirations and insights.

The contents of this volume can best be described by excerpts from the original Call for Papers:

AgeS 2002
Over the past decade, software agents and multi-agent systems have grown into one of the most active areas of research and development activity in computing generally. There are many reasons for the current intensity of interest, but certainly one of the most important is that the concept of an agent as an autonomous system, capable of interacting with other agents in order to satisfy its design objectives, is a natural one for software designers. Just as we can understand many systems as being composed of essentially passive objects, which have state, and upon which we can perform operations, so we can understand many others as being made up of interacting, semi-autonomous agents. Recently, there has been a growth of interest in the potential of agent technology in the context of software engineering. Some researchers (mainly in the Agent-Oriented Software Engineering community) point out that agents can be looked upon as a new paradigm for software engineering, a different way of looking at and modeling complex and dynamic systems in terms of concepts such as collaboration, coordination, and negotiation; others rather stress the need to explore the usage of existing software engineering processes, instruments, and methodologies in the design of multi-agent systems; still other researchers look for an incremental way of identifying and adopting proven concepts from agent technology into existing software engineering approaches. The goal of the AgeS workshop is to foster interaction between the agents and software engineering communities, to gain a better understanding of the requirements from software engineering and the possible roles that agent technology can play in the contexts of software engineering, including but not necessarily restricted to agent-oriented software engineering. In this workshop we will seek to examine the credentials of agent-based approaches as a software engineering paradigm, and to gain an insight into what agent-oriented software engineering will look like. By colocating the workshop with a major software engineering event such as NODe, we hope to attract a strong software engineering audience, and hence to implement a fruitful forum for discussion and the identification of research needs and collaboration possibilities between the agents and software engineering communities.

ATES 2002
The workshop on Agent Technologies for e-Services (ATES 2002) was held in conjunction with Net.ObjectDays 2002 (NODe 2002) in Erfurt, Germany on 9 October 2002. It aimed at exploring and promoting the use of software agent technologies for electronic services (e-services) that can deliver information, knowledge, and decision support, perform tasks and conduct transactions, control and monitor operations, and interact and integrate with other e-services in the global, dynamic, and open environment of the Internet. Typical examples of e-services are Web-accessible resources and applications, Web-enabled business processes and relationships, and networked devices and portable information appliances. Software agents with the capabilities of autonomous reasoning, lear-

ning, adaptation, social interactions, cooperation, and mobility is a very promising technology for e-services. In particular there is growing success in a wide range of related applications including agent-based e-commerce, e-business, and mobile applications, and it is envisaged that agent technology can also be very useful in the context of e-services.

As editors of this volume, we would like to thank once again all programme committee members and all external referees for their excellent work in evaluating the submitted papers. Moreover, we would like to thank Mr. Hofmann from Springer-Verlag for his cooperation and help in putting this volume together.

December 2002

Ryszard Kowalczyk
Jörg P. Müller
Huaglory Tianfield
Rainer Unland

3rd International Symposium on Multi-Agent Systems, Large Complex Systems, and E-Businesses (MALCEB 2002)

Programme Co-chairs

Prof. Dr. Huaglory Tianfield
Department of Computing
Glasgow Caledonian University
City Campus
70 Cowcaddens Road
Glasgow G4 0BA
Scotland, UK
Tel: 0044 141 331 8025
Fax: 0044 141 331 8445
E-mail: lcsr@gcal.ac.uk

Prof. Dr. Hans Czap
Universität Trier
Wirtschaftsinformatik
54286 Trier
Germany
Tel: 0049 651 201 2859
Fax: 0049 651 201 3959
E-mail:
cz@wiinfo.uni-trier.de

Members of the International Programme Committee

Paul Alpar, Philipps Universität Marburg, Germany
Joseph Barjis, Delft University of Technology, The Netherlands
Michael Berger, Siemens AG, Germany
Hans-Dieter Burkhard, Humboldt Universität, Berlin, Germany
X.Q. Cai, Chinese University of Hong Kong, China
Peter Chamoni, Gerhard Mercator Universität Duisburg, Germany
Haoxun Chen, Université de Technologie de Troyes, France
Armin B. Cremers, Universität Bonn, Germany
Rohan de Silva, University of New South Wales, Australia
Philippe De Wilde, Imperial College of Science, UK
Torsten Eymann, Albert-Ludwigs-Universität Freiburg, Germany
Liping Fang, Ryerson University, Canada
Baogang Hu, Chinese Academy of Sciences, China
Horace H.S. Ip, City University of Hong Kong, China
Hermann Krallmann, Technische Universität Berlin, Germany
Karl Kurbel, Europa Universität Viadrina, Germany
Markus Lemmen, Ford Motor Company, Cologne, Germany
Jiming Liu, Hong Kong Baptist University, China
Julian Liu, Oxford University, UK
Kecheng Liu, Reading University, UK

Workshop on "Agent Technology and Software Engineering" (AgeS 2002)

Organizing Committee

Bernhard Bauer
Siemens AG, CT IC 6, Munich, Germany
E-mail: bernhard.bauer@mchp.siemens.de

Klaus Fischer
DFKI GmbH, Saarbrücken, Germany
http://www.dfki.de/~kuf
E-mail: klaus.fischer@dfki.de

Jörg P. Müller
Siemens AG, CT IC 6, Munich, Germany
E-mail: joerg.mueller@mchp.siemens.de

Bernhard Rumpe
Munich University of Technology, Munich, Germany
http://www.in.tum.de/~rumpe/
E-mail: Bernhard.Rumpe@in.tum.de

Workshop on "Agent Technologies for e-Services" (ATES 2002)

Organizing Committee

Ryszard Kowalczyk
CSIRO, Mathematical and Information Sciences, Australia
ryszard.kowalczyk@csiro.au

Table of Contents

Reuse

Negotiation and Communication

Large Complex Systems

E-business

Applications

Software Agents: The Future of Web Services

Michael N. Huhns

University of South Carolina, Department of Computer Science and Engineering,
Columbia, SC 29208, USA
Huhns@engr.sc.edu
http://www.cse.sc.edu/~huhns

Abstract. The World-Wide Web is evolving from an environment for
people to obtain information to an environment for computers to accomplish tasks on behalf of people. The resultant Semantic Web will be
computer-friendly through the introduction of standardized Web services. This paper describes how Web services will become more agent-like,
and how the envisioned capabilities and uses for the Semantic Web will
require implementations in the form of multiagent systems. It also describes how the resultant agent-based Web services will yield unprecedented
levels of software robustness.

1 Introduction

I recently transfered the title for my daughter's car from my name to hers.
This straightforward transaction involved the following flurry of documents and
interactions with various agencies: I needed tax receipts from my county and my
state, a cancelled check for taxes paid, a driver's license verification from the
motor vehicle department, a new license plate for the car from the license plate
bureau, a proof of disposal of the old license plate, the old title, and, finally,
the new title. The county tax agency has strict rules about issuing duplicate
receipts, but these rules are difficult to understand and not fully known by my
state motor vehicle department.

What does this have to do with Web services? Well, the organizations participating in my tax-and-title transaction might have implemented their capabilities as on-line Web services. By invoking each other's functionalities, the Web
services could have determined the necessary forms, the required fees, and the
verification of the identities of myself and my daughter. A single visit to exchange
license plates physically would have been sufficient, I would have received a new
title on-line, and I could have authorized payment via an automatic debit from
my bank account.

Because of the potential illustrated above, Web services are the hottest trend
in information technology: it is hard to find a computer magazine today that
doesn't feature them. Web services are XML-based, work through firewalls, are
lightweight, and are supported by all software companies. They are a key component of Microsoft's .NET initiative, and are deemed essential to the business
directions being taken by IBM, Sun, and SAP.

R. Kowalczyk et al. (Eds.): Agent Technology Workshops 2002, LNAI 2592, pp. 1–18, 2003.

Web services are also central to the envisioned *Semantic Web* [1], which is what the World Wide Web is evolving into. But the Semantic Web is also seen as a friendly environment for software agents, who will add capabilities and functionality to the Web. What will be the relationship between agents and Web services?

1.1   The Semantic Web

The World-Wide Web was designed for humans. It is based on a simple concept: information consists of pages of text and graphics that contain links, and each link leads to another page of information, with all of the pages meant to be viewed by a person. The constructs used to describe and encode a page, the Hypertext Markup Language (html), describe the appearance of the page, but not its contents. Software agents don't care about appearance, but rather the contents.

There are, however, some agents that make use of the Web as it is now. A typical kind of such agent is a *shopbot*, an agent that visits the on-line catalogs of retailers and returns the prices being charged for an item that a user might want to buy. The shopbots operate by a form of "screen-scraping," in which they download catalog pages and search for the name of the item of interest, and then the nearest set of characters that has a dollar-sign, which presumably is the item's price. The shopbots also might submit the same forms that a human might submit and then parse the returned pages that merchants expect are being viewed by humans. The Semantic Web will make the Web more accessible to agents by making use of semantic constructs, such as ontologies represented in DAML, RDF, and XML, so that agents can *understand* what is on a page.

The World-Wide Web was designed for people to get information, such as finding out about books; the Web also supports people getting work done, such as buying a book. In its current form, the Web has the following characteristics:

- HTML describes how things appear
- HTTP is stateless
- Sources are independent and heterogeneous
- Processing is asynchronous and client-server
- There is no support for integrating information
- There is no support for meaning and understanding

The envisioned Web services of the Semantic Web are expected to be

- Robust
- Composable
- Dynamic
- Distributed
- Aware of client's needs so that they can volunteer their services.

1.2   Current Standards for Web Services

There are a number of definitions for Web services. For example, a Web service is said to be

- ...a piece of business logic accessible via the Internet using open standards..."
  (Microsoft)
- Loosely coupled software components that interact with one another dynamically via standard Internet technologies (Gartner)
- A software application identified by a URI, whose interfaces and binding are capable of being defined, described, and discovered by XML artifacts and supports direct interactions with other software applications using XML-based messages via Internet-based protocols (W3C)

My working definition is: A Web service is functionality that can be engaged over the Web.

Web services are currently based on the triad of functionalities depicted in Fig. 1. The architecture for Web services is founded on principles and standards for connection, communication, description, and discovery. For providers and requestors of services to be connected and exchange information, there must be a common language. This is provided by the eXtensible Modeling Language (XML).

A common protocol is required for systems to communicate with each other, so that they can request services, such as to schedule appointments, order parts,

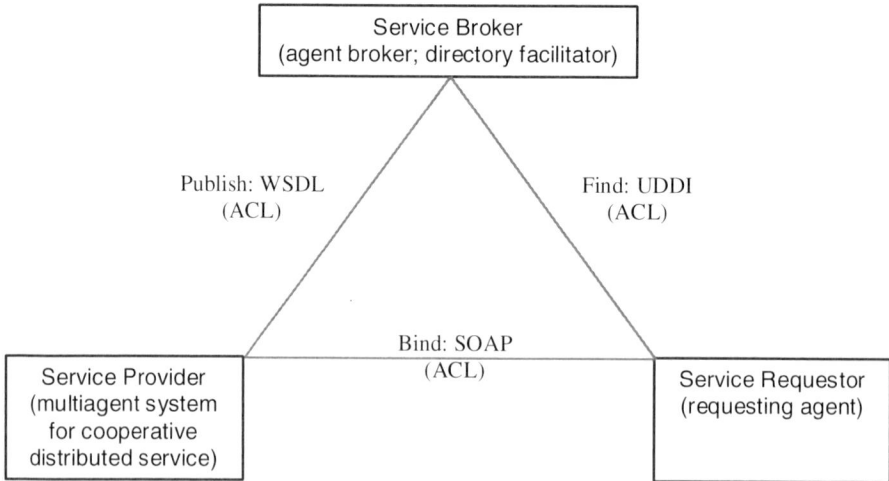

Fig. 1. The general architectural model for Web services. Web services rely on the functionalities of publish, find, and bind. The equivalent agent-based functionalities are shown in parentheses, and all interactions are via an agent-communication language (ACL)

and deliver information. This is provided by the Simple Object Access Protocol
(SOAP) [4].

The services must be described in a machine-readable form, where the names
of functions, their required parameters, and their results can be specified. This
is provided by the Web Services Description Language (WSDL).

Finally, clients—users and businesses—need a way to find the services they
need. This is provided by Universal Description, Discovery, and Integration
(UDDI), which specifies a registry or "yellow pages" of services.

Besides standards for XML, SOAP, WSDL, and UDDI, there is a need for
broad agreement on the semantics of specific domains. This is provided by the
Resource Description Framework (RDF) [7,8], the DARPA Agent Modeling Language (DAML), and, more generally, ontologies [12].

1.3   Directory Services

The purpose of a directory service is for components and participants to be able
to locate each other, where the components and participants might be applications, agents, Web service providers, Web service requestors, people, objects,
and procedures. There are two general types of directories, determined by how
entries are found in the directory: (1) name severs or *white pages*, where entries
are found by their name, and (2) *yellow pages*, where entries are found by their
characteristics and capabilities.

The implementation of a basic directory is a simple database-like mechanism
that allows participants to insert descriptions of the services they offer and query
for services offered by other participants. A more advanced directory might be
more active than others, in that it might provide not only a search service,
but also a brokering or facilitating service. For example, a participant might
request a brokerage service to recruit one or more agents who can answer a
query. The brokerage service would use knowledge about the requirements and
capabilities of registered service providers to determine the appropriate providers
to which to forward a query. It would then send the query to those providers,
relay their answers back to the original requestor, and learn about the properties
of the responses it passes on (e.g., the brokerage service might determine that
advertised results from provider X are incomplete, and so seek out a substitute
for provider X).

UDDI is itself a Web service that is based on XML and SOAP. It provides
both a white-pages and a yellow-pages service, but not a brokering or facilitating
service.

The DARPA DAML effort has also specified a syntax and semantics for
describing services, known as DAML-S. This service description provides

- Declarative ads for properties and capabilities, used for discovery
- Declarative APIs, used for execution
- Declarative prerequisites and consequences, used for composition and intero-
  peration.

2   Foundations for Web Services

Current Web services have either a database or a programming basis. Both are unsatisfactory. To illustrate the database basis and its shortcomings, consider the following simple business-to-customer Web service example: suppose a business wants a software application to sell cameras over the Web, debit a credit card, and guarantee next-day delivery. The application must

- Record a sale in a sales database
- Debit the credit card
- Send an order to the shipping department
- Receive an OK from the shipping department for next-day delivery
- Update an inventory database

What if the order is shipped, but the debit fails? What if the debit succeeds, but the order was never entered or shipped? A traditional database approach works only for a closed environment:

- Transaction processing monitors (such as IBM's CICS, Transarc's Encina, BEA System's Tuxedo) can ensure that all or none of the steps are completed, and that systems eventually reach a consistent state
- But what if the user's modem is disconnected right after he clicks on OK? Did the order succeed? What if the line went dead before the acknowledgement arrives? Will the user order again?

The essential problem is that the transaction processing monitor cannot get the user into a consistent state! The user is part of the software system's environment, which is *open* because it can accommodate any user. In more modern approaches designed for open environments, a server application could send email about credit problems, or detect duplicate transactions. A downloaded applet could synchronize with the server after a broken connection was restored and then recover the transaction; the applet could communicate using http, or directly with server objects via CORBA/IIOP or RMI.

If there are too many orders to process synchronously, they could be put in a message queue, managed by a Message Oriented Middleware server (which guarantees message delivery or failure notification), and customers would be notified by email when the transaction is complete. In essense, *the server behaves like an agent!*

With a programming basis for Web services, software is partitioned into composable services, which are invoked by an application using, for example, RMI. This is illustrated in Fig. 2. In this figure, suppose application A invokes service B, but B is busy and delegates the request to service C. When service C sends a response to A, A *fails* because it expected a response from B.

3   Composing Cooperative Web Services

Imagine that a merchant would like to enable a customer to be able to track the shipping of a sold item. Currently, the best the merchant can do is to point the

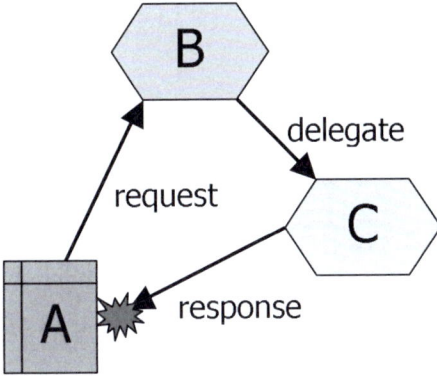

Fig. 2. An illustration of a programming basis for Web services

customer to the shipper's Web site, and the customer can then go there to check on delivery status. If the merchant could compose its own production notification system with the shipper's Web services, the result would be a customized delivery notification service by which the customer—or the customer's agents—could find the status of a purchase in real time.

As Web uses (and thus Web interactions) become more complex, it will be increasingly difficult for one server to provide a total solution and increasingly difficult for one client to integrate solutions from many servers. Web services currently involve a single client accessing a single server, but soon applications will demand federated servers with multiple clients sharing results. Cooperative peer-to-peer solutions will have to be managed, and this is an area where agents have excelled. In doing so, agents can balance cooperation with the interests of their owner.

Composing Web services requires capturing patterns of semantic and pragmatic constraints on how services may participate in different compositions. It also requires tools to help reject unsuitable compositions so that only acceptable systems are built. The following challenges have not yet been met by current implementations and standards for Web services:

Information Semantics. The composer and the member services must agree on the semantics of the information that they exchange.
Collaboration. To perform even simple protocols reliably, service providers must ensure that the parties to an interaction agree on its current state and where they desire to take it. This requires elements of teamwork through (1) persistence of the computations, (2) ability to manage context, and (3) retrying.
Autonomous Interests. Services should be able to participate in automated markets, where various mechanisms are required for effective participation. This requires abilities to (1) set prices, (2) place bids, (3) accept or reject bids, and (4) accommodate risks.

Personalization. Effective usage of services often requires customization of
the compositions in a manner that is context sensitive, especially with respect to user needs. This requires (1) learning a customer's preferences, (2)
mixed-initiative interactions, offering guidance to a customer (best if it is not
intrusive) and letting a user interrupt the composed service, and (3) acting
on behalf of a user, which is limited to ensure that a user's autonomy is not
violated.

Exception Conditions. To construct virtual enterprises dynamically in order
to provide more appropriate, packaged goods and services to common customers requires the ability to (1) construct teams, (2) enter into multiparty
deals, (3) handle authorizations and commitments, and (4) accommodate
exceptions.

Service Location. Recommendations must be provided to help customers find
relevant, high quality, and trustworthy services. This requires a means to (1)
obtain evaluations, (2) aggregate evaluations, and (3) find evaluations.

Distributed Decision-Making. Decision-making will be distributed across
the composed services, which requires intelligent decisions by each service
so the composed services can collaborate and compete appropriately. The
objective is to achieve the desired composition while accommodating exceptions.

With these concerns being addressd by various research efforts, the Web will
evolve from being passive to active, client-server to peer-to-peer to cooperative,
services to processes, and semantics to mutual understanding to pragmatics and
cognition. The result, as indicated in Fig. 3, will be a Semantic Web that enables
work to get done and better decisions to be made.

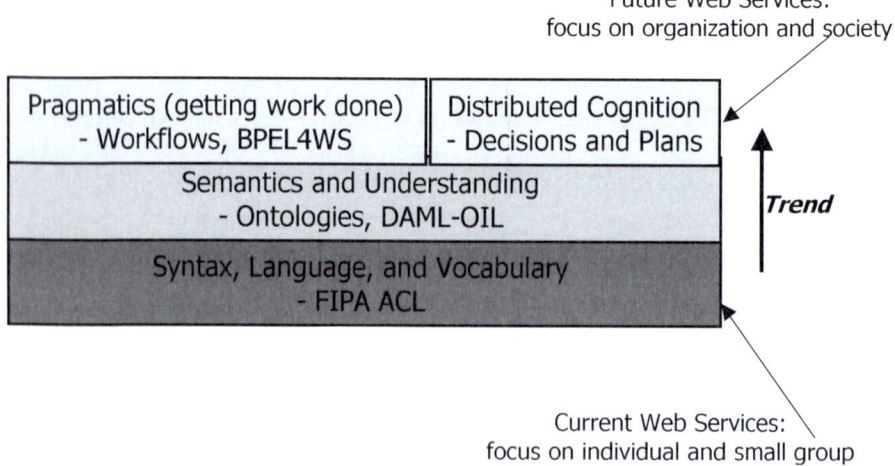

Fig. 3. Beyond the Semantic Web

4   Agents versus Web Services

Typical agent architectures have many of the same features as Web services. Agent architectures provide yellow-page and white-page directories, where agents advertise their distinct functionalities and where other agents search to locate the agents in order to request those functionalities. However, agents extend Web services in several important ways:

- A Web service knows only about itself, but not about its users/clients/ customers. Agents are often self-aware at a metalevel, and through learning and model building gain awareness of other agents and their capabilities as interactions among the agents occur. This is important, because without such awareness a Web service would be unable to take advantage of new capabilities in its environment, and could not customize its service to a client, such as by providing improved services to repeat customers.
- Web services, unlike agents, are not designed to use and reconcile ontologies. If the client and provider of the service happen to use different ontologies, then the result of invoking the Web service would be incomprehensible to the client.
- Agents are inherently communicative, whereas Web services are passive until invoked. Agents can provide alerts and updates when new information becomes available. Current standards and protocols make no provision for even subscribing to a service to receive periodic updates.
- A Web service, as currently defined and used, is not autonomous. Autonomy is a characteristic of agents, and it is also a characteristic of many envisioned Internet-based applications. Among agents, autonomy generally refers to social autonomy, where an agent is aware of its colleagues and is sociable, but nevertheless exercises its independence in certain circumstances. Autonomy is in natural tension with coordination or with the higher-level notion of a commitment. To be coordinated with other agents or to keep its commitments, an agent must relinquish some of its autonomy. However, an agent that is sociable and responsible can still be autonomous. It would attempt to coordinate with others where appropriate and to keep its commitments as much as possible, but it would exercise its autonomy in entering into those commitments in the first place.
- Agents are cooperative, and by forming teams and coalitions can provide higher-level and more comprehensive services. Current standards for Web services do not provide for composing functionalities.

4.1   Benefits of an Agent-Oriented Approach

Multiagent systems can form the fundamental building blocks for software systems, even if the software systems do not themselves require any agent-like behaviors [19]. When a conventional software system is constructed with agents as its modules, it can exhibit the following characteristics:

- Agent-based modules, because they are active, more closely represent real-world things
- Modules can hold beliefs about the world, especially about themselves and others
- Modules can negotiate with each other, enter into social commitments to collaborate, and can change their mind about their results
- Modules can volunteer to be part of a software system.

The benefits of building software out of agents are [5,15]

1. Agents enable dynamic composibility, where the components of a system can be unknown until runtime
2. Agents allow interaction abstractions, where interactions can be unknown until runtime
3. Because agents can be added to a system one-at-a-time, software can continue to be customized over its lifetime, even potentially by end-users
4. Because agents can represent multiple viewpoints and can use different decision procedures, they can produce more robust systems. The essence of multiple viewpoints and multiple decision procedures is redundancy, which is the basis for error detection and correction.

4.2   Advanced Composition

Suppose an application needs simply to sort some data items, and suppose there are 5 Web sites that offer sorting services described by their input data types, output date type, time complexity, space complexity, and quality:

1. One is faster
2. One handles more data types
3. One is often busy
4. One returns a stream of results, while another returns a batch
5. One costs less

An application could take one of the following possible approaches:

- Application invokes services randomly until one succeeds
- Application ranks services and invokes them in order until one succeeds
- Application invokes all services and reconciles the results
- Application contracts with one service after requesting bids
- Services self-organize into a team of sorting services and route requests to the best one

The last two require that the services behave like agents. Furthermore, the last two are scalable and robust, because they take advantage of the redundancy that is available.

5 Redundancy and Robustness

Redundancy is the basis for most forms of robustness. It can be provided by replication of hardware, software, and information, and by repetition of communication messages. For years, NASA has made its satellites more robust by duplicating critical subsystems. If a hardware subsystem fails, there is an identical replacement ready to begin operating. The space shuttle has quadruple redundancy, and won't leave the ground without all copies functioning. However, software redundancy has to be provided in a different way. Identical software subsystems will fail in identical ways, so extra copies do not provide any benefit.

Moreover, code cannot be added arbitrarily to a software system, just as steel cannot be added arbitrarily to a bridge. When we make a bridge stronger, we do it by adding beams that are not identical to ones already there, but that have equivalent functionality. This turns out to be the basis for robustness in software systems as well: there must be software components with equivalent functionality, so that if one fails to perform properly, another can provide what is needed. The challenge is to design the software system so that it can accommodate the additional components and take advantage of their redundant functionality.

We hypothesize that agents are a convenient level of granularity at which to add redundancy and that the software environment that takes advantage of them is akin to a society of such agents, where there can be multiple agents filling each societal role [13]. Agents by design know how to deal with other agents, so they can accommodate additional or alternative agents naturally. They are also designed to reconcile different viewpoints.

Fundamentally, the amount of redundancy required is well specified by information and coding theory. Assume each software module in a system can behave either correctly or incorrectly. Then two modules with the same intended functionality are sufficient to detect an error in one of them, and three modules are sufficient to correct the incorrect behavior (by voting, or choosing the best two-out-of-three). This is exactly how parity bits work in code words. Unlike parity bits, and unlike bricks and steel bridge beams, however, the software modules can't be identical, or else they would not be able to correct each other's errors.

If we want a system to provide n functionalities robustly, we must introduce $m \times n$ agents, so that there will be m ways of producing each functionality. Each group of m agents must understand how to detect and correct inconsistencies in each other's behavior, without a fixed leader or centralized controller. If we consider an agent's behavior to be either correct or incorrect (binary), then, based on a notion of Hamming distance for error-correcting codes, $4m$ agents can detect $m - 1$ errors in their behavior and can correct $(m - 1)/2$ errors.

Fundamentally, redundancy must be balanced with complexity, which is determined by the number and size of the components chosen for building a system. That is, adding more components increases redundancy, but might also increase the complexity of the system. This is just another form of the common software engineering problem of choosing the proper size of the modules used to implement a system. Smaller modules are simpler, but their interactions are more complicated because there are more modules.

An agent-based system can cope with a growing application domain by increasing the number of agents, each agent's capability, the computational resources available to each agent, or the infrastructure services needed by the agents to make them more productive. That is, either the agents or their interactions can be enhanced, but to maintain the same degree of redundancy n, they would have to be enhanced by a factor of n.

To underscore the importance being given to redundancy and robustness, several initiatives are underway around the world to investigate them. IBM has a major initiative to develop autonomic computing—"a systemic view of computing modeled after the self-regulating autonomic nervous system." Systems that can run themselves incorporate many biological characteristics, such as self-healing (redundancy), adaptability to changing environments (reconfigurability), identity (awareness of their own resources), and immunity (automatic defense against viruses). An autonomic computing system will adhere to self-healing, not by "cellular regrowth," but by making use of redundant elements to act as replenishment parts. By taking advantage of redundant services located around the world, a better range of services can be provided for customers in business transactions.

5.1   N-Version Programming

N-version programming, also called dissimilar software, is a technique for achieving robustness first considered in the 1970's. It consists of N disparate and separately developed implementations of the same functionality. Although it has been used to produce several robust systems, it has had limited applicability, because (1) N independent implementations have N times the cost, (2) N implementations based on the same flawed specification might still result in a flawed system, and (3) the resultant system might have N times the maintenance cost (e.g., each change to the specification will have to be made in all N implementations).

5.2   Transaction Checkpointing, Rollback, and Recovery

Database systems have exploited the idea of transactions for maintaining the consistency of their data. A transaction is an atomic unit of processing that moves a database from one consistent state to another. Consistent transactions are achievable for databases because the types of processing done are very regular and limited.

Applying this idea to general software execution requires that the state of a software system be saved periodically (a checkpoint) so that the system can return to that state if an error occurs. The system then returns to that state and processes other transactions or alternative software modules. This is depicted in Fig. 4.

There are two ways of returning to a previous state: (1) reloading a saved image of the system before the recently failed computation, or (2) rolling back,

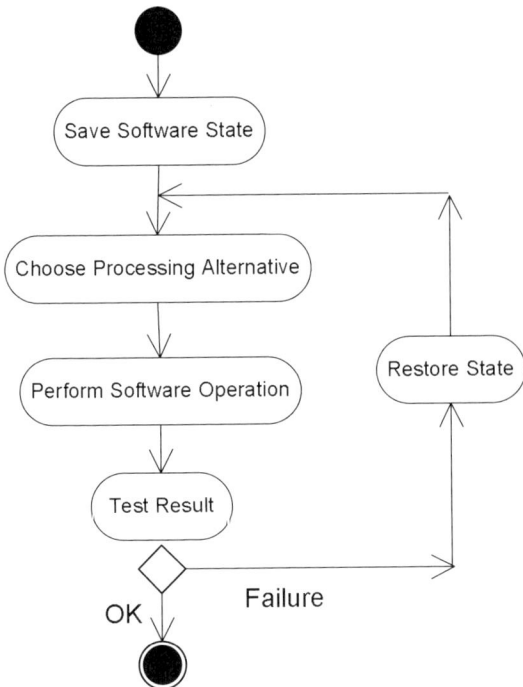

Fig. 4. A transaction approach to correcting for the occurrence of errors in a software system

i.e., reversing and undoing, each step of the failed computation. Both of the ways suffer from major difficulties:

1. The state of a software system might be very large, necessitating the saving of very large images
2. Many operations cannot be undone, such as those that have side-effects. Examples of these are sending a message, which cannot be un-sent, and spending resources, which cannot be un-spent. Rollback is successful in database systems, because most database operations do not have side-effects.

5.3   Compensation

Because of this, compensation is often a better alternative for software systems. Figure 5 depicts the architecture of a robust software system that relies on compensation of failed operations.

6   Architecture and Process

Suppose there are a number of sorting algorithms available. Each might have strengths, weaknesses, and possibly errors. One might work only for integers,

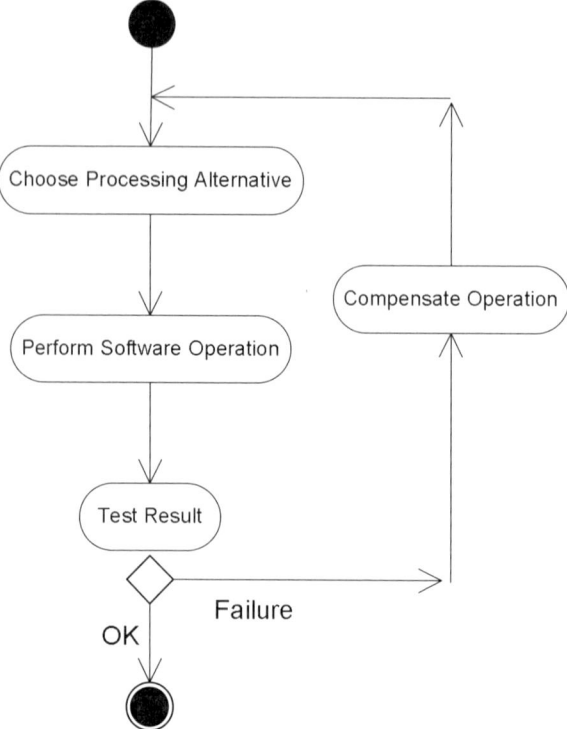

Fig. 5. An architecture for software robustness based on compensating operations

while another might be slower but be able to sort strings as well as integers. How can the algorithms be combined so that the strengths of each are exploited and the weaknesses or flaws of each are compensated or covered? In solving this in a general way, we hypothesize that the end result is an "agentizing" of each algorithm.

A centralized approach, as shown in Fig. 6, would use an omniscient preprocessing algorithm to receive the data to be sorted and would then choose the best algorithm to perform the sorting. Each module's characteristics have to be encoded into the central unit. The central unit can use a simplistic algorithm for determining the best, based on known facts about each of the modules. The difficulties with this approach are (1) the preprocessing algorithm might be flawed and (2) it is difficult to maintain such a preprocessing algorithm as new algorithms are added and existing algorithms become unavailable. Also, only one module at-a-time executes, there is low CPU usage, and results are taken as-is when completed.

An improvement might be a postprocessing algorithm, as shown in Fig. 7, that receives the results of all sorting algorithms and chooses the best result to be the output. Results have to be compared and voted on to determine the best. This approach is also centralized and suffers from a waste of CPU resources,

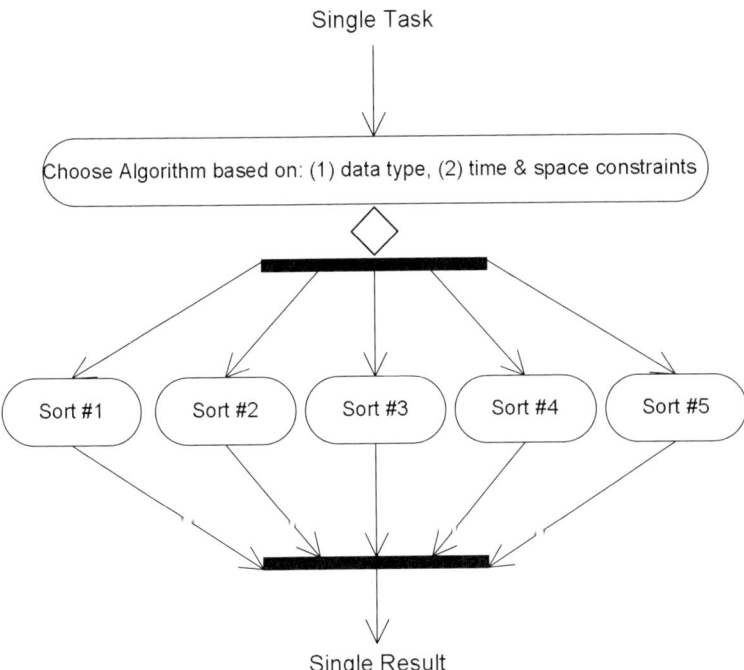

Fig. 6. Centralized architecture for combining N versions of a sorting algorithm into a single, more robust system for sorting, where a preprocessing algorithm chooses which sorting algorithm will execute

because all algorithms work on the data. However, due to the comparison of outcomes, it is likely to produce better results.

A combination of the preprocessing and postprocessing centralized systems could also be used. Since the characteristics of each module are known, a subgroup could be selected to perform the desired task based on known factors such as speed, time, and space. This subgroup would then have its results compared to determine the best results as above. Because certain modules will be selected every time the same set of circumstances come up, a better way of developing a conventional system would be to hard-wire these sets of circumstances to eliminate the need for a central intelligent filtering unit.

A fourth approach is a distributed solution, where the algorithms jointly decide which one(s) should perform the sorting, and if there is more than one qualified algorithm, they jointly decide on the best result. Conventional algorithms do not typically have such a distributed decision-making ability, so we investigated whether there is a generic capability that can be added to an algorithm to enable it to participate in a distributed decision. We discovered that the result has the characteristics of a software agent.

We collected a number of sorting algorithms, each written by different people and therefore having different characteristics (such as input data type, output

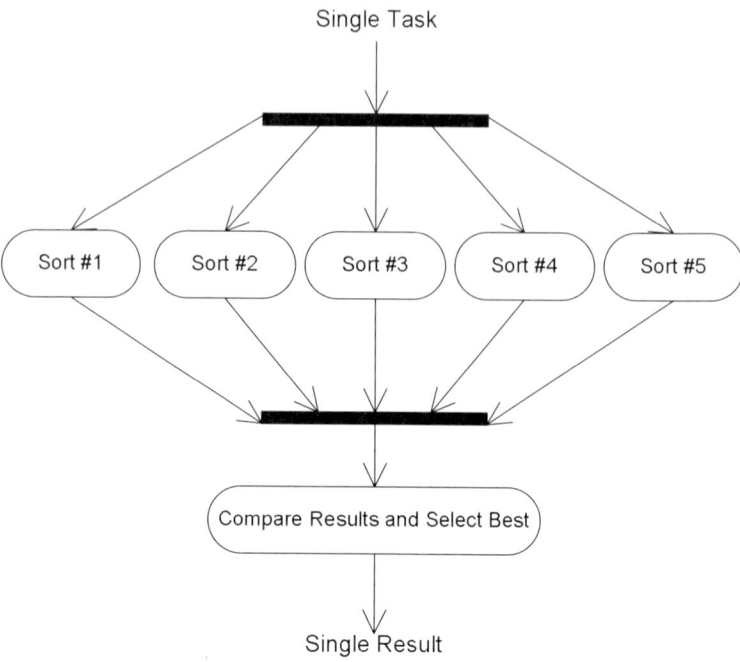

Fig. 7. Centralized architecture for combining N versions of a sorting algorithm into a single, more robust system for sorting, where a postprocessing algorithm chooses one result to be the output

data type, and time and space complexity). For our experiments, we converted each algorithm into a sorting agent.

Each sorting agent is composed of a sorting algorithm and a wrapper for that algorithm. The wrapper knows nothing about the inner workings of the algorithm with which it is associated. It has knowledge only about the characteristics of its algorithm, such as the data type(s) it can sort, the data type it produces, its time complexity, and its space complexity. The sorting algorithms were written in Java and the wrappers in JADE [18].

The system sends data to be sorted to all the sorting agents. Their responsibility (as a group) is the sorting of the data, and they should be able to do this better than any one of them alone. Upon receiving data to be sorted, each agent determines whether or not it can sort it successfully (based on the type of the data and its own knowledge of what types it can sort) and if the agent can, it broadcasts a message to every other agent specifying its intention, along with a measure of performance for its algorithm (based on time and space complexity).

The decision of which agent (i.e., algorithm) to choose (among those that are capable of sorting the input data) is made in a distributed manner: upon receiving the estimates from the other agents, each agent compares its own performance measure against those received in the messages. If the agent has the best performance measure, it will run its algorithm and send the results back to the system. If it does not have the best performance measure, it will do nothing. The results, not surprisingly, showed that the agent-based composition of sorting algorithms performed better than any individual algorithm.

7 Conclusion: Challenges and Implications for Developers

Producing robust software has never been easy, distributing it across the Web makes it much more difficult, and the approach recommended here would have major effects on the way that developers construct software systems:

- It is difficult enough to write one algorithm to solve a problem, let alone n algorithms. However, algorithms, in the form of agents, are easier to reuse than when coded conventionally and easier to add to an existing system, because agents are designed to interact with an arbitrary number of other agents,
- Agent organizational specifications need to be developed to take full advantage of redundancy.
- Agents will need to understand how to detect and correct inconsistencies in each other's behavior, without a fixed leader or centralized controller.
- There are problems when the agents either represent or use nonrenewable resources, such as CPU cycles, power, and bandwidth, because they will use it n times as fast.
- Although error-free code will always be important, developers will spend more time on algorithm development and less on debugging, because different algorithms will likely have errors in different places and can cover for each other.
- In some organizations, software development is competitive in that several people might write an algorithm to yield a given functionality, and the "best" algorithm will be selected. Under the approach suggested here, all algorithms would be selected.

Web services are extremely flexible, and a major advantage is that a developer of Web services does not have to know who or what will be using the services being provided. They can be used to tie together the internal information systems of a single company or the interoperational systems of virtual enterprises. But how Web services tie the systems together will be based on technologies being developed for multiagent systems.

Acknowledgements. The US National Science Foundation supported this work under grant number IIS-0083362.

References

1. Berners-Lee, Tim, James Hendler, and Ora Lassila: "The Semantic Web," *Scientific American*, vol. 284, no. 5, May 2001, pp. 34–43.

2. Beugnard, Antoine, Jean-Marc Jezequel, Noel Plouzeau, and Damien Watkins: "Making Components Contract Aware," *IEEE Computer*, Vol. 32, No. 7, July 1999, pp. 38–45.

3. Booch, Grady, James Rumbaugh, and Ivar Jacobson: *The Unified Modeling Language User Guide*. Addison-Wesley, Reading, MA, 1999.

4. Box, D., et al.: "Simple Object Access Protocol (SOAP) 1.1," 2000.
   http://www.w3.org/TR/SOAP

5. Coelho, Helder, Luis Antunes, and Luis Moniz: "On Agent Design Rationale." In *Proceedings of the XI Simposio Brasileiro de Inteligencia Artificial (SBIA)*, Fortaleza (Brasil), October 17–21, 1994, pp. 43–58.

6. Cox, Brad J.: "Planning the Software Industrial Revolution." *IEEE Software*, November 1990, pp. 25–33.

7. Decker, Stefan, et al.: "The Semantic Web: The Roles of XML and RDF," *IEEE Internet Computing*, vol. 4, no. 5, September-October 2000, pp. 63–74.

8. Decker, Stefan, P. Mitra, and S. Melnik: "Framework for the Semantic Web: An RDF Tutorial," *IEEE Internet Computing*, vol. 4, no. 6, November-December 2000, pp. 68–73.

9. DeLoach, S.: "Analysis and Design using MaSe and agentTool." In *Proceedings of the 12th Midwest Artificial Intelligence and Cognitive Science Conference (MAICS 2001)*, 2001.

10. Dignum, Frank, Barbara Dunin-Keplicz, and Rineke Verbrugge: "Dialogue in team formation: a formal approach" In van der Hoek, W., Meyer, J. J., and Wittenveen, C., Editors, *ESSLLI99 Workshop: Foundations and applications of collective agent based systems*, (1999).

11. Hasling, John: *Group Discussion and Decision Making*, Thomas Y. Crowell Company, Inc. (1975).

12. Heflin, Jeff and James A. Hendler: "Dynamic Ontologies on the Web," In *Proceedings American Association for Artificial Intelligence (AAAI)*, AAAI Press, Menlo Park, CA, 2000, pp. 443–449.

13. Holderfield, Vance T. and Michael N. Huhns: "A Foundational Analysis of Software Robustness Using Redundant Agent Collaboration." In *Proceedings International Workshop on Agent Technology and Software Engineering*, Erfurt, Germany, October 2002.

14. Huhns, Michael N. and Vance T. Holderfield: "Robust Software," *IEEE Internet Computing*, vol. 6, no. 2, March-April 2002, pp. 80–82.

15. Huhns, Michael N.: "Interaction-Oriented Programming." In *Agent-Oriented Software Engineering*, Paulo Ciancarini and Michael Wooldridge, editors, Springer Verlag, Lecture Notes in AI, Volume 1957, Berlin, pp. 29-44 (2001).

16. Iglesias, C. A., M. Garijo, J. C. Gonzales, and R. Velasco: "Analysis and Design of Multi-Agent Systems using MAS-CommonKADS." In *Proceedings of the AAAI'97 Workshop on agent Theories, Architectures and Languages*, Providence, USA, 1997.

17. Iglesias, C. A., M. Garijo, and J. Gonzalez: "A survey of agent-oriented methodologies." In J. Muller, M. P. Singh, and A. S. Rao, editors, *Proceedings of the 5th International Workshop on Intelligent Agents V: Agent Theories, Architectures, and Languages (ATAL-98)*. Springer-Verlag: Heidelberg, Germany, 1999.

18. *JADE: Java Agent Development Environment*,
    http://sharon.cselt.it/projects/jade.

19. Jennings, Nick R.: "On Agent-Based Software Engineering" *Artificial Intelligence*, 117 (2) 277–296 (2000).

20. Juan, T., A. Pearce, and L. Sterling: "Extending the Gaia Methodology for Complex Open Systems." In *Proceedings of the 2002 Autonomous Agents and Multi-Agent Systems*, Bologna, Italy, July 2002.

21. Kalinsky, David: "Design Patterns for High Availability." *Embedded Systems Programming* (August 2002) 24–33.

22. Kinny, David and Michael Georgeff: "Modelling and Design of Multi-Agent Systems," in J.P. Muller, M.J. Wooldridge, and N.R. Jennings, eds., *Intelligent Agents III — Proceedings of the Third International Workshop on Agent Theories, Architectures, and Languages*, Springer-Verlag, Berlin, 1997, pp. 1–20.

23. Kinny, D., M. Georgeff, and A. Rao: "A Methodology and Modeling technique for systems of BDI agents." In *Proceedings of the 7th European workshop on modeling autonomous agents in a multi-agent world*, LNCS 1038, pp. 56–71, Springer-Verlag, Berlin Germany, 1996.

24. Laddaga, Robert: "Creating Robust Software through Self-Adaptation," *IEEE Intelligent Systems*, Vol. 14, No. 3, May/June 1999, pp. 26–29.

25. Light, Donald, Suzanne Keller, and Craig Calhoun: *Sociology* Alfred A. Knopf/New York (1989).

26. Lorge, I. and H. Solomon: "Two models of group behavior in the solution of Eureka-type problems." *Psychometrika* (1955).

27. Nwana, Hyacinth S. and Michael Wooldridge: "Software Agent Technologies." *BT Technology Journal*, 14(4):68-78 (1996).

28. Odell, J., H. Van Dyke Parunak, and Bernhard Bauer: "Extending UML for Agents." In *Proceedings of the Agent-Oriented Information Systems Workshop*, Gerd Wagner, Yves Lesperance, and Eric Yu eds., Austin, TX, 2000.

29. Padgham, L. and M. Winikoff: "Prometheus: A Methodology for Developing Intelligent Agents." In *Proceedings of the Third International Workshop on Agent-Oriented Software Engineering*, at AAMAS 2002. July, 2002, Bologna, Italy.

30. Paulson, Linda Dailey: "Computer System, Heal Thyself," *IEEE Computer*, (August 2002) 20–22.

31. Perini, A., P. Bresciani, F. Giunchiglia, P. Giorgini, and J. Mylopoulos: "A knowledge level software engineering methodology for agent oriented programming." In *Proceedings of Autonomous Agents*, Montreal CA, 2001.

32. Schreiber, A. T., B. J. Wielinga, and J. M. A. W. Van de Velde: "CommonKADS: A comprehensive methodology for KBS development," 1994.

33. Shapley, L. S. and B. Grofman: "Optimizing group judgmental accuracy in the presence of interdependence" *Public Choice*, 43: 329-343 (1984).

34. Swap, Walter C., et al.: *Group Decision Making*, SAGE Publications, Inc., Beverly Hills, London, New York (1984).

35. Tambe, Milind, David V. Pynadath, and Nicolas Chauvat: "Building Dynamic Agent Organizations in Cyberspace," *IEEE Internet Computing*, Vol. 4, No. 2, March/April 2000.

36. Wooldridge, M., N. R. Jennings, and D. Kinny: "The Gaia Methodology for Agent-Oriented Analysis and Design." *Journal of Autonomous Agents and Multi-Agent Systems*, 2000.

Building Automated Negotiators

Nick Jennings

Department of Electronics and Computer Science
University of Southampton
Highfield
Southampton
SO17 1BJ
United Kingdom
nrj@ecs.soton.ac.uk
http://www.ecs.soton.ac.uk/~nrj/

Abstract. Computer systems in which autonomous software agents negotiate with one another in order to come to mutually acceptable agreements are likely to become pervasive in the next generation of networked systems (e.g., Semantic Web, Grid Computing, Pervasive Computing). In such systems, the agents will be required to participate in a range of negotiation scenarios and exhibit a range of negotiation behaviors (depending on the context). To this end, this talk explores the issues involved in designing and implementing a number of automated negotiators for auctions, bi-lateral negotiations and persuasive negotiations for real-world applications.

R. Kowalczyk et al. (Eds.): Agent Technology Workshops 2002, LNAI 2592, p. 19, 2003.
© Springer-Verlag Berlin Heidelberg 2003

Emergence in Cyberspace: Towards the Evolutionary Self-Organizing Enterprise

Mihaela Ulieru

University of Calgary, Department of Electrical and Computer Engineering

Abstract. As result of the process of evolution driven by the law of synergy, emergence endows the dynamics of composite systems with properties unidentifiable in their individual parts. The phenomenon of emergence involves: – self-organization of the dynamical systems such that the synergetic effects can occur; – interaction with other systems from which the synergetic properties can evolve in a new context. Multi-agent Systems enable cloning of real-life systems into autonomous software entities with a 'life' of their own in the dynamic information environment offered by today's Cyberspace. After introducing the concept of Holonic Enterprise (HE) as a paradigm for the networked world that enables virtual representation of real-life organizations as multi-agent systems I will present a fuzzy-evolutionary approach which mimics emergence in Cyberspace as follows: – it induces self-organizing properties by minimizing the entropy measuring the information spread across the virtual system/organization such that equilibrium is reached in an optimal interaction between the system's parts to reach the system's objectives most efficiently; – it enables system's evolution into a better one by enabling interaction with external systems found via genetic search strategies (mimicking mating with most fit partners in natural evolution) such that the new system's optimal organizational structure (reached by minimizing the entropy) is better then the one before evolution. The holonic enterprise paradigm provides a framework for information and resource management in global virtual organizations by modeling enterprise entities as software agents linked through the internet. Applying the proposed fuzzy-evolutionary approach to the virtual societies 'living' on the dynamic Web endows them with behavioral properties characteristic to natural systems. In this parallel universe of information, enterprises enabled with the proposed emergence mechanism can evolve towards better and better structures while at the same time self-organizing their resources to optimally accomplish the desired objectives.

R. Kowalczyk et al. (Eds.): Agent Technology Workshops 2002, LNAI 2592, p. 20, 2003.
© Springer-Verlag Berlin Heidelberg 2003

Requirements Analysis in Tropos: A Self-Referencing Example

Paolo Bresciani and Fabrizio Sannicolò

ITC-Irst
Via Sommarive, 18, I-38050 Trento-Povo, Italy
{bresciani,sannico}@irst.itc.it

Abstract. *Tropos*, a novel agent-oriented software engineering methodology, is heavily characterized, among other features, by the fact that it pays great attention to the activities that precede the specification of the prescriptive requirements, such as understanding how the intended system would meet the organizational goals. This is obtained by means of the two requirement phases: the *early requirements analysis* and the *late requirements analysis*. Moreover, Tropos uses, along these phases, a uniform notation and an homogeneous, smooth, incremental, and iterative process, based on a set of progressive transformational steps.
This paper will take into account the application of the Tropos methodology to a *self-motivating* case study: the definition of a support tool for the Tropos methodology itself. The focus here is on the early requirements and on how to manage the transition from them to the late requirement analysis.

1 Introduction

Tropos [19,13,11] is a novel agent-oriented software engineering methodology characterized by three key aspects [18]. First, it pays attention to the activities that precede the specification of the prescriptive requirements, like understanding *how* and *why* the intended system would meet the organizational goals[1]. Second, it deals with all the phases of system requirement analysis and all the phases of system design and implementation in a uniform and homogeneous way, based on common mentalistic notions as those of *actors*, *goals*, *softgoals*, *plans*, *resources*, and *intentional dependencies*. Third, the methodology rests on the idea of building a model of the system-to-be that is incrementally refined and extended from a conceptual level to executable artifacts, by means of a sequence of transformational steps [3].

One of the main advantages of the Tropos methodology is that it allows to capture not only the *what* or the *how*, but also the *why* a piece of software is developed. This, in turn, allows for a more refined analysis of the system dependencies and, in particular, for a much better and uniform treatment not only of

---

[1] In this, Tropos is largely inspired by Eric Yu's framework for requirements engineering, called *i**, which offers actors, goals, and actor dependencies as primitive concepts [23,24,26].

R. Kowalczyk et al. (Eds.): Agent Technology Workshops 2002, LNAI 2592, pp. 21–35, 2003.
© Springer-Verlag Berlin Heidelberg 2003

the system functional requirements, but also of its non-functional requirements. Tropos, although not exclusively, addresses particularly well the Agent Oriented Programming [18]. In fact, the decision of using mentalistic notions in all the analysis phases has important consequences. In particular, agent oriented specifications and programs use the same notions and abstractions used to describe the behavior of human agents and the processes involving them; thus, the conceptual gap between users' specifications (in terms of *why* and *what*) and system realization (in terms of *what* and *how*), is reduced to a minimum.

Tropos supports five phases of software development. The **early requirements analysis** is concerned with the understanding of a problem by studying an existing organizational setting. The output of this phase is an organizational model which includes relevant actors and their respective dependencies. Actors in the organizational setting are characterized by having goals that each single actor, in isolation, would be unable – or not as well or as easily – to achieve. The goals are achievable in virtue of reciprocal means-end knowledge and dependencies. During the **late requirements analysis**, the system-to-be is described within its operational environment, along with relevant functions and qualities. This description models the system as a (relatively small) number of actors, which have a number of social dependencies with other actors in their environment. The **architectural design** phase deals with the definition of the system global architecture in terms of subsystems, that are represented as actors, and their data dependencies, that are represented as actor dependencies. The **detailed design** phase aims at specifying each architectural component in further detail (adopting a subset of the AUML diagrams [17,1]) in terms of inputs, outputs, control and other relevant information. Finally, during the **implementation** phase, the actual implementation of the system is carried out, consistently with the detailed design. More details and motivation about these five phases can be found in [12,11,13,19].

The present paper mainly focuses on the analysis of the early requirement analysis phase and, partially, on the late requirement analysis phase. In particular, concerning the early requirements analysis, the task of encoding initial informal requirements into the diagrammatic format used in the methodology, as well as the incremental transformational process that is at the basis of the construction of the complete model, will be addressed. With respect to previous papers [3], this revision mechanism will be motivated in its high level aspects and applied to a case study, rather than analyzed in its fundamental details. Another aspect that will be addressed in this paper is the transition process from the early to the late requirements. The example that will be used along all the present paper to illustrate the above aspects is the definition of a support tool for the Tropos methodology itself, called, since now on, the *Tropos tool*.

The rest of the paper is structured as follows. Section 2 describes the Tropos tool problem and some background motivating some early choices. Section 3 shows how these choices can be embodied into actor and goal diagrams. Section 4 develops a preliminary actor diagram for late requirements with the only aim to provide a glance on the transition from early to late requirements. Conclusions and directions for further research are presented in Sect. 5.

2   The Problem

As of today, the Tropos methodology lacks a tool which supports both the analyst during the process of acquiring, modeling, and analyzing requirements and the designer during the process of software engineering. Of course, the availability of a tool that supports along the whole development process would be of great impact on the applicability of the methodology. Although relevant efforts have already been made in order to provide a tool for managing diagrams in i^* [15], the realization of an integrated tool that supports all the phases of Tropos (or at least the first four) in a uniform and integrate way has not been analyzed, yet.

In order to start with the definition of the Tropos tool, some observations may be useful. First, let us recall that early and late requirement analysis and architectural design in Tropos are aimed at producing domain and system conceptual and architectural models. These conceptual models must conform to a precise Tropos modeling language [20]. Of course, the Tropos tool shall support for syntax checking during the conceptual model construction.

Another point to be taken under consideration is that, during the process of building actor and dependencies models, resulting from the analysis of social and system actors, the analyst may need to look at more diagrams from many perspectives at the same time. For example, she may want to look back why she introduced some subgoals and how she did it. Therefore, the Tropos tool shall allow us to analyze the conceptual model from several points of view at the same time.

One important check that has to be performed during the analysis and before it may be considered completed, is the closure of all the alternative or conjunctive decompositions (of goals, softgoals, and plans) and the evaluations of the goal and softgoal contributions. This process may be graphically visualized by putting and propagating ticks from the leafs up to the root (goal, softgoal, or plan) of the analysis tree, also using weighted propagation mechanisms for the qualitative (NFR) analysis [6]. This process will be later referred as the capability of *managing ticks*.

In addition, it may also be the case that the project manager wants to use a common graphical interface for viewing and analyzing documents (feasibility study, requirements model, capability diagrams, and so on) generated during different phases, such as the analysis phase and the design phase. Thus, it is desirable that the Tropos tool adopts a common interface for different phases of the methodology, specially when adopting similar graphical notations, which is a pervasive characteristic of Tropos.

One of the main advantages of the Tropos methodology is that the software engineer can also capture the *why* a sort of analysis is carried on or has been made. This important feature gives an extraordinary, although not yet fully explored, value to the notion of *traceability*. Traditionally, traceability is the property that a methodology or a CASE system exhibits when artifacts produced during later phases can be clearly referred back to artifacts or requirements produced earlier. In Tropos, this feature assumes an extra value due to two aspects. First, Tropos is aimed at uniformly covering development activities ranging from early requirements analysis down to the actual implementation; thus, traceability in Tropos may be thought as spanning over several phases and very distant points in the development process. Second, the early and late requirements of Tropos provides an intentional analysis of the problem, facing the description of the *why*, together with that of the *what* and the *how*. In this context, tracing late artifacts back to requirements – both early and late – provides a powerful method for giving strong motivations to all the developed artifacts, virtually even to each single line of the produced code, and justifying them with respect to the original requirements. Of course, the Tropos tool must give full support for traceability.

Finally, as always desired, the Tropos tool has also to be friendly, understandable and, above all, useful for all the users, like the analyst, the formal engineer, the designer, the developer, and so on. Useful means that the user has not to waste time in order to understand how the tool works, giving her more time for depicting and investigating the best way to model the actors, their goals, and their strategic dependencies.

In the following section, these preliminary requirements will be detailed and further analyzed, by means of an early requirement analysis.

3   Early Requirement Analysis

In Tropos, during the early requirement analysis, the requirement engineer models and analyzes the intentions of the stakeholders. These are modeled as goals that, through some form of analysis [16,8], such as AND-OR decomposition, means-ends analysis, and contribution analysis, eventually lead to the functional and non-functional requirements of the system-to-be. To this end, *actor diagrams* for describing the network of social dependency relationships among actors, as well as *goal diagrams* for analyzing and trying to fulfill goals, are produced.

Starting from our common understanding of the Tropos tool problem, as described in the previous section, the following list of more detailed issues to be addressed may be provided:

- Fundamental features of Tropos:
  - software engineer support for all the phases of the software development process (at least the first four);
  - support for the transformational approach as in [3];
  - goal/plan analysis [16,8] and non-functional analysis [6].
- System management:
  - model and views storage;
  - multi-users management.
- Extensibility:
  - integration with tools for formal analysis (e.g., NuSMV [10,7]) and version management (e.g., CVS [9]);
  - integration with debugging tools.

– User interaction:
  - visualization of different possible views on the conceptual model, which may be needed both because of the different roles played by the user, and because of her changing design focus along the different phases;
  - subgraph visualization and zooming in order to manage visual complexity;
  - support for multiple decomposition of the same goal from the perspective of one actor, thus allowing for more goal diagrams of the same goal;
  - verification that the functional and non-functional (or quality) requirements are achieved (manage ticks).
– Help and documentation:
  - user guide for the Tropos language and its concepts (ontology, metamodel for the informal language [20], formal language [10], and so on);
  - automatic generation of the documentation about a particular view on the model or a set of diagrams (i.e., static diagrams, dynamic diagrams);
  - export of the conceptual model or some views in several formats (e.g., Scalable Vector Graphics (SVG) [21]).

3.1   The Analysis

After the definitions of the high level requirements, the process proceeds by identifying the most relevant stakeholders of the environment. In particular, the potential users of the tool, their goals, and the respective dependencies are identified and then modeled by means of a first actor diagram.

In order to proceed with a deeper analysis, it is worth to recall that an actor may be embodied by the more specific notions of *agent, role* and *position*. An *agent* represents a concrete instantial actor like a *physical agent* (e.g., a person, an hardware system, and so on), or a *software agent*; a *role* corresponds to an abstract characterization of one specific behavior of an actor within some specialized context; a *position* represents a set of roles, typically played by a single agent. An agent can occupy a position, while a position is said to cover one or more roles. Also, an agent is said to play one or more roles. For more detailed distinctions and examples see [24,25].

In the actor diagram depicted in Fig. 1, Software Professional is depicted as an agent who may occupy different positions: PRG Manager, Analyst, Formal Engineer, Architecture Designer, Details Designer, and Developer.

Although in general it is assumed that each position covers several roles, in Fig. 1, for lack of space, only one role is present as an example: Customer relations keeper. The interesting aspect to be noticed here is that this role can be covered both by the Analyst position and by the PRG Manager position, that is, each agent that occupies one of these position, or possibly both, must play the role Customer relations keeper.

After the introduction of the actors, the dependency analysis focuses on the dependencies of each single different actor. For example, Customer depends on the PRG Manager for achieving the goal system provided. Also, the PRG

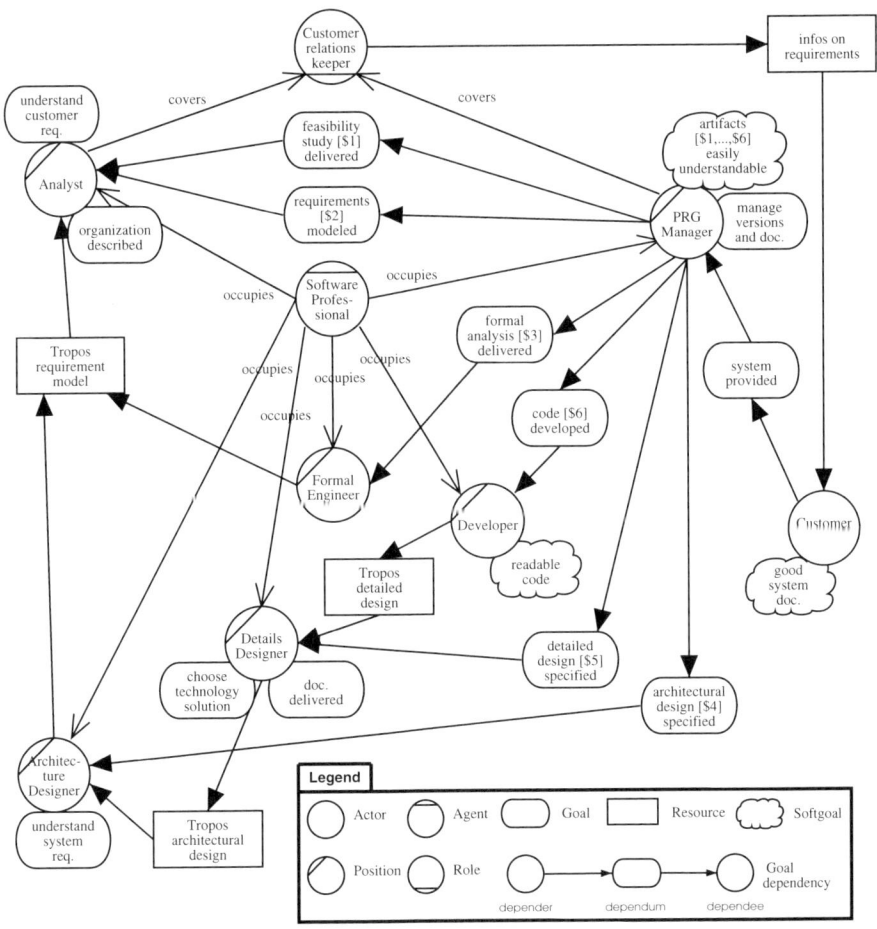

Fig. 1. An actor diagram specifying the stakeholders of the Tropos tool project

Manager wants to manage the different versions of artifacts that other actors deliver to her (manage versions and doc.); in addition, she has the softgoal artifacts [$1...$6] easily understandable[2].

PRG Manager delegates two goals to the Analyst: feasibility study [$1] delivered and requirements [$2] modeled. The PRG Manager depends on the Formal Engineer for the goal formal analysis [$3] delivered; in order to fulfill formal analysis [$3] delivered the Formal Engineer relies on the Analyst for the resource Tropos requirement model. In a resource depen-

---

[2] Note that the place holders $1...$6 refer to objects mentioned in particular goals of the PRG Manager, namely feasibility study [$1] delivered, requirements [$2] modeled, formal analysis [$3] delivered, architectural design [$4] specified, detailed design [$5] specified, and code [$6] developed.

dency, the depender depends on the dependee for the availability of a physical or informational entity (dependum). Along a similar path, the PRG Manager depends on the Architecture Designer for achieving the architectural design [$4] specified, and Architecture Designer depends on the Analyst for the Tropos requirement model. Another relevant position corresponds to the Details Designer, who has to produce a detailed design (detailed design [$5] specified); also, she depends on the Architecture Designer for the Tropos architectural design. Finally, PRG Manager depends on the Developer in order to develop the source code (code [$6] developed).

The next step in the analysis is the decomposition of each goal from the point of view of the actor who committed for its fulfillment. Goals and plans (plan represents a set of actions which allow for the satisfaction of one or more goals) are analyzed from the perspective of each specific actor in turn, by using three basic analysis techniques: *means-ends analysis, contribution analysis,* and *AND-OR decomposition* [16,6,20]. For goals, means-ends analysis aims at identifying goals, plans, resources and softgoals that provide means for achieving the goal (the end). Contribution analysis allows the designer to point out goals, softgoals, and plans that can contribute positively or negatively at reaching the goal under analysis. AND-OR decomposition allows for a combination of AND and OR decompositions of a root goal into subgoals, thereby refining a goal structure.

As depicted in Fig. 2, the goal requirements [$2] modeled delegated from the PRG Manager to the Analyst is the "end" in a means-ends analysis; of course, the chosen "mean" is the goal build a requirements Tropos model, that makes explicit our implicit "strong" requirement of building a Tropos oriented tool. The goal build a requirement Tropos model is then AND decomposed into the four subgoals: learn Tropos language, model managed, model documented, and model reasoned.

The softgoal exhaustive guide contributes positively at satisfying the goal learn Tropos language. In order to build a requirements Tropos model (model managed), it is indispensable to create a new model, manage an exist model, add/edit/remove some elements of the model and so on. All this corresponds to the goal diagrams edited[3]. In addition, some goals concerning the most important features of Tropos, already introduced in Sect. 3, are also depicted here, that are: retrotraceability managed, in order to deal with backward and multiple phases traceability, ticks marked, in order to deal with the ticks mechanism for the check of the completeness of model analysis, and syntax of model checked, to verify the syntax of the model with respect to the metamodel as described in [20]. Finally, also multiple views management is considered (multiple views visualized).

For the ticks mechanism, it is useful to take into account the softgoals contribution, represented by the softgoal evaluable softgoal. A first proposal for partially fulfilling model managed relied on the use of DIA[4] graphic tool (DIA

---

[3] Further decomposition of this and other goals are not provided because of space limits. See [5,2] for further details.

[4] Dia is a graphic tool that supports many language like UML, ER, and so on [14].

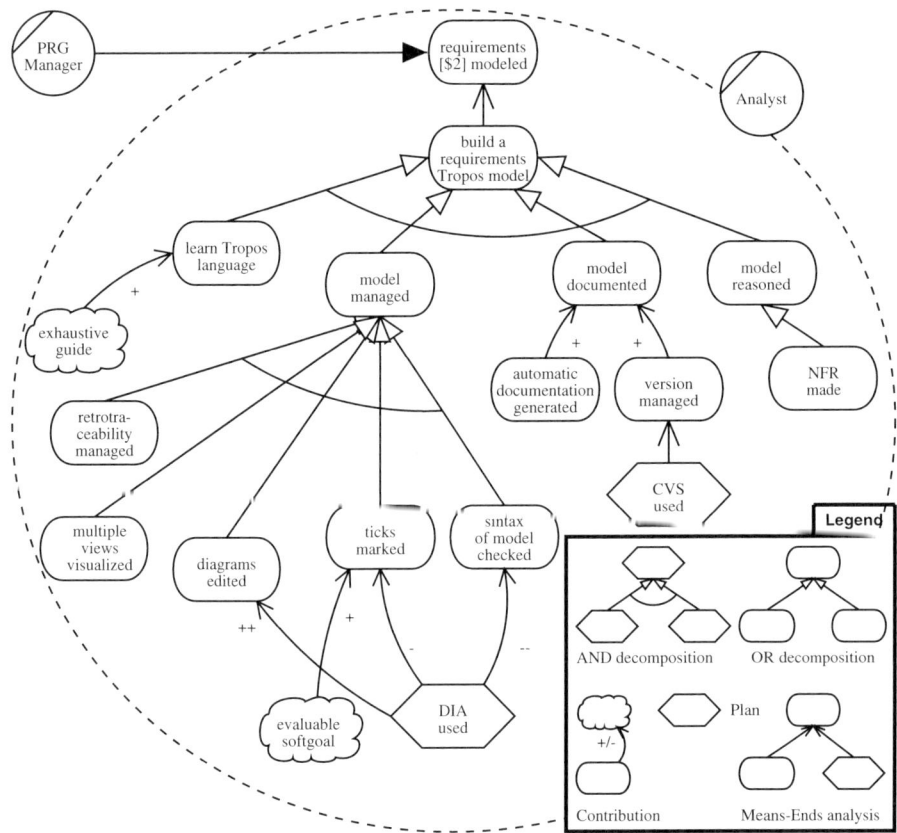

Fig. 2. Goal analysis from the perspective of the `Analyst`

used). Although this solution could satisfy `diagrams edited` (note the positive contribution ++), it does not satisfy `ticks marked` and `syntax of model checked` (note the negative contribution − and −−, respectively).

The subgoals `automatic documentation generated` and `version managed` contribute positively to the achievement of the goal `model documented`. The first subgoal concerns with the automatic generation of the documentation about the conceptual model and/or some views on it, while the second regards the management of different versions for each analyst. A mean for the fulfillment of the last subgoal is to integrate the CVS software system in the Tropos tool. Finally, the last subgoal `model reasoned` can be satisfied with a *Non-Functional Analysis*, as proposed in [6] (`NFR made`).

Figure 3 shows the goal diagram from the point of view of the position `Architecture Designer`. The goal delegated by `PRG Manager` (see Figure 1) is labeled with `architectural design [$4] specified`. This is AND decomposed into three subgoals: `choose pattern`, `choose paradigm`, and `requirements`

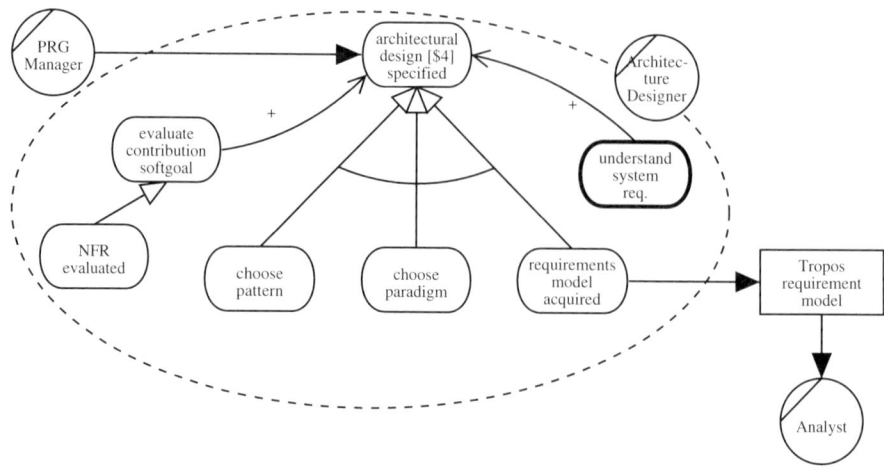

Fig. 3. A goal diagram including the `Architecture Designer`

`model acquired`. This last subgoal may be considered as depending on the
resource `Tropos requirements model` that, by further analysis, can be assumed as to be delegated to the `Analyst`. `Choose paradigm` concerns with
the selection among several architectural paradigms (e.g., BDI architecture).
Aside the AND decomposition, the root goal `architectural design [$4]`
`specified` also receives positive contributions from two goals: `understand`
`system requirements` and `evaluate contribution softgoal`. The first goal
was originally foreseen as a proper goal of `Architecture Designer` (see Fig. 1).
This fact is evidenced here, as well as in analogous situations in other figures,
by using a thicker borderline.

The second contributing goal is further enabled by `NFR evaluated`. Due to
lack of space, the goal diagrams of `Details Design`, `Formal Engineer`, and
`Developer` are not reported here, but can be found in [5,2].

3.2   Revising the Analysis

The previous section presented a first result of the activity of requirements definition and analysis. It is obvious that the diagrams developed are not sufficiently
detailed. For this reason, iterative steps of incremental refinement of the model
have to be performed. As already mentioned in the introduction, this way of
proceeding is a typical feature of the Tropos methodology. The final setting of
each Tropos phase may be reached possibly after several refinements, in each of
which not only new details may be added, but, also, already present elements
and dependencies can be revised or even deleted [3]. This iterative process not
only may require intra-phase refinements, but, possibly, also revisions of artifacts produced during earlier phases (inter-phases refinements). The importance
of retrotraceability is, here, evident.

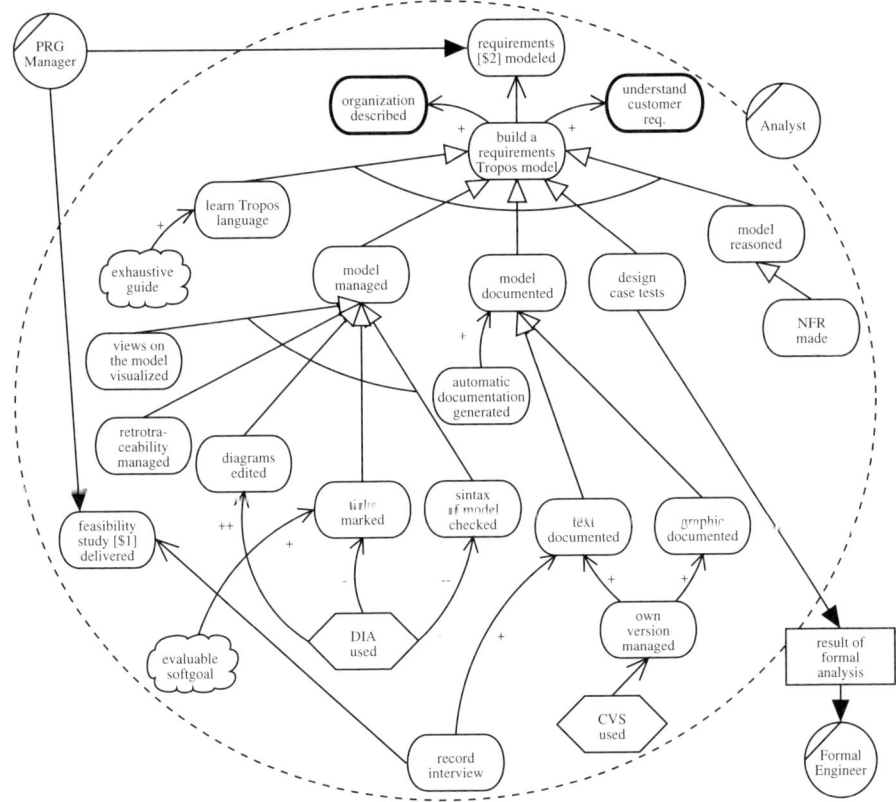

Fig. 4. Revising Goal analysis from the point of view of the Analyst

Just as an example of the intra-phase refinement activity, the revision of the goal diagram in Fig. 2 is presented in Fig. 4.

The positive contributions from build a requirements Tropos model to the goals organization described and understand customer requirements are the first relevant differences. The two positively contributed goals – namely, organization described and understand customer requirements – were initially considered as original Analyst's goal (see Fig. 1), but later not further analyzed (in fact, they are not taken into account in Fig. 2), until the revision depicted in Fig. 4. Thus, this revision is necessary to complete the analysis of the requirements initially introduced in Fig. 1.

Among other new elements added by the revision, let us note the subgoal design case tests, considered necessary because the analyst has also to design the cases to be tested in order to validate and verify the functionalities of the software system with respect to the functional and non-functional requirements. Another source of revision derives from the observation that the evaluation of the formal analysis conducted by Formal Engineer can provide more hints to the Analyst; thus, Fig. 4 also introduces a dependence between the Analyst and the Formal Engineer, upon the resource result of formal analysis, motivated by the fulfillment of design case tests. Another sub-goal of build a requirements Tropos model, namely model documented, has been OR decomposed into text documented and graphic documented. Finally, the goal feasibility study [$1] delivered (see Fig. 1), became the end in a means-ends analysis where the "mean" is the goal record interview, which, also, contributes positively for the fulfillment of the text documented.

4   Late Requirements

During late requirement analysis the system-to-be (the Tropos tool in our case) is described within its operating environment, along with relevant functions and qualities. The system is represented as one actor which have a number of dependencies with the other actors of the organization. These dependencies define all the functional and the non-functional requirements for the system-to-be.

A very important feature of Tropos is that the type of representations adopted for the early requirements (that are, the actor and the goal diagrams) are used in the same way also during the late requirement analysis. The only different element is that, here, the system-to-be is introduced in the analysis as one of the actors of the global environment. Of course, the goal of the engineer, since now on, is to decompose and analyze in details the system goals. To do this, a sequence of more and more precise and refined goal diagrams for the actor Tropos tool, in our case, have to be produced, applying the iterative refinement process already introduced in the previous section.

Of course, during the goal analysis of the system-to-be some differences with respect to the early analysis may be taken into account, as, for example, the fact that the system-to-be is characterized by a much different level of intentionality and autonomy if compared with the social actors (it may be assumed that the system is much more prone to fulfill the delegated goals and tasks – it is built for this – than a social actor, and that it has no backward dependency on the social environment, unless for resources representing I/O flow)[5]. Also, the analysis of the system goals should be carried on within the scope of the system as far as possible, trying to minimize revisions of decisions already taken during early requirements analysis. Finally, the system-to-be is characterized only by delegated goals, which are, then, exclusively generated by the early requirement analysis in this simplified setting.

Due to lack of space, and considering, apart the principled differences listed above, that there is no technical difference in the process of late requirement

---

[5] Actually, some more dependencies could be introduced, especially if we consider the system in relations with other systems or its subsystems. This possibility, indeed, allow us for an interesting level of analysis for, e.g., the so called *autonomous systems*. In any case, the choice of relating the system with other systems, or decompose it into subsystems, comes later in the process, and falls out of the scope of the present paper.

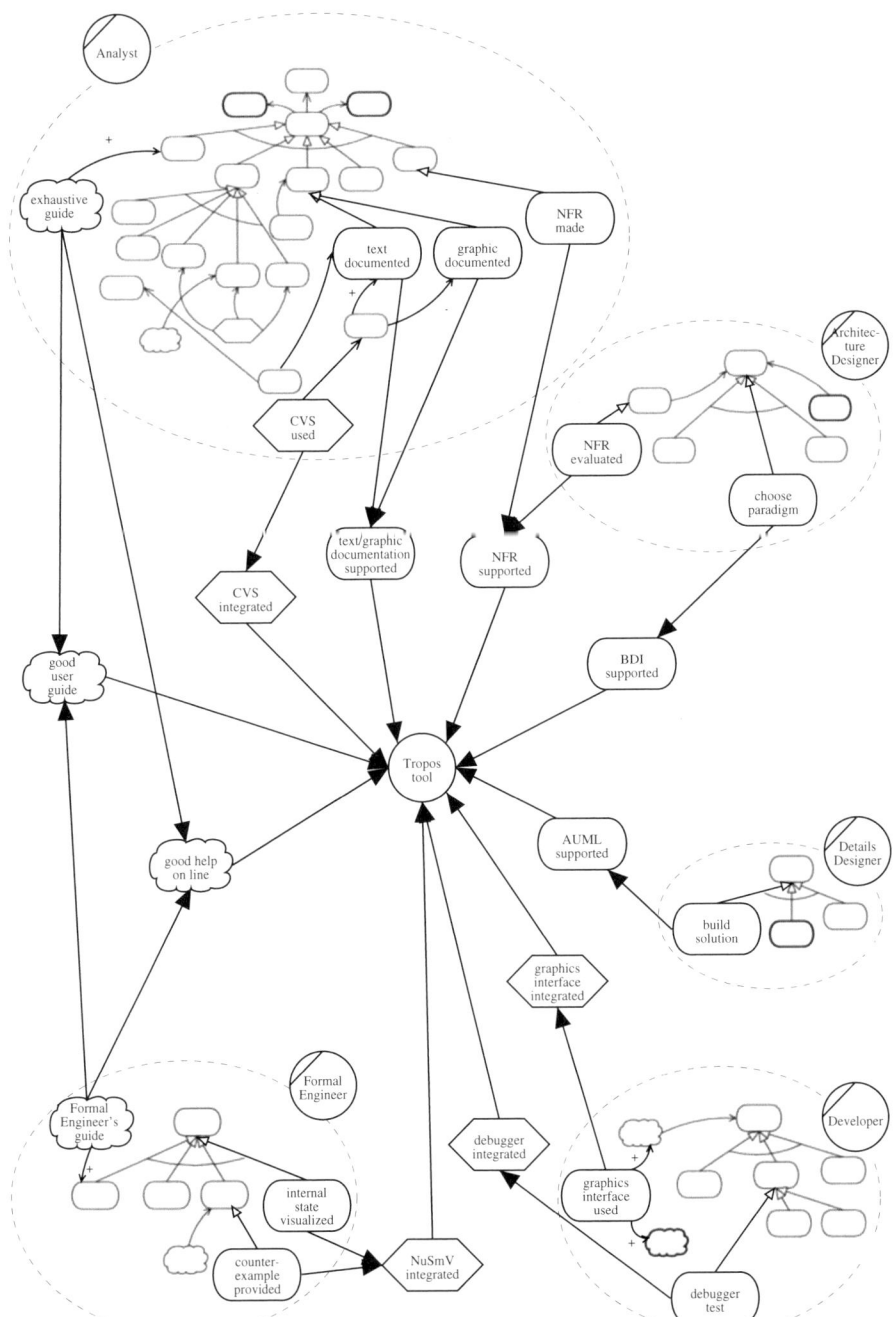

Fig. 5. Actor diagram: focus on the system actor `Tropos tool`

analysis with respect to the process of early requirement analysis, only a short description of the first steps is introduced here. In particular, it is relevant to show the very beginning of the late requirement analysis, and how a set of goals can be assigned to the system-to-be starting from the previous analyses. In fact, as mentioned above, the system goals, softgoals and plans have to be motivated by the unresolved goals, softgoals and plans elicited in the early requirements. At this end, Fig. 5 highlights which goals, softgoals and plans in the social actors goal diagrams may be satisfied by means of strategic dependencies on the `Tropos tool`. It is worth noticing that not necessarily an unresolved social actor goal has to be resolved through a goal dependency on the system; instead, it may generate for example, a plan dependency, as happens for `debugger integrated`. The same may apply to all the kinds of dependum.

Referring in detail at Fig. 5, it can be seen that `Analyst` depends on the `Tropos tool` for `NFR supported`, as well as the `Architecture Designer` does, although for fulfilling a different internal goal (`NFR evaluated` instead of `NFR made`). `Analyst` also depends on the goal `text/graphic documentation supported` in order to contribute to `text documented` and to `graphic documented`, on the plan `CVS integrated` in order to fulfill the plan `CVS used`, and, finally, on the softgoals `good user guide` and `good help on line`. As another example, let us consider the position `Formal Engineer`, whose diagram had not been presented in the previous pages due to lack of space. She depends on the `Tropos tool` for the plan `NuSMV integrated`, with the motivations of `internal state visualized` and `counter-example provided`. Other examples, not commented here, are shown in Fig. 5, which, of course, has the only aim of exemplifying the process, and has not to be considered exhaustive.

5   Conclusion

In the present paper the definition process of a Tropos tool, which is currently in phase of analysis and design at IRST, has been used as a case study for presenting some features of the Tropos methodology itself.

First, a stronger emphasis with respect to previous papers [11,12,18,19,4] has been put on the process that lead from the very informal elicitations of requirements to their descriptions in terms of actor and goal diagrams. Also, the process of goals definition, analysis, and revision has been presented, pointing out, in particular, how the goal diagram construction has to be considered as an incremental process, based on a sequence of revision steps.

Finally, the last figure shown in the paper is aimed at giving an introductive hint on how the transition from the early requirement analysis to the late requirement analysis can be seen as part of a smooth and natural process.

The management of traceability has been raised as a crucial point for correctly dealing with the revision (specially the inter-phase revision) process. In future works, we aim at further developing this issue with other specifically focused case studies and examples. Nevertheless, we believe that clearly showing the connecting items (dependums) between early and late requirements, as done in Fig. 5, already provides for a first step towards the solution. Of course, the best support can be reached only through the development of an appropriate tool: the Tropos tool.

Acknowledgments. Our acknowledgments are addressed to all the IRST people working at the Tropos tool project. A special thanks is due to Eric Yu, who provided precious suggestions and comments during very preliminary discussions on the idea. We would also like to thank Angelo Susi and Paolo Busetta for comments on drafts of this paper.

References

1. B. Bauer, J. P. Müller, and J. Odell. Agent UML: A formalism for specifying multiagent software system. In P. Ciancarini and M. Wooldridge, editors, *Agent-Oriented Software Engineering - Proceedings of the First International Workshop (AOSE2000)*, volume 1957, pages 91–103, Limerick, Ireland, June 2000. Springer-Verlag Lecture Notes in Computer Science.
2. D. Bertolini, P. Bresciani, A. Daprà, A. Perini, and F. Sannicolò. Requirement Specification of a CASE tool supporting the Tropos methodology. Technical Report 0203-01, ITC-irst, via Sommarive, Povo, Trento, January 2002.
3. P. Bresciani, A. Perini, P. Giorgini, F. Giunchiglia, and J. Mylopoulos. Modelling early requirements in Tropos: a transformation based approach. In Wooldridge et al. [22], pages 151–168.
4. P. Bresciani and F. Sannicolò. Applying Tropos Requirements Analysis for defining a Tropos tool. In P. Giorgini, Y. Lespérance, G. Wagner, and E. Yu, editors, *Agent-Oriented Information System. Proceedings of AOIS-2002: Fourth International Bi-Conference Workshop*, pages 135–138, Toronto, Canada, May 2002.
5. P. Bresciani and F. Sannicolò. Applying Tropos to requirement analysis for a Tropos tool. Technical Report 0204-01, ITC-irst, via Sommarive, Povo, Trento, April 2002.
6. L. K. Chung, B. A. Nixon, E. Yu, and J. Mylopoulos. *Non-Functional Requirements in Software Engineering*. Kluwer Publishing, 2000.
7. A. Cimatti, E. M. Clarke, F. Giunchiglia, and M. Roveri. NuSMV: a new symbolic model checker. *International Journal on Software Tools for Technology Transfer (STTT)*, 2(4):410–425, March 2000.
8. A. Dardenne, A. van Lamsweerde, and S. Fickas. Goal-directed requirements acquisition. *Science of Computer Programming*, 20(1–2):3–50, 1993.
9. P. Cederqvist et al. *Version Management with CVS*. http://www.cvshome.org/docs/manual/.
10. A. Fuxman, M. Pistore, J. Mylopoulos, and P. Traverso. Model Checking Early Requirements Specifications in Tropos. In *Proceedings Fifth IEEE International Symposium on Requirements Engineering (RE01)*, pages 174–181, Toronto, Canada, August 2001.
11. P. Giorgini, A. Perini, J. Mylopoulos, F. Giunchiglia, and P. Bresciani. Agent-oriented software development: A case study. In *Proceedings of the Thirteenth International Conference on Software Engineering - Knowledge Engineering (SEKE01)*, pages 283–290, Buenos Aires - ARGENTINA, June 2001.

12. F. Giunchiglia, A. Perini, and J. Mylopoulus. The Tropos Software Development Methodology: Processes, Models and Diagrams. In C. Castelfranchi and W.L. Johnson, editors, *Proceedings of the first international joint conference on autonomous agents and multiagent systems*, pages 63–74, palazzo Re Enzo, Bologna, Italy, July 2002. ACM press. Featuring: 6th International Conference on Autonomous Agents, 5th International Conference on MultiAgents System, and 9th International Workshop on Agent Theory, Architectures, and Languages.

13. F. Giunchiglia, A. Perini, and F. Sannicolò. Knowledge level software engineering. In J.-J.C. Meyer and M. Tambe, editors, *Intelligent Agents VIII*, LNCS 2333, pages 6–20, Seattle, WA, USA, August 2001. Springer-Verlag.

14. Gnome. *Dia Tutorial*.
    http://www.lysator.liu.se/~alla/dia/diatut/all/all.html.

15. Knowledge Management Lab at the University of Toronto. *OME3 Documentation*.
    http://www.cs.toronto.edu/km/ome/docs/manual/manual.html.

16. J. Mylopoulos, L. Chung, S. Liao, H. Wang, and E. Yu. Exploring Alternatives during Requirements Analysis. *IEEE Software*, 18(1):92–96, February 2001.

17. J. Odell, H. V. D. Parunak, and B. Bauer. Extending UML for Agents. In G. Wagner, Y. Lesperance, and E. Yu, editors, *Proc. of Agent-Oriented Information System Workshop at the 17th National conference on Artificial Intelligence*, pages 3–17, Austin, TX, 2000.

18. A. Perini, P. Bresciani, P. Giorgini, F. Giunchiglia, and J. Mylopoulos. Towards an Agent Oriented approach to Software Engineering. In A. Omicini and M. Viroli, editors, *WOA 2001 – Dagli oggetti agli agenti: tendenze evolutive dei sistemi software*, Modena, Italy, 4–5 September 2001. Pitagora Editrice Bologna.

19. A. Perini, P. Bresciani, F. Giunchiglia, P. Giorgini, and J. Mylopoulos. A Knowledge Level Software Engineering Methodology for Agent Oriented Programming. In J. P. Müller, E. Andre, S. Sen, and C. Frasson, editors, *Proceedings of the Fifth International Conference on Autonomous Agents*, pages 648–655, Montreal CA, May 2001.

20. F. Sannicolò, A. Perini, and F. Giunchiglia. The Tropos modeling language. A User Guide. Technical Report 0204-13, ITC-irst, January 2002.

21. A. H. Watt. *Designing SVG web graphics*. D. Dwyer, 2002.

22. M.J. Wooldridge, G. Weiß, and P. Ciancarini, editors. *Agent-Oriented Software Engineering II*. LNCS 2222. Springer-Verlag, Montreal, Canada, Second International Workshop, AOSE2001 edition, May 2001.

23. E. Yu. Modeling Organizations for Information Systems Requirements Engineering. In *Proceedings First IEEE International Symposium on Requirements Engineering*, pages 34–41, San Jose, January 1993.

24. E. Yu. *Modelling Strategic Relationships for Process Reengineering*. PhD thesis, University of Toronto, Department of Computer Science, University of Toronto, 1995.

25. E. Yu. Software Versus the World. In Wooldridge et al. [22], pages 206–225.

26. E. Yu and J. Mylopoulos. Understanding 'why' in software process modeling, analysis and design. In *Proceedings Sixteenth International Conference on Software Engineering*, pages 159–168, Sorrento, Italy, May 1994.

A Mechanism for Dynamic Role Playing

V. Hilaire, A. Koukam, and P. Gruer

UTBM
Systems and Transports Laboratory
90010 Belfort Cedex
FRANCE
vincent.hilaire@utbm.fr
tel: +33 384 583 009 - fax +33 384 583 342

Abstract. The work presented in this article is based upon a methodological approach for building Multi-Agent Systems specifications. The basic idea is to define such systems as a set of entities playing roles which have interactions between them. We present a mechanism for dynamic role-playing specification within a formal framework. The framework use a formalism which can express Multi-Agent Systems aspects. This formalism composes Object-Z and statecharts. The main features of this approach are: enough expressive power to obtain unbiased specifications, tools for specification analysis and refinement mechanisms allowing the refinement of a high level specification into a low level specification which can be easily implemented.

Keywords: Multi-agent systems, formal specification, methodology

1 Introduction

Software agents and multiagent systems have become an appealing paradigm for the design of computer systems composed of autonomous cooperating software entities. This paradigm consists in new ways for analyzing, designing and implementing such systems based upon the central notions of agents, their interactions and the environment which they perceive and in which they act. There exists several software engineering approaches [24] which try to fill the gap of MAS analysis and development life-cycle.

These approaches differ in many ways like for example: the notation employed, the basic concepts used and the presence or absence of a methodology. The notation may be formal or not, in this paper we are interested in formal notations which allow formal specification of MAS and which guide implementation phase. Indeed, formal specification approaches may authorize validation and verification of the specification. The process of validation and verification provides a support for incremental specification leading to an executable model of the system being built.

Due to their complexity, MAS have reactive and transformational features. A formalism which: specifies easily and naturally both aspects, enables validation and verification and guides the implementation phase is to be defined yet.

R. Kowalczyk et al. (Eds.): Agent Technology Workshops 2002, LNAI 2592, pp. 36–48, 2003.

We have thus chosen to use a multi-formalisms approach that results of the composition of Object-Z [11] and statecharts [19]. This formalism enables the specification of reactive and transformational aspects of MAS.

A specification method is essential to manage MAS complexity by decomposition and abstraction. Some approaches use organizational concepts to model MAS [12,31,22]. The use of such primitive concepts enable to go from the requirements to detailed design and helps to decompose a MAS in terms of roles and organizations. In fact, it is a three step approach. The first step views the system as an organization or a society defined by a set of roles and their interactions. The second introduces the agents and assigns roles to them according to some design criteria. The third focus on the design of the internal architecture of agents. With these concepts we have defined a framework, named RIO which stands for Role-Interaction-Organization, composed of Object-Z classes which specify each informal concept. In doing so we have methodological guidelines formally grounded for the analysis and design of MAS. We have used the RIO framework on several problems. We have, for example, specified a MAS for solving a radiomobile network field problem [27]. We have also animated a RIO model of foot-and-mouth disease simulation [22].

One drawback of organizational based methodology is that they frequently restrict to static role-playing relationship. The purpose of this paper is to extend our approach to deal with dynamic role-playing relationship.

The idea that we suggest for dynamic role-playing relationship is inspired of viewpoint specification. Viewpoint techniques have been widely used in requirement analysis [1,8]. Indeed, viewpoint specifications consist in separating partial specifications of large and complex systems. There has been some work on combination of specifications. The problem of combining dynamically roles is similar to combining viewpoints. We have to merge several specifications and check their consistency. In this paper, due to lack of place, we deal only with combination of the Object-Z part of the specification.

The rest of the paper is organized as follows. Section 2 introduces the extended RIO framework, Sect. 3 illustrates the dynamic role-playing relationship on an example, Sect. 4 presents related works and eventually Sect. 5 concludes.

2   RIO Framework

This section presents the RIO framework concepts and their specification in Object-Z. Indeed, the RIO framework is composed of two parts. The first part is semi-formal and allows the representation of MAS as systems of agents playing interacting roles in several organizations by the way of graphic diagrams. The second part gives formal meaning and notations for the concepts depicted by the diagrams. In this paper we insist mainly on the latter part.

The first subsection introduces the organizational level which roughly correspond to the analysis level and the second subsection introduces agentification level which belongs to the design level.

Each concept is specified by an Object-Z class which takes the form of a graphic named box. The specification unit of Object-Z is the schema represented by inner boxes in a class. Each class owns several schemas which are for example state schema and operation schemas. A schema is divided in two parts by a short line. The upper part consists of declaration in a set theory fashion. The part below the short line consists of constraints, in first order predicate logic.

2.1   Organizational Level

We base our work on the RIO framework defined in [22]. This framework had three main concepts which was: role, interaction and organization. We have modified these concepts to take into account dynamic role-playing aspects. In fact we have modified relationships between classes of the meta-model. We have generalized attributes, stimulus and actions in a generic entity named *Actor*. *Role* and *Agent* classes inherit from *Actor*. We have also introduced a specific class *RoleContainer* which represents set of roles played simultaneously. The Fig. 1 describes with UML notation the relationships between these classes. This meta-model is similar to the one used by Yu [32] for his i^* framework which is used in many MAS methodology like the one presented in [4].

An actor is an abstraction of a behavior or a status. We have chosen to specify it by the *Actor* class. This class represents characteristic set of attributes whose elements are of [*Attribute*] type. These elements belong to the *attributes* set. An actor is also defined by stimulus it can react to and actions it can execute. They are respectively specified by *stimulus* set and *actions* set. The [*Attribute*], [*Event*] and [*Action*] types are defined as given types which are not defined further.

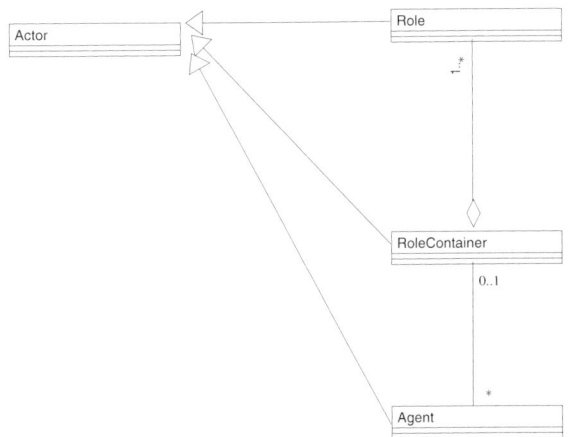

Fig. 1. RIO meta-model

Actor
> attributes : $\mathbb{P}\ Attribute$
> stimulus : $\mathbb{P}\ Event$
> actions : $\mathbb{P}\ Action$

The *Role* class inherits from *Actor*, This is stated by the first line of the class. The reactive aspect of a role is specified by the sub-schema *behavior* which includes a statechart. It is to say that *behavior* specifies the different states of the role and transitions among these states. It is in the *behavior* sub-schema that we include statecharts. The *obtainConditions* and *leaveConditions* attributes specify conditions requested to obtain and leave the role. These conditions require specific capabilities or features to be present in order to play or leave the role. Stimulus which trigger a reaction in the role behavior must appear in one transition at least. The action belonging to the statechart transitions must belong to the *actions* set. The constraints specified in the *Role* class use the heterogeneous basis concepts in order to ensure coherence between Object-Z and statechart parts.

Role
> Actor
>
> **behavior**
>
> obtainConditions, leaveConditions : Condition
>
> $\forall\ s \in stimulus, \exists\ e \in behavior.\rho \bullet$
> $\quad (\exists\ t \in e.transitions \bullet t.label.event = s)$
> $\forall\ e \in behavior.\rho \bullet$
> $\quad (\forall\ t \in e.transitions \bullet t.label.action \subseteq actions)$

An interaction is specified by a couple of role which are the origin and the destination of the interaction. The role *orig* and *dest* interacts by the way of operations op_1 and op_2. These operations are combined by the $\|$ operator which equates output of op_1 and input of op_2. In order to extend interaction to take into account more than two roles or more complex interactions involving plan exchange one has to inherit from *Interaction*.

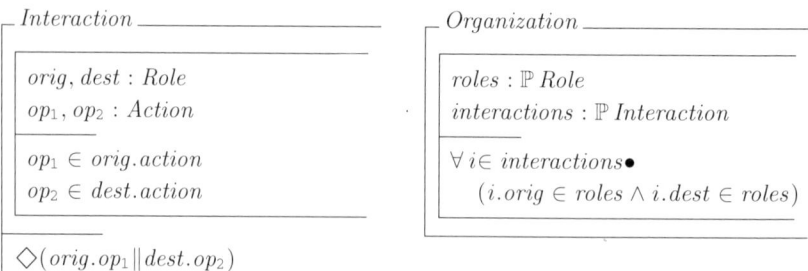

Interaction
> orig, dest : Role
> op_1, op_2 : Action
>
> $op_1 \in orig.action$
> $op_2 \in dest.action$
>
> $\Diamond(orig.op_1 \| dest.op_2)$

Organization
> roles : $\mathbb{P}\ Role$
> interactions : $\mathbb{P}\ Interaction$
>
> $\forall\ i \in interactions \bullet$
> $\quad (i.orig \in roles \wedge i.dest \in roles)$

An organization is specified by a set of roles and their interactions. Interactions happen between roles of the concerned organization. It is to say that for each interaction of the *interactions* set the roles of the interaction must belong to *roles* set of the organization. Moreover, each role must be part of at least one interaction.

A *RoleContainer* specifies an entity which plays a set of roles. The role-playing relationship is static. In other words the set of played roles does not change. In *RoleContainer* the link between entities which play roles and roles specification is precised. Indeed, three functions f_1, f_2 et f_3 map respectively attributes, stimulus and actions of roles and attributes, stimulus and actions of the entity playing the roles. These functions impose that for each attribute, stimulus or action of a role there exists an attribute, stimulus or action which refines it in the *RoleContainer*. The refinement relationship is denoted by the $\preccurlyeq$ symbol. An object o' is said to refine an object o is represented as $o \preccurlyeq o'$.

__ *RoleContainer* _____

  Actor

    ―――――――――――――――――――――――――――――

    playing : $\mathbb{P}$ *Role*

    ―――――――――――――――――――――――――――――

    $\exists f_1$: *Attribute* $\mapsto$ *Attribute* $\bullet$
        $\forall r \in playing \bullet$
        $\forall a \in r.attributes \, \exists a' \in attributes \bullet a' = f_1(a) \land a \preccurlyeq a'$
    $\exists f_2$: *Action* $\mapsto$ *Action* $\bullet$
        $\forall r \in playing \bullet$
        $\forall a \in r.actions \, \exists a' \in actions \bullet a' = f_2(a) \land a \preccurlyeq a'$
    $\exists f_3$: *Stimulus* $\mapsto$ *Stimulus* $\bullet$
        $\forall r \in playing \bullet$
        $\forall a \in r.stimulus \, \exists a' \in actions \bullet a' = f_3(a) \land a \preccurlyeq a'$

This mechanism allows the specification of several, possibly overlapping, roles to be merged in one entity. There are three different combination cases. We take the attributes example to illustrate them. The first case is when attributes are completely distincts the mapping is then the identity function. The second case is when two attributes specifying the same *RoleContainer* attribute are defined with different definition domains of the same type. The function must map these attributes to a *RoleContainer* attribute which is defined over the union of the two definition domains. The third case is when two attributes specifying the same *RoleContainer* attribute are defined with different definition domains but these domains aren't of the same type. The function must map the two attributes to a subset of the cartesian product of the definition domains.

```
┌─ Agent ─────────────────────────────────────────────────────────
│ Actor
│ ┌──────────────────────────────────────────────────────────────
│ │ acquaintances : ℙ Agent
│ │ position : RoleContainer
│ ├──────────────────────────────────────────────────────────────
│ │ ∀ a ∈ acquaintances, ∃ r₁, r₂ : Role,
│ │        ∃ i : Interaction •
│ │        r₁ ∈ position.playing
│ │        ∧ r₂ ∈ a.position.playing
│ │        ∧ (r₁, r₂) ∈ i.roles
│ │ ∀ r ∈ position • r.behavior ⊆ behavior
│ └──────────────────────────────────────────────────────────────
│
│ ┌─ addRole ────────────────────────────────────────────────────
│ │ Δ(position)
│ │ r? : Role
│ ├──────────────────────────────────────────────────────────────
│ │ position.playing′ = position.playing ∪ r?
│ │ r.obtainConditions
│ │ ∀ a ∈ r.attributes • (position.f₁′ = position.f₁ ⊕ {r.a, a′})
│ │        ∧ ((a′ ∈ attributes ∧ a ≼ a′)
│ │          ∨ (attributes′ = attributes ∪ {a′} ∧ a ≼ a′))
│ │ ∀ a ∈ r.actions • (position.f₂′ = position.f₁ ⊕ {r.a, a′})
│ │        ∧ ((a′ ∈ actions ∧ a ≼ a′)
│ │          ∨ (actions′ = actions ∪ {a′} ∧ a ≼ a′))
│ │ ∀ a ∈ r.stimulus • (position.f₃′ = position.f₃ ⊕ {r.a, a′})
│ │        ∧ ((a′ ∈ stimulus ∧ a ≼ a′)
│ │          ∨ (stimulus′ = stimulus ∪ {a′} ∧ a ≼ a′))
│ └──────────────────────────────────────────────────────────────
│
│ ─────────
│ □(∀ r ∈ position.playing, ∀ e ∈ r.behavior.ϱ •
│                     e.instate
│                     ∧ (∃ t : Transition • t.source = e) ⇒
│                     (t.event
│                       ∧ t.conditions ⇒
│                       (∀ f ∈ t.destinations • ○f.instate))
│ □(∀ r : Role • (r ∉ playing) ∧ ○(r ∈ playing) ⇒ r.obtainConditions)
│ □(∀ r : Role • (r ∈ playing) ∧ ○(r ∉ playing) ⇒ r.leaveConditions)
└──────────────────────────────────────────────────────────────────
```

Where the math renders as:

$\forall a \in acquaintances, \exists\, r_1, r_2 : Role,$
$\quad \exists\, i : Interaction \bullet$
$\quad r_1 \in position.playing$
$\quad \wedge\, r_2 \in a.position.playing$
$\quad \wedge\, (r_1, r_2) \in i.roles$
$\forall r \in position \bullet r.behavior \subseteq behavior$

$position.playing' = position.playing \cup r?$
$r.obtainConditions$
$\forall a \in r.attributes \bullet (position.f_1' = position.f_1 \oplus \{r.a, a'\})$
$\quad \wedge ((a' \in attributes \wedge a \preccurlyeq a')$
$\quad\quad \vee (attributes' = attributes \cup \{a'\} \wedge a \preccurlyeq a'))$
$\forall a \in r.actions \bullet (position.f_2' = position.f_1 \oplus \{r.a, a'\})$
$\quad \wedge ((a' \in actions \wedge a \preccurlyeq a')$
$\quad\quad \vee (actions' = actions \cup \{a'\} \wedge a \preccurlyeq a'))$
$\forall a \in r.stimulus \bullet (position.f_3' = position.f_3 \oplus \{r.a, a'\})$
$\quad \wedge ((a' \in stimulus \wedge a \preccurlyeq a')$
$\quad\quad \vee (stimulus' = stimulus \cup \{a'\} \wedge a \preccurlyeq a'))$

$\Box(\forall r \in position.playing, \forall e \in r.behavior.\varrho \bullet$
$\quad\quad\quad\quad\quad e.instate$
$\quad\quad\quad\quad\quad \wedge (\exists\, t : Transition \bullet t.source = e) \Rightarrow$
$\quad\quad\quad\quad\quad (t.event$
$\quad\quad\quad\quad\quad\quad \wedge t.conditions \Rightarrow$
$\quad\quad\quad\quad\quad\quad (\forall f \in t.destinations \bullet \bigcirc f.instate))$
$\Box(\forall r : Role \bullet (r \notin playing) \wedge \bigcirc(r \in playing) \Rightarrow r.obtainConditions)$
$\Box(\forall r : Role \bullet (r \in playing) \wedge \bigcirc(r \notin playing) \Rightarrow r.leaveConditions)$

2.2   Agentification Level

The *Agent* class inherits from *Actor*. This class is defined by a *position* which is an instance of *RoleContainer*. The agent position is the set of roles it plays at a moment in time. These roles define the agent status and behavior in all context it may intervene. The position of an agent may change during it lifetime. This mechanism allows dynamic role-playing. The *addRole* (resp *leaveRole*) operation add (resp substract) a role to an agent. The preconditions of this operation impose that *obtainConditions* (resp *leaveConditions*) must be true. These operations modify the functions which map attributes, stimulus and actions from roles to agent. These functions defined statically in the *RoleContainer* class are modified dynamically whenever an agent change one of his role.

An agent *A* is also defined by an *acquaintances* set. This set represents the other agents which are currently interacting with *A*.

In this context, an agent is only specified as an active communicative entity which plays roles [12]. In fact agents instantiate an organization (roles and interactions) when they exhibit behaviors defined by the organization's roles and when they interact following the organization interactions. The main reason for this choice is that one can study agent behaviours and agent architecture separately. Indeed, the different roles an agent plays define it behaviour. The architectures used by agents may be different for the same behaviour and so it is sound to study it apart from the core agent behaviour. In [15] we have specified a specific agent architecture by extending the RIO framework classes.

3   Example

In this section we present a simplified example of a virtual university which allows office automation. The aim of this example is to illustrate the dynamic role playing mechanism. As the refinement mechanism for the statechart part is not presented in this paper we present only the Object-Z specification of the roles.

The first role we specify is the *Researcher* role. It is described by a set of *publications*, a set of *contracts*, the *diplomas* it owns and eventually the *institute* it works in.

The team leader is a researcher which leads a team composed of researchers.

Researcher	TeamLeader
Role	Researcher
$publications : \mathbb{P}\,Publication$	$members : \mathbb{P}\,Researcher$
$contracts : \mathbb{P}\,Contract$	$name : \mathbb{P}\,Team$
$diplomas : \mathbb{P}\{DEA, PhD, Hdr\}$	$(Hdr \in diplomas) \in obtainConditions$
$institute : \mathbb{P}\,Laboratory$	

A teacher may be involved in different courses described by the set *courses*. He is responsible of courses among the one he is involved in. He owns *diplomas* and he is attached to a teaching department.

$\boxed{\begin{array}{l} \text{\textit{Teacher}} \\ \hline \\ \textit{courses} : \mathbb{P}\ \textit{Courses} \\ \textit{responsible} : \mathbb{P}\ \textit{Courses} \\ \textit{diplomas} : \mathbb{P}\{\textit{ing}, \textit{DEA}, \textit{PhD}, \textit{Hdr}\} \\ \textit{institute} : \mathbb{P}\ \textit{Department} \\ \hline \\ \textit{responsible} \subseteq \textit{courses} \end{array}}$

The *TeacherResearcher* class specifies a *roleContainer* which plays the roles *Teacher* and *Researcher*. The attributes *diplomas* and institute *institute* are refined. In the first case the definition domain is extended by union of the two definition domain. In the second case the definition domain is the cartesian product of the definition domains of role attributes. All other attributes are mapped without change.

Eventually, the *TRVirtualUniversityAgent* specify an agent which can play the roles defined above. Each agent of this type owns several categories which enable, by the way of the *access* function, to determine which role he can play.

$\boxed{\begin{array}{l} \text{\textit{TeacherResearcher}} \\ \text{\textit{RoleContainer}} \\ \hline \\ \textit{diplomas} : \mathbb{P}(\{\textit{DEA}, \textit{PhD}, \textit{Hdr}\} \\ \qquad\qquad \cup \{\textit{ing}, \textit{DEA}, \textit{PhD}, \textit{Hdr}\}) \\ \textit{institute} : \textit{Department} \times \textit{Laboratory} \\ \hline \\ \textit{playing} = \{\textit{Teacher}, \textit{Researcher}\} \\ f_1(\textit{Teacher.diplomas}) \\ \quad = f_1(\textit{Researcher.diplomas}) \\ \quad = \textit{diplomas} \\ f_1(\textit{Teacher.institute}) = 1 \mid \textit{institute} \\ f_1(\textit{Researcher.institute}) = 2 \mid \textit{institute} \end{array}}$

$\boxed{\begin{array}{l} \text{\textit{TRVirtualUniversityAgent}} \\ \text{\textit{Agent}} \\ \hline \\ \textit{categories} : \mathbb{P}\ \textit{Category} \\ \textit{access} : \textit{Category} \mapsto \mathbb{P}\ \textit{Role} \\ \hline \\ \text{\_\_\textit{INIT}\_\_} \\ \forall\, r \in \textit{access}(\textit{categories}) \bullet \textit{addRole}(r) \\ \hline \\ \{\textit{Teacher}, \textit{Researcher}\} \in \textit{position.playing} \\ \hline \\ \text{\_\textit{addCategories}\_} \\ \Delta(\textit{categories}) \\ \textit{new?} : \textit{Category} \\ \hline \\ \textit{categories}' = \textit{categories} \cup \textit{new?} \\ \forall\, r \in \textit{access}(\textit{new?}) \bullet \textit{addRole}(r) \end{array}}$

4   Related Works

This section describes approaches for Agent-Oriented Software Engineering. We have divided these approaches in two parts: the semi-formal approaches and the formal ones.

4.1   Semi-formal Approaches

The i^* framework of Yu [32] is based upon a requirements analysis by mean of goals, dependencies, roles, actors, positions and agents. These notions are similar to the one we use. They are graphically presented on a same schema. It is difficult to read such schemas where all concepts are on the same level. We think that a schema which separates analysis and design level is better.

Kendall [25] suggests the use of extended Object Oriented methodologies like design patterns and CRC. CRC are extended to Role Responsibilities and Collaborators. In [26] a seven layered architectural pattern for agent is presented. The seven layers are: mobility, translation, collaboration, actions, reasoning, beliefs and sensory. This architecture is dedicated for mobile agents.

Cassiopeia [6] and Andromeda [10] use role notion and propose a step by step methodology in order to design MAS. Andromeda is an enrichment of Cassiopeia. Indeed, it deals with machine learning techniques for MAS. These methodologies are oriented towards the design of reactive MAS.

The MaSE methodology [7] insist on the necessity of software tools for software engineering, specifically code generation tools. This methodology has seven steps. These steps are structured in sequence. It begins with the identification of overall goals and their structuration. From goals with use case diagrams and sequence diagrams one can identify roles and define them as a set of tasks. Agents are then introduced with class diagrams with a specific semantics. The last steps consist in defining precisely: high level protocols, agent architectures and deployment diagram. For each step a different diagram is introduced. Moreover, MaSE methodology suffers of limited one-to-one agent interactions.

Bergenti and Poggi [2] suggest the use of four UML-like diagrams, like for example the class diagram [23]. These diagrams are modified in order to take into account MAS specific aspects. Among these specific MAS aspects there are conceptual ontology description, MAS architecture, interaction protocols and agent functionalities.

In [29], authors present an approach extending UML for represent agent relative notions. In particular, authors insist on role concept and suggest the use of modified sequence diagrams to deal with roles.

The problem with the two latter notations is the UML starting point. Indeed, using an object oriented notation in order to describe MAS lead specifiers to use object oriented concepts. We think that a MAS methodology must insist on agent oriented concepts first.

4.2   Formal Approaches

Formal approaches for MAS specification are numerous but they are often abstract and unrelated to concrete computational models [9]. Temporal modal logic, for example, have been widely used [30]. Despite the important contribution of these works to a solid underlying foundation for MAS, no methodological guidelines are provided concerning the specification process and how an implementation can be derived.

Another type of approach consists in using traditional software engineering or knowledge based formalisms [28]. One advantage of using such approaches is that they are more widely used and expertise concerning notations is greater then newer ones and there are tools which help the specification process.

For example, the approach proposed in [20] is based upon the refinement of informal requirement specifications to semi-formal and then formal specifications. The system structuration is based on a hierarchy of components [3]. These components are defined in term of input/ouput and temporal constraints. With this approach it seems difficult to refine down specifications to an implementation language. Moreover, the verification technique is limited to model checking.

Luck and d'Inverno [28] propose a formal framework which use the Z language. This framework is the starting point of any specification. It is composed of concepts to be refined in order to obtain a MAS specification. However, this approach has two drawbacks. First, the specifications unit is the schema. State spaces and operations of agents are separated. This drawback is avoided in our approach as we specify structure, properties and operations of an entity in a same Object-Z class. Second, the Luck and d'Inverno framework does not allow to specify temporal and reactive properties of MAS [13]. In our framework these aspects are specified by temporal invariants and statecharts.

Wooldridge, Jennings and Kinny [31] propose the Gaia methodology for agent oriented analysis and design. This methodology is composed of two abstraction levels: agent level and structural organizational level. The role concept exist in Gaia however the relationship between agent and role is static.

5   Conclusion

In this paper, we have presented a formal specification approach for MAS based upon an organizational model. The organizational model describes interaction patterns which are composed of roles. When playing these roles, agents instantiate interaction patterns. An agent can play several roles and can change the roles it plays. This model is well suited for describing complex interactions which are among MAS main features. The language used by the specification framework can describe reactive and functional aspects. It is structured as a classes hierarchy so one can inherit from these classes to produce its own specification. As an example we have specified the Virtual University Helper Agent. Several MAS have already been specified with the RIO framework with or without using a specific agent architecture [21,22,5,15]. These MAS have been applied to complex problems like radio-mobile network field [27], foot-and-mouth disease simulation [22] or office automation [14].

The used specification language allows prototyping of specification [22]. Prototyping is not the only means of analysis, indeed, in another work [16], we have introduced a formal verification approach. Moreover, the specification structure enables incremental and modular validation and verification through its decomposition. Eventually, such a specification can be refined to an implementation with multi-agent development platform like MadKit [18] which is based upon an organizational model [12].

Despite the encouraging results already achieved, we are aware that our approach still has some limitations. Indeed, it doesn't tackle all problems raised by a MAS development methodology. Among issues remaining for future work our organizational model needs more work to do ahead. The Object-Z part of the specification is not yet executable. However a preliminary work [17] has shown that it is possible to give an operational semantic to Object-Z but it must be strengthened. We are also working on the automation of the methodology steps. Indeed, in order to spread this methodology must be supported by software tools like code generation or specifications animation.

References

1. M. Ainsworth, A. H. Cruickshank, P. J. L. Wallis, and L. J. Groves. Viewpoint specification and Z. *Information and Software Technology*, 36(1):43–51, 1994.
2. Federico Bergenti and Agostino Poggi. Exploiting uml in the design of multi-agent systems. In Andrea Omicini, Robert Tolksdorf, and Franco Zambonelli, editors, *Engineering Societies in the Agents' World*, Lecture Notes in Artificial Intelligence. Springer Verlag, 2000.
3. F.M.T. Brazier, B. Dunin Keplicz, N. Jennings, and J. Treur. Desire: Modelling multi-agent systems in a compositional formal framework. *International Journal of Cooperative Information Systems*, 6:67–94, 1997.
4. P. Bresciani and F. Sannicolo. Requirements analysis in tropos: A self referencing example. 2002. In this volume.
5. R. Campero, P. Gruer, V. Hilaire, and P. Rovarini. Modeling and simulation of agent-oriented systems: an approach based on object-z and the statecharts. In Christoph Urban, editor, *Agent Based Simulation*, 2000.
6. Anne Collinot, Alexis Drogoul, and Philippe Benhamou. Agent oriented design of a soccer robot team. In Victor Lesser, editor, *ICMAS*. Springer Verlag, 1995.
7. Scoot DeLoach. Multiagent systems engineering: a methodology and language for designing agent systems. In *Agent Oriented Information Systems '99*, 1999.
8. J. Derrick, H. Bowman, and M. Steen. Viewpoints and objects. In J. P. Bowen and M. G. Hinchey, editors, *Ninth Annual Z User Workshop*, volume 967 of *Lecture Notes in Computer Science*, pages 449–468, Limerick, September 1995. Springer-Verlag.
9. M. d'Inverno, M. Fisher, A. Lomuscio, M. Luck, M. de Rijke, M. Ryan, and M. Wooldridge. Formalisms for multi-agent systems. *Knowledge Engineering Review*, 12(3), 1997.
10. A. Drogoul and J. Zucker. Methodological issues for designing multi-agent systems with machine learning techniques: Capitalizing experiences from the robocup challenge, 1998.
11. Roger Duke, Paul King, Gordon Rose, and Graeme Smith. The Object-Z specification language. Technical report, Software Verification Research Center, Departement of Computer Science, University of Queensland, AUSTRALIA, 1991.
12. Jacques Ferber and Olivier Gutknecht. A meta-model for the analysis and design of organizations in multi-agent systems. In Y. Demazeau, E. Durfee, and N.R. Jennings, editors, *ICMAS'98*, july 1998.

13. M. Fisher. if Z is the answer, what could the question possibly be? In *Intelligent Agents III*, number 1193 in Lecture Note of Artificial Intelligence, 1997.

14. P. Gruer, V. Hilaire, and Abder Koukam. approche multi-formalismes pour la spécification des systèmes multi-agents. Technical report, UTBM-SeT, 2001. to appear in "Système Multi-Agents : des Théories Organisationnelles aux Applications Industrielles" Hermés.

15. P. Gruer, V. Hilaire, Abder Koukam, and Krzysztof Cetnarowicz. A formal framework for multi-agent systems analysis and design. *Expert Systems with Applications*, 23, December 2002.

16. Pablo Gruer, Vincent Hilaire, and Abder Koukam. an Approach to the Verification of Multi-Agent Systems. In *International Conference on Multi Agent Systems*. IEEE Computer Society Press, 2000.

17. Pablo Gruer, Vincent Hilaire, and Abder Koukam. Verification of Object-Z Specifications by using Transition Systems. In T. S. E. Maibaum, editor, *Fundamental Aspects of Software Engineering*, number 1783 in Lecture Notes in Computer Science. Springer Verlag, 2000.

18. Olivier Gutknecht and Jacques Ferber. The madkit agent platform architecture. In *1st Workshop on Infrastructure for Scalable Multi-Agent Systems*, june 2000.

19. David Harel. Statecharts: A visual formalism for complex systems. *Science of Computer Programming*, 8(3):231–274, June 1987.

20. D. E. Herlea, C. M. Jonker, J. Treur, and N. J. E. Wijngaards. Specification of behavioural requirements within compositional multi-agent system design. *Lecture Notes in Computer Science*, 1647:8–27, 1999.

21. V. Hilaire, T. Lissajoux, and A. Koukam. Towards an executable specification of Multi-Agent Systems. In Joaquim Filipe and José Cordeiro, editors, *International Conference on Enterprise Information Systems'99*. Kluwer Academic Publisher, 1999.

22. Vincent Hilaire, Abder Koukam, Pablo Gruer, and Jean-Pierre Müller. Formal specification and prototyping of multi-agent systems. In Andrea Omicini, Robert Tolksdorf, and Franco Zambonelli, editors, *Engineering Societies in the Agents' World*, number 1972 in Lecture Notes in Artificial Intelligence. Springer Verlag, 2000.

23. M.P. Huget. Agent uml class diagrams revisited. 2002. In this volume.

24. Carlos Iglesias, Mercedes Garrijo, and José Gonzalez. A survey of agent-oriented methodologies. In Jörg Müller, Munindar P. Singh, and Anand S. Rao, editors, *Proceedings of the 5th International Workshop on Intelligent Agents V : Agent Theories, Architectures, and Languages (ATAL-98)*, volume 1555 of *LNAI*, pages 317–330, Berlin, July 04–07 1999. Springer.

25. Elizabeth A. Kendall. Role modeling for agent system analysis, design, and implementation. *IEEE Concurrency*, 8(2):34–41, 2000.

26. Elizabeth A. Kendall, P. V. Murali Krishna, C. B. Suresh, and Chira G. V. Pathak. An application framework for intelligent and mobile agents. *ACM Computing Surveys*, 32(1), 2000.

27. T. Lissajoux, V. Hilaire, A. Koukam, and A. Caminada. Genetic Algorithms as Prototyping Tools for Multi-Agent Systems: Application to the Antenna Parameter Setting Problem. In S. Albayrak and F. J. Garijo, editors, *Lecture Notes in Artificial Intelligence*, number 1437 in LNAI. Springer Verlag, 1998.

28. Michael Luck and Mark d'Inverno. A formal framework for agency and autonomy. In Victor Lesser and Les Gasser, editors, *Proceedings of the First International Conference on Multi-Agent Systems*, pages 254–260. AAAI Press, 1995.

29. J. Odell, H. Parunak, and B. Bauer. Extending uml for agents. In Yves Lesperance E. Y. Gerd Wagner, editor, *Information Systems Workshop at the 17th National conference on Artificial Intelligence*, pages 3–17, 2000.

30. M. Wooldridge and N. R. Jennings. Intelligent agents: Theory and practice. *The Knowledge Engineering Review*, 10(2):115–152, 1995.

31. Michael Wooldridge, Nicholas R. Jennings, and David Kinny. A methodology for agent-oriented analysis and design. In *Proceedings of the Third International Conference on Autonomous Agents (Agents'99)*, pages 69–76, Seattle, WA, USA, 1999. ACM Press.

32. E. Yu. Towards modelling and reasoning support for early-phase requirements engineering. In *3rd IEEE Int. Symp. on Requirements Engineering*, pages 226–235, 1997.

Agent UML Class Diagrams Revisited

Marc-Philippe Huget

Agent ART Group
University of Liverpool
LIVERPOOL L69 7ZF
United Kingdom
M.P.Huget@csc.liv.ac.uk

Abstract. Agent UML is a graphical modeling language based on UML. Like UML, Agent UML provides several types of representation covering the description of the system, the components, the dynamics of the system and the deployment. Multiagent system designers already use Agent UML to represent interaction protocols [13,2]. Since agents and objects differ on several points, UML class diagram has to be modified for describing agents. The aim of this paper is to present how to extend UML class diagrams in order to represent agents. We then compare our approach to Bauer's approach [1].

1 Introduction

Multiagent system designers like object-oriented system designers need methodologies and tools to design their systems. Several methodologies are available for multiagent system design (see [10] for a helpful survey). We are particularly interested in the graphical modeling language Agent UML [13,2]. This language is used for the analysis and design of multiagent systems and is an extension of the well known UML [3]. As Odell and Bauer noted it [13,2], it is important to make profit of previous work (here UML) since it steeps the learning curve for software designers and several industrial-strength tools are available for UML. UML is not directly used since agents differ from objects. They present some features that are not available in objects such as autonomy, enriched communication or rationality [12,16]. When agents match objects, multiagent system designers use UML diagrams and when it is not the case, specific Agent UML diagrams are provided. Sequence diagrams and class diagrams are two examples of specific diagrams tailored to agents.

A first extension of Agent UML class diagrams was proposed by Bauer in [1]. This paper presents a prolongation of this work.

The remainder of this paper is as follows. Section 2 presents our proposal of class diagrams. Section 3 gives a comparison between our approach and Bauer's approach [1]. Section 4 concludes the paper and gives some information about future work. We do not give a long example but several small examples are in Section 2.

R. Kowalczyk et al. (Eds.): Agent Technology Workshops 2002, LNAI 2592, pp. 49–60, 2003.

2   Our Proposal of Agent UML Class Diagrams

Class diagrams in UML are used in order to represent the relationships between classes and to define attributes and operations for these ones. It is possible, in UML, to consider different levels of abstraction for the description of class diagrams. Actually, at some levels, it is not interesting to have an accurate view of all classes but just the name of classes. These different levels of abstraction help designers to tackle the complexity of the system. For instance, in software engineering, a particular view can only consider the connection to a data server. Description of UML class diagrams is in [3]. Most of these elements are present in our proposal.

Agent UML class diagrams may contain both agents and classes. Actually, agents could be defined as a set of classes or could use classes. For instance, when one considers ants and pheromones, ants are defined as agents and pheromones as instances of classes. Thus, we have to distinguish agents from classes on diagrams. We propose to prepend the agent name by the stereotype <<agent>> as shown on Fig. 3.

UML considers several levels of abstraction: from low level of abstraction to high level of abstraction [3]. This range of levels of abstraction allows designers to have different views of the system. They are particularly interesting when the system is complex. Two levels of abstraction are usually in use in UML: conceptual level and implementation level. The conceptual level corresponds to high level of abstraction. The implementation level corresponds to low level of abstraction. Section 2.1 presents the conceptual view of the system. Section 2.2 describes the implementation level.

2.1   Conceptual Level

The conceptual level allows designers to have a bird's eye view of both the agents and the classes. The conceptual level is especially used to represent the different agents and classes used in the system and the relationships between these elements. Since agents and classes are mixed on the same diagrams, we have to consider what the four relationships in UML (association, generalization, aggregation and dependency) mean when they are applied between agents or between an agent and objects.

In UML, an *association* is a structural relationship, specifying that objects of one thing are connected to objects of another. Given an association connecting two classes, it is possible to navigate from an object of one class to an object of the other class, and vice-versa. An association that connects exactly two classes is called a binary association. For instance, if an association is defined between one class *company* and one class *employee*, it is then possible to know the company while one employee is considered and it is possible to know the employees while the company is selected. An association is rendered as a solid line connecting the same or different classes (see Fig. 1). There are four basic adornments that apply to an association: a name, the role at each end of the association, the multiplicity at each end of the association and aggregation.

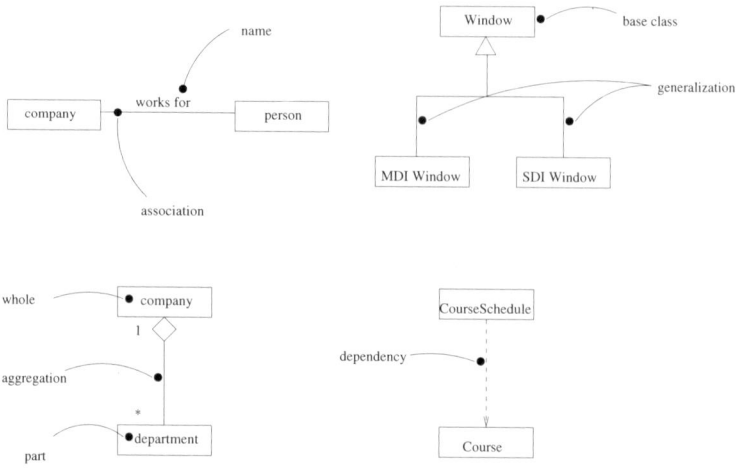

Fig. 1. Relationships in UML

In UML, a *generalization* is a relationship between a general thing (called the superclass or parent) and a more specific kind of that thing (called the subclass or child). Generalization is sometimes called "is-a-kind-of" relationship: a thing is-a-kind-of a more general thing. The *child* inherits the properties of its parents: attributes and operations. Graphically, a generalization is rendered as a solid directed line with a large open arrowhead pointing to its parent as shown on Fig. 1.

In UML, an *aggregation* is a plain association between two classes that represents a structural relationship between peers, meaning that both classes are conceptually at the same level, no one more important than the other. An aggregation is a "whole/part" relationship, in which one class represents a larger thing (the "whole"), which consists of smaller things (the "parts"). It is for instance, the case for the class *car* which is composed of wheels, bodywork, doors, etc. Graphically, an aggregation is rendered a solid directed line with an open diamond as shown in Fig. 1. The diamond points to the class which merges parts together.

In UML, a *dependency* is a using relationship, specifying that a change in the specification of one thing may affect another thing that uses it but not necessarily the reverse. Graphically, a dependency is rendered as a dashed directed line, directed to the class being depended on as shown in Fig. 1.

The relationships as defined above correspond to the relationships between classes. Some modifications have to be done when these relationships deal with agents and classes.

The relationships have the following meaning while they are used between agents:

association: the *association* means that there is a relation of acquaintance between the linked agents. An acquaintance corresponds to a relationship where two agents know each other. This relationship is used when agents are not in a context of cooperation or coordination. As a consequence, agents may exchange messages.

generalization: the *generalization* has the same meaning as for UML classes. The definition of an agent can be derived from another one.

aggregation: the *aggregation* seems to be only possible in one situation: recursion [7]. For instance, if the agent has a recursive architecture, i.e. an agent is an aggregation of several agents.

dependency: the *dependency* relationship in UML corresponds to a unilateral dependency in multiagent theory. We propose to extend this relationship to mutual dependency. If A has a dependency over B, B has also a dependency over A. It is the case if A needs a result that B can provide to it but B needs information that A has.

A new relationship between agents has to be considered: the relationship of *order*. This relationship is particularly used in hierarchies to exhibit that an agent higher in the hierarchy commands an agent lower in the hierarchy to do something.

The relationships have the following meaning while they are used between agents and classes:

association: the *association* relationship connects an agents and several classes. It means that this agent use the connected classes for their execution. It might be a specific class for messages, for goals, for plans. We see further the example of beliefs which are defined as an associated class to agents. This relationship is unidirectional from agents to classes.

generalization: since agents and objects present many differences, it is then impossible that the agents are derived from objects.

aggregation: the *aggregation* between agents and classes implies that agents are defined as an aggregation of several classes. It is frequently the case when designing agent architectures: an agent architecture contains a reasoning part, an interaction part and a perception part such as InteRRaP [11]. All these parts are implemented as objects.

dependency: the *dependency* between agents and classes is possible and means that one agent needs this class either in its code or during its execution since an instance of this class corresponds to an object that the agent can use.

Such an example of conceptual level is given in Fig. 3. This is the example of supply chain management [14] where it is represented clients, the company and providers as shown on Fig. 2.

Briefly, we consider nine types of agents: one type for the client, one type for the provider and seven types for the company. The agent *Order Acquisition* who receives the orders and negotiates the delays and the prices. Orders are negotiated with the agent *Logistics*. As soon as orders are accepted, these orders are sent to the agent *Logistics*. It asks the agent *Scheduler* to generate a plan for this order to allocate materials, workers and transportation. As soon as the plan is available, it is sent to the agents *Transporter*, *Dispatcher* and *Resource*. The agent *Transporter* is responsible for the transportation of the materials and

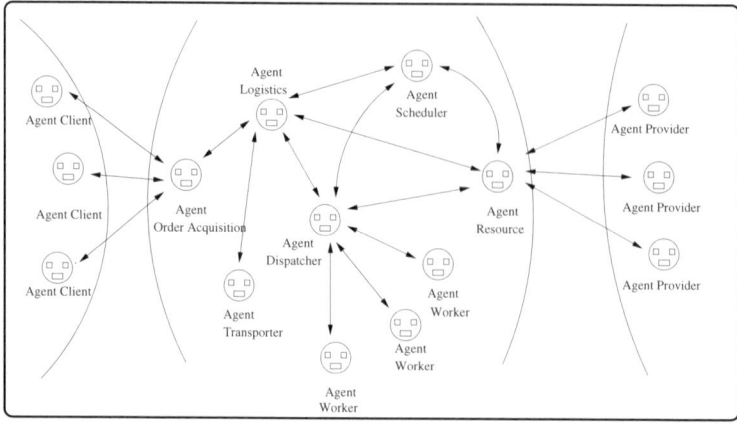

Fig. 2. A multiagent approach of the supply chain management

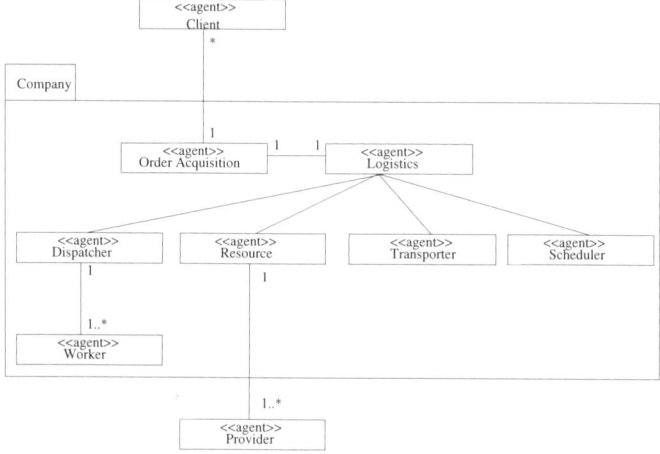

Fig. 3. Conceptual level of the supply chain management

the final product to the client. The agent *Dispatcher* is responsible for finding agents *Workers* who have to do the allocated tasks. The agent *Resource* has to provided materials when needed. If the materials are not available, it has to order these materials to providers.

This example points out that no attributes nor operations are defined in this diagram. It allows designers not to be drowned by numerous information which are not important at this level of abstraction. Figure 3 gives one package called *Company* containing all the agents working in the company. It is interesting to merge together agents working in the same entity. A package is mechanism in UML which organizes objects into groups [3].

It is clear in this example that several levels of abstraction are possible. Actually, in Fig. 3, no classes are given even if these classes are used. One of the great advantage of UML class diagrams is that designers are not required to represent all the information. They can decide what kind of data is important on a particular diagram.

Generalization is not considered in this example. However, it could be used if we consider different kinds of orders. In this case, we have one agent broker who receives the orders and dispatches them to the relevant *Order Acquisition* agent. These agents have the same behavior but different properties according to the different lines of products of this company.

2.2   Implementation Level

The conceptual level defined in the previous section is too abstract to be considered for an implementation. As soon as designers want to implement their systems, they need to have an accurate view of their systems. The implementation level is given for this purpose. Obviously, designers can define as much diagrams as they want in order to point out different views of the system.

This level is the most difficult one. Actually, it is required to give all the elements which define the agents and the classes. The description of agents might be really difficult if the application domain is complex. In order to help designers, we propose to follow the Vowels approach [5,6] for the definition of agents. The Vowels model considers four elements in multiagent systems: agent, environment, interaction and organization. These four elements represent the four different views that designers could have when designing agents: (1) an *agency* view where it is given agents' knowledge, beliefs, intentions, plans and behaviors, (2) an *environmental* view where it is given how agents react to the events coming from the environment and other agents, (3) an *interactional* view where it is given how agents interact with other agents and (4) an *organizational* view where it is given the organizations in which the agents evolve and what are their roles in these organizations.

We add several new compartments to UML class diagrams as shown on Fig. 4. UML class diagrams only consider attributes and operations. These two compartments are insufficient to represent the agent complexity. Figure 4 summarizes the information defined below.

Agency View. Several information have to be supplied according to the agency view. An agent is represented by a name which could be a generic name if this agent class is a pattern for agents. An agent class is denoted by the stereotype `<<agent>>` and a unique agent name. Agent name is insufficient to represent the role of the agent in the multiagent system. We add the compartment *Role* for this purpose. This compartment describes what roles are played by this agent. This information is then used for interaction protocols and organizations.

Like UML classes, agents are characterized by several attributes. These attributes correspond to data used during the agent execution. Common attributes are for instance, the agent identifier, its physical address, its profile if it acts on behalf of users, etc. These attributes are stored in the compartment *Attribute*.

We propose to store beliefs, desires, intentions and goals not as attributes but as objects. An object in UML is an instance of a class with well defined boundary.

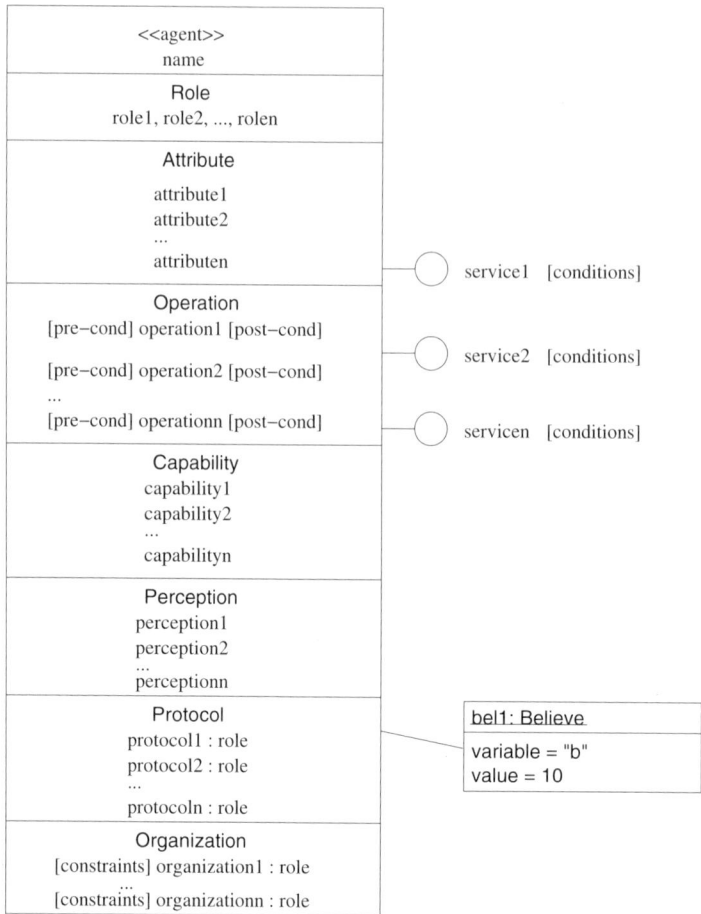

Fig. 4. Agent class diagram

This approach allows designers to represent concrete data within agents. For instance, it is difficult to represent a belief such as *believe(Melbourne sunny)* directly within agent class particularly if we need to modify it after. As a consequence, it is easier to handle these information. When considering a goal, it is then possible to retrieve a part of the goal or to modify it. As far as we are concerned this approach brings flexibility to the management of BDI information. Moreover, it helps designers in the context of mutual beliefs and joint intentions [15] since it is easier to share beliefs or intentions if they are defined outside agents.

Operations are stored in the compartment *Operation*. Operations correspond to actions performed by agents. We divide operations in three categories: internal operations, pro-active operations and reactive operations. Internal operations correspond to operations that are not visible by other agents. For instance, if agents have a specific interaction module for communication [9], such operations could be used to retrieve messages. Pro-active operations correspond to operations that are fired given some conditions such as timer, exceptions or conditions. An usual example is a server which informs agents of a specific value every t units of time. Reactive operations are actions linked to modifications of the environment or in reaction of other agents' actions. These operations are associated to the statecharts corresponding to agent behaviors.

These three kinds of operations are distinguished on class diagrams by a keyword: `int` for internal operations, `pro` for pro-active operations and `reac` for reactive operations.

Environmental View. The environmental view deals with what agents do as soon as the environment is modified. In order to represent the modifications and actions, we use statecharts [8]. These statecharts allow designers to represent the agent states and actions needed for going from state to state. Statechart diagram is the solution that we follow even if statecharts are not the only possibility. We can also use activity diagrams or state machines.

Here follows a small example of an agent monitoring the temperature room. The resulting statechart is given in Fig. 5.

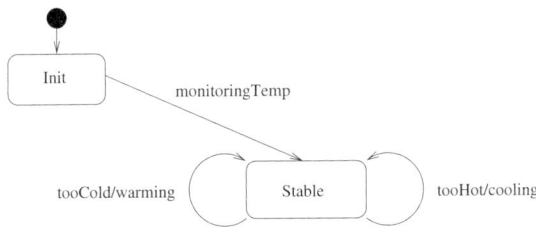

Fig. 5. Statechart diagram for agent monitoring room temperature

Statecharts are not defined within class diagrams but outside, only the name of statecharts is given in class diagrams.

Interactional View. The interactional view gives the interaction protocols used by this agent. Interaction protocols are represented by sequence diagrams [13,2] and extends UML sequence diagrams. Sequence diagrams allow designers to represent agents by their role in this interaction. The messages are sent between roles. The example of the English auction protocol is given in Fig. 6.

Sequence diagrams are not defined within class diagrams but outside, only the name of the protocol diagrams are given in class diagrams. The roles played by the agent are given for each sequence diagram.

Organizational View. Agents are represented in multiagent systems by their roles and their actions. Several information are provided according to the organizational view: capabilities, services and organizations in which agents involve.

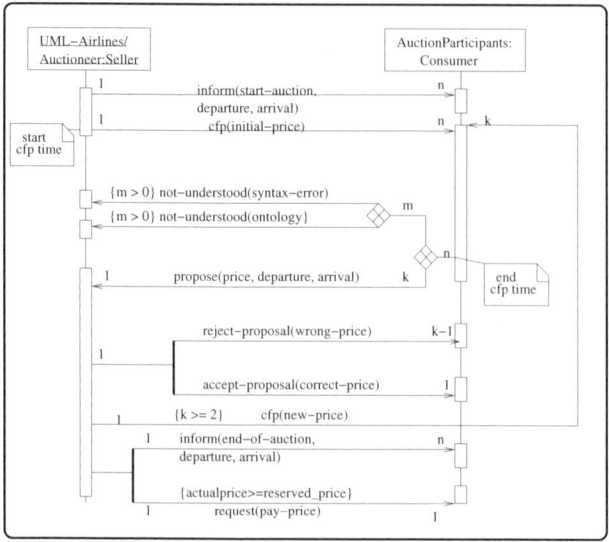

Fig. 6. English auction protocol in agent UML

Capabilities describe what agents are able to do. Capabilities are written as a free-format text. Capabilities are stored in the compartment *Capability*. Capabilities are derived as services to agents. These services are not rendered within agent class diagrams but as lollipops linked to agent class diagrams. We add the name of the service and the conditions under which the service is available.

The last information in the organizational view is the list of organizations, the agents belong to. This list of organizations is stored in the compartment *Organization*. Agents play roles in organizations. Designers insert the roles played for each organization. We also add the constraints imposed for entering into, belonging to and leaving this organization.

3   Comparison with Bauer's Approach

Only one work (except ours) exists in the domain of Agent UML class diagrams [1]. In this section, we compare our approach to Bauer's approach.

Several information are considered in Bauer's approach:

1. Agent class description and roles
2. State description
3. Actions
4. Methods
5. Capabilities, services and supported protocols
6. Organization
7. Agent head automata

Agent class description and roles merges two information that we have in our proposal: the agent class name and the roles played by this agent.

State description corresponds to our compartment *Attribute*. The difference between our compartment *Attribute* and this compartment *State description* is for the computation of beliefs, desires, intentions and goals. Bauer considers these information as well formed formulas. As far as we are concerned, we think that it is better to represent beliefs, desires and intentions as objects linked to the agent class. As a consequence, it is easier to modifiy the belief or to handle information. For instance, if the agent learnt that the value of b is no longer 5 but 10. It is easier to modify the associated belief since the agent has just to modify the attribute value of the belief. This approach is also more efficient for retrieving parts of goals or to merge different sub-goals. Last advantage of such a method is in the context of mutual beliefs, joint intentions and shared plans. It is easier to share these elements since they are defined outside agents.

Actions corresponds to two kinds of actions in Bauer's approach: pro-active actions and reactive actions. Pro active actions correspond to actions fired by the agent itself such as exceptions or triggered actions. Reactive actions are llnked to the modification of the environment. These two kinds of actions are equivalent to ours. We also propose internal actions.

Methods corresponds to our compartment *Operation* when operations are internal ones.

Capabilities corresponds to our compartment *Capabilities*. *Services* are described within agent class diagrams. It is not explained how these services are described. In our proposal, we follow the approach used in UML for interfaces for rendering services.

Supported protocols corresponds to our compartment *Protocol*. However, we add the roles played by the agents. Actually, it is possible that agents play several roles in the same protocol.

Organization corresponds to our compartment *Organization*. One more time, we add the notion of roles played by the agents. Moreover, it is not clear what the piece of information *constraint* means. In our case, we have the information about the conditions to enter the organization, the conditions to stay in it and the conditions to leave it.

The main difference between Bauer's approach and ours remain in the *Agent head automata*. Bauer proposes to render on agent class diagrams, the matching of communicative acts. This approach seems to be unrealistic as soon as agents encompass several complex protocols. This agent head automata contains communicative acts as well as actions triggered by the receipt of messages. In our opinion, this rendering must not be done within agent class diagrams but outside. It is better to use sequence diagrams if designers want to represent protocols or activity diagrams if they want to represent the actions triggered by the messages. Moreover, if designers insert an agent head automata within agent class diagrams, they break one principle of UML saying that one diagram is only performing one kind of view of the system: agent class diagrams are used to represent the content of agents and not how they perform actions linked to messages.

To Sum Up

Several elements are present both in Bauer's approach and in our approach even if sometimes, the term used is not the same: roles, attributes, operations, actions, capabilities, protocols or organizations. The main differences between our two approaches are located in the management of beliefs, desires, intentions and goals; and in the agent head automata. Beliefs, desires, intentions and goals are stored as objects. This approach makes the handle ease. The second main difference is the agent head automata. In our opinion, this agent head automata is inefficient as soon as agents encompass complex protocols and that several actions have to be done between the receipt of messages and the sending of messages. Moreover, this approach is contrary to one principle in UML saying that one diagram is designed to perform only one view of the system.

4   Conclusion

Following the example of software engineering where designers have modeling languages for describing elements used in their software, it is important that multiagent system designers have modeling languages too. The modeling language Agent UML [13] seems to be an answer to this need. This modeling language extends UML in order to consider special features encompassed in agents such as autonomy, cooperation and richer interactions.

Since objects and agents are not the same, some UML representations have to be modified. The aim of this paper is to rethink the class diagrams which allow designers to define the structure of the system and the elements. In our proposal, class diagrams contain agents as well as classes. Several compartments are inserted in these diagrams in order to represent agent features such as behaviors, capabilities, services or protocols.

Bauer has also considered how to augment class diagrams in order to represent agents [1]. His approach presents several similarities with ours but interaction is managed according to a different approach. Bauer proposes to render incoming messages, resulting actions and outgoing messages on class diagrams. This approach tends to reduce the readability of the class diagrams if designers have to represent numerous messages and actions.

Several different future directions are possible for future work: (1) until now, designers use UML tools in order to represent their Agent UML diagrams. Since we have modified deeply the class diagrams, we have to design specific tools for Agent UML, (2) the design of class diagrams is really difficult since designers have to consider a really accurate view of the agents and multiagent systems. It seems to be useful to propose a methodology or recipes in order to help designers in managing this task. (3) it might be interesting to consider how to generate code for class diagrams giving the skeleton of a multiagent systems. Finally, (4) we need to link agent class design to agents and multiagent system analysis. We can make profit of Tropos [4] to this purpose and its two analysis stages.

References

1. B. Bauer. UML class diagrams revisited in the context of agent-based systems. In M. Wooldridge, P. Ciancarini, and G. Weiss, editors, *Proceedings of Agent-Oriented Software Engineering (AOSE 01)*, number 2222 in LNCS, pages 1–8, Montreal, Canada, May 2001. Springer-Verlag.
2. B. Bauer, J. P. Müller, and J. Odell. An extension of UML by protocols for multiagent interaction. In *International Conference on MultiAgent Systems (ICMAS'00)*, pages 207–214, Boston, Massachussetts, july, 10-12 2000.
3. G. Booch, J. Rumbaugh, and I. Jacobson. *The Unified Modeling Language User Guide*. Addison-Wesley, Reading, Massachusetts, USA, 1999.
4. P. Bresciani and F. Sannicolo. Requirements analysis in tropos: A self referencing example. In *In this volume*.
5. Y. Demazeau. Steps towards multi agent oriented programming. slides Workshop, 1st International Workshop on Multi-Agent Systems, IWMAS '97, October 1997.
6. Y. Demazeau. *VOYELLES*. Habilitation à diriger les recherches, Institut National Polytechnique de Grenoble, Grenoble, avril 2001.
7. K. Fernandes and M. Occello. A recursive approach to build hybrid multi-agent systems. In *III Iberoamerican Workshop on Distributed Artificial Intelligence and Multi-Agent Systems, SBIA/IBERAMIA*, Sao Paulo, Brazil, November 2000.
8. D. Harel. Statecharts: A visual formalism for complex systems. *Science of Computer Programming*, 8:231–274, 1987.
9. M.-P. Huget. Design agent interaction as a service to agents. In M.-P. Huget, F. Dignum, and J.-L. Koning, editors, *AAMAS Workshop on Agent Communication Languages and Conversation Policies (ACL2002)*, Bologna, Italy, July 2002.
10. C. A. Iglesias, M. Garijo, and J. C. Gonzalez. A survey of agent-oriented methodologies. 1999.
11. J. Müller. *The Design of Intelligent Agents - a layered approach*. Number LNAI 1177 in Lecture Notes in Artificial Intelligence. Springer-Verlag, Berlin, 1996.
12. J. Odell. Objects and agents compared. *Journal of Object Computing*, 1(1), May 2002.
13. J. Odell, H. V. D. Parunak, and B. Bauer. Extending UML for agents. In G. Wagner, Y. Lesperance, and E. Yu, editors, *Proceedings of the Agent-Oriented Information Systems Workshop at the 17th National conference on Artificial Intelligence*, Austin, Texas, july, 30 2000. ICue Publishing.
14. J. Swaminathan, S. Smith, and N. Sadeh-Koniecpol. Modeling supply chain dynamics: A multiagent approach. *Decision Sciences*, April 1997.
15. M. Wooldridge. *Reasoning about Rational Agents*. MIT Press, 2000.
16. M. Wooldridge, N. R. Jennings, and D. Kinny. The Gaia methodology for agent-oriented analysis and design. *Journal of Autonomous Agents and Multi-Agent Systems*, 3(3):285–312, 2000.

The Behavior-Oriented Design of Modular Agent Intelligence

Joanna J. Bryson

University of Bath, Department of Computer Science
Bath BA2 7AY, United Kingdom
jjb@cs.bath.ac.uk [+44] (0)1225 38 6811

Abstract. Behavior-Oriented Design (BOD) is a development methodology for creating complex, complete agents such as virtual-reality characters, autonomous robots, intelligent tutors or intelligent environments. BOD agents are modular, but not multi-agent systems. They use hierarchical reactive plans to perform arbitration between their component modules. BOD provides not only architectural specifications for modules and plans, but a methodology for building them. The BOD methodology is cyclic, consisting of rules for an initial decomposition and heuristics for revising the specification over the process of development.

1 Introduction

This chapter examines how to build complete, complex agents (CCA). A *complete agent* is an agent that can function naturally on its own, rather than being a dependent part of a Multi-Agent System (MAS). A *complex agent* is one that has multiple, conflicting goals, and multiple, mutually-exclusive means of achieving those goals. Examples of complete, complex agents are autonomous robots, virtual reality (VR) characters, personified intelligent tutors or psychologically plausible artificial life (ALife) [2, 8, 25]. Being able to reliably program CCA is of great practical value both commercially, for industrial, educational and entertainment products, and scientifically, for developing AI models for the cognitive and behavioural sciences.

All of the methodologies I describe in this chapter are modular. The advantages of treating a software system as modular rather than monolithic are well understood. Modularity allows for the problem to be decomposed into simpler components which are easier to build, maintain and understand. In particular, the object-oriented approach to software engineering has shown that bundling behavior with the state it depends on simplifies both development and maintenance.

The problem of when to treat a module as an individual actor, an *agent*, is currently less well understood [3]. Unlike most chapters in this volume, this chapter separates the issues of agency from the issues of MAS. This chapter addresses how to develop and maintain a system that pursues a set of goals relatively autonomously, responding to the challenges and opportunities of dynamic environments that are not fully predictable. However, it does this without agent communication languages, negotiation or brokering. The assumption is that the system will be run on a single platform where no module is

R. Kowalczyk et al. (Eds.): Agent Technology Workshops 2002, LNAI 2592, pp. 61–76, 2003.

likely to die without the entire system crashing, and that the system is being developed by a relatively small team who can share code and interfaces. In other words, the assumption is that this is in many respects a normal software engineering project, except that it is producing an intelligent, proactive system.

2   The Previous State of the Art

The last decade of research has shown impressive convergence on the gross characteristics of software architectures for CCA. The field is now dominated by 'hybrid', three-layer architectures [19, 22]. These hybrids combine the following:

1. *behavior-based AI* (BBAI), the decomposition of intelligence into simple, robust, reliable modules,
2. *reactive planning*, the ordering of expressed actions via carefully specified program structures, and
3. (optionally) *deliberative planning*, which may inform or create new reactive plans, or, in principle, even learn new behaviors.

In this section I will discuss these systems and their history in more detail. The remainder of this chapter presents an improvement to three-layer architectures, called Behavior-Oriented Design.

2.1   Behavior-Based Artificial Intelligence (BBAI)

BBAI was first developed by Brooks [6], at a time when there were several prominent modular theories of natural intelligence being discussed [13, 17, 29]. In BBAI, intelligence is composed of a large number of modular elements that are relatively simple to design. Each element operates only in a particular context, which the module itself recognizes. In Brooks' original proposal, these modules are finite state machines organized into interacting *layers*, which are themselves organized in a linear hierarchy or stack. The behaviors have no access to each other's internal state, but can monitor and/or alter each other's inputs and outputs. Layers are an organizational abstraction: the behaviors of each layer achieve one of the agent's goals, but a higher layer may subsume the goal of a lower layer through the mechanism of affecting inputs and outputs. This is the *subsumption architecture* [6].

BBAI has proved powerful because of the robustness and simplicity of the programs. Each behavior module is straight-forward enough to program reliably. BBAI is also strongly associated with *reactive intelligence*, because the subsumption architecture was both behavior-based and reactive. Reactive intelligence operates with no deliberation or search, thus eliciting the good response times critical for the successful operation of real-time systems such as robots or interactive virtual reality [34].

The cost of reactive intelligence is engineering. The agent's intelligence must be designed by hand since it performs no search (including learning or planning) itself. Although there have been some efforts made to learn or evolve BBAI programs off-line [e.g. 23, 33], these efforts themselves take immense amounts of design, and have not proved superior to hand-designed systems [32, 36]. Although behavior modules are themselves definitionally easy to create, engineering the interactions *between* behaviors has proved difficult. Some authors have taken the approach of limiting behaviors to representations that are easily combined [e.g. 1, 24], but this in turn limits the complexity of the agent that can be created by limiting its most powerful building blocks.

2.2   Reactive Plans and Three-Layer Architectures

At roughly the same time as BBAI was emerging, so were reactive plans [16, 20]. Reactive plans are powerful plan representations that provide for robust execution. A single plan will work under many different contingencies given a sufficiently amenable context. An agent can store a number of such plans in a library, then use context-based preconditions to select one plan that should meet its current goals in the current environment. The alternative – constructing a plan on demand – is a form of search and consequently costly [12]. In keeping with the goal of all reactive intelligence, reactive plans provide a way to avoid search during real-time execution.

A hybrid behavior-based system takes advantage of behaviors to give a planning system very powerful primitives. This in turn allows the plan to be relatively high-level and simple, a benefit to conventional planners as well as reactive plans [28]. Most three-layer hybrid architectures (as described above) have a bottom layer of behaviors, which serve as primitives to a second layer of reactive plans. They may then optionally has a third 'deliberative' (searching) layer either to create new plans or to choose between existing ones.

Consider this description of the ontology underlying three-layered architectures:

> The three-layer architecture arises from the empirical observation that effective algorithms for controlling mobile robots tend to fall into three distinct categories:
> 1. reactive control algorithms which map sensors directly onto actuators with little or no internal state;
> 2. algorithms for governing routine sequences of activity which rely extensively on internal state but perform no search; and
> 3. time-consuming (relative to the rate of change of the environment) search-based algorithms such as planners.
>
> Gat [19, p. 209]

In this description, behaviors are the simple stateless algorithms and reactive plans serve as state- or context-keeping devices for ordering the activity of the behaviors. In Gat's own architecture, ATLANTIS, [18] the second, reactive-plan layer dominates the agent: it monitors the agent's goals and selects its actions. If the second layer becomes stuck or uncertain, it can reduce the agent's activity while consulting a third-level planner, while still monitoring the environment for indications of newer, more urgent goals.

3   Behavior-Oriented Design (BOD)

Despite the development of three-layer hybrid architectures, programming CCA is still hard. Consequently, more approaches are being tried, including using MAS as architectures for single CCA [e.g. 35, 37]. I believe that the problem with three-layer architectures is that they tend to trivialise the behavior modules. The primary advantage of the behavior-based approach and modular approaches in general is that they simplify coding the agent by allowing the programming task to be decomposed. Unfortunately, the drive to simplify the planning aspects of module coordination has lead many developers to tightly constrain what a behavior module can look like [24, fip]. Consequently, the engineering problem has become hard again.

To address these issues, I have developed Behavior-Oriented Design (BOD) [8, 11]. BOD has two major engineering advantages over three-layer architectures:

1. Perception, action, learning and memory are *all* encapsulated in behavior modules. They are expressed in standard programming languages, using object-oriented design (OOD) techniques.
2. The reactive plan structures used to arbitrate between these behavior modules are also designed to be easily engineered by programmers familiar with conventional programming languages.

One way to look at BOD is that the behaviors are used to determine *how* an agent acts, while the plans are used to largely determine *when* those actions are expressed.

At first glance, it may appear that BOD differs from the three-layer approach in that it has dropped the top, deliberative layer, but this is not the case. The real difference between my approach and Gat's is the loss of the bottom layer of purely reactive modules. I don't believe that there *are* many elements of intelligence that don't require some state. Perception requires memory: everything from very recent sensory information, which can help disambiguate a sight or sound, to life-long learning, which can establish expectations in the form of semantic knowledge about word meanings or a building's layout.

Under BOD, the behaviors are semi-autonomous modules, optionally with their own processes. A behavior can store sensory experience and apply complex processes of learning or deliberation across them. Or it can simply encapsulate clever algorithms for performing tasks. Nevertheless, a BOD system is reactive. This is both because the individual behaviors can respond immediately to salient sensory information, and because arbitration between modules is controlled by reactive plans. The primitive elements of the reactive plans are *an interface* to the behaviors; they are implemented as methods on the objects that encode the behaviors. These methods should produce immediate (or at least very rapid) results. Reactive plan primitives require a behavior to provide an anytime response [14] on demand, but the behaviors are free to perform the sorts of longer-term computations described by Gat continuously in the background. Thus all of an agent's behaviors (including behavior arbitration) can run continuously in parallel. Only when actions are expressed externally to the agent are they likely to be subject to action selection through behavior arbitration. Action selection is forced by competition for resources, such as the location of visual attention or the position of the agent's body [4].

4 Building an Agent with Good BOD

Behavior-Oriented Design is not just an architecture, but a design methodology, partially inspired by Object-Oriented Design (OOD) [e.g. 26, 31]. The analogy between BOD and OOD is not limited to the metaphor of the behavior and the object, nor to the use of methods on the behavior objects as primitives to the reactive plans. The most critical aspect of BOD is its emphasis on the design process itself.

The fundamental problem of using a modular approach is deciding what belongs in a module – how many modules should there be, how powerful should they be, and so on. In BBAI this problem is called *behavior decomposition*; obviously analogous problems exist for OOD and MAS. BOD adopts the current accepted OOD practise for solving object decomposition: it focuses on the agent's adaptive state requirements, then uses iterative design and a set of heuristics to refine the original decomposition.

BOD emphasizes cyclic design with rapid prototyping. The process of developing an agent alternates between developing libraries of behaviors and the reactive plans to control the expression of those behaviors, and the process of clarifying and simplifying the agent by re-examining its behavior decomposition. The following sections explain the BOD guidelines for both the initial decomposition and for recognizing and correcting problems in the decomposition during the development process.

4.1 The Initial Decomposition

The initial decomposition is a set of steps. Executing them correctly is not critical, since the main development strategy includes correcting assumptions from this stage of the process. Nevertheless, good work at this stage greatly facilitates the rest of the process.

1. Specify at a high level what the agent is intended to do.
2. Describe likely activities in terms of sequences of actions. These sequences are the the basis of the initial reactive plans.
3. Identify an initial list of sensory and action primitives from the previous list of actions.
4. Identify the state necessary to enable the described primitives and drives. Cluster related state elements and their dependent primitives into specifications for behaviors. This is the basis of the behavior library.
5. Identify and prioritize goals or drives that the agent may need to attend to. This describes the initial roots for the reactive plan hierarchy (described below).
6. Select a first behavior to implement.

The lists compiled during this process should be kept, since they are an important part of the documentation of the agent. The process of documenting BOD agents is described below in Sect. 8.

4.2 Iterative Development

The heart of the BOD methodology is an iterative development process:

1. Select a part of the specification to implement next.
2. Extend the agent with that implementation:
   - code behaviors and reactive plans, and
   - test and debug that code.
3. Revise the current specification.

BOD's iterative development cycle can be thought of as sort of a hand-cranked version of the Expectation Maximization (EM) algorithm [15]. The first step is to elaborate the current model, then the second is to revise the model to find the new optimum representation. Of course, regardless of the optimizing process, the agent will continue to grow in complexity. But if that growth is carefully monitored, guided and pruned, then the resulting agent will be more elegant, easier to maintain, and easier to further adapt.

Unlike behaviors, which are simply coded directly in a standard object-oriented language, reactive plans are stored in script files. The plan is normally read when the agent is initialized, or "comes to life," though in theory new plans could be added during execution. The reactive plans for an agent grow in complexity over the course of development. Also, multiple reactive plans may be developed for a single AI platform (and set of behavior modules), each creating agents with different overall characteristics, such as goals or personality.

Even when there are radically different plan scripts for the same platform or domain, there will generally only be one behavior library – one set of code. Each agent will have its own instance or instances of behavior objects when it is running, and may potentially save run-time state in its own persistent object storage. But it is worth making an effort to support all scripts for a single platform or domain in a single library of behavior code.

Testing should be done as frequently as possible. Using languages that do not require compiling or strong typing, such as lisp or perl, significantly speeds the development process, though they may slow program execution time. "Optimize later", one of the modern mantras of software engineering, applies to programming languages too. In my experience, the time spent developing an AI agent generally far outweighs the time spent watching the agent run. Particularly for interactive real-time agents like robots and VR characters, the bottle-necks are much more likely to be caused by motor constraints or speech-recognition than by the intelligent control architecture.

The most interesting part of BOD's iterative design cycle is the set of rules for revising the specifications. However, understanding these rules requires understanding BOD reactive plans. The following section explains the details of BOD action selection. Section 6 returns to the question of knowing exactly how to optimize the agent.

5   BOD Reactive Plans

Reactive plans support action selection. At any given time step, most agents have a number of actions which could potentially be expressed, at least some of which cannot be expressed simultaneously, for example sitting and walking. In architectures without centralized action selection, such as the Subsumption Architecture [6] or the Agent Network Architecture (ANA) [27], the developer must fully characterize *for each action* how to determine when it should be expressed. This task grows in complexity with the number of new behaviors. For engineers, it is generally easier to describe the desired behavior in terms of sequences of events.

Of course, action-selection sequences can seldom be specified precisely in advance, due to the non-determinism of environments, including the unreliability of the agent's own sensing or actuation. Several types of events may interrupt the completion of an intended action sequence. These events fall into two categories:

1. some combination of alarms, requests or opportunities may make pursuing a different plan more relevant, and
2. some combination of opportunities or difficulties may require the current 'sequence' to be reordered.

Thus the problems of action selection can be broken into three categories: things that need to be checked regularly, things that only need to be checked in a particular context, and things that do not strictly need to be checked at all.

BOD uses reactive plans to perform action selection through behavior arbitration. Individual behavior modules should be simple enough to be programmed to only recommend one action at any particular instant. BOD reactive plans provide three types of plan elements corresponding (respectively) to the three categories of action selection above: drive collections, competences, and action patterns.

The rest of this section explains these three types of elements, in reverse of the above order. BOD reactive plans are described more formally elsewhere [7, 10, 11]. This section gives a quick, informal introduction through examples.

5.1   Action Patterns: Some Things Always Follow

The *action pattern* is a simple sequence of primitives. Primitives are either actions or sensory predicates, and supported directly by the behaviors. Including the sequence as an element type is useful for two reasons. First, it allows an agent designer to keep the system as simple as possible, which both makes it more likely to succeed and communicates more clearly to a subsequent designer the expected behavior of that plan segment. Second, it allows for speed optimization of elements that are reliably run in order, which can be particularly useful in sequences of preconditions or in fine motor control. Here's an example that might be useful to an artificial monkey:

$$\langle \text{get a banana} \rightarrow \text{peel a banana} \rightarrow \text{eat a banana} \rangle \tag{1}$$

5.2   Competences: Some Things Depend on Context

A sequence is *not* equivalent to the process of *chaining* a set of productions, where each element's precondition is set to the fire its action as a consequence of the outcome of the prior element. Besides the possibility of optimizing away the step of checking preconditions, a sequence includes an additional piece of control state. Its elements may also occur in different orders in other contexts, and there is no ambiguity if more than one sequence's element fires in a perceptually equivalent context.

The advantage of productions of course is that, within the confines of a particular context, they allow for flexible behavior. A *competence* combines the advantages of both productions and sequences. Here is an example of the above sequence rewritten as a competence:

$$
\text{(have hunger)} \Rightarrow \left.\middle|\begin{array}{c} \text{(full)} \Rightarrow goal \\ \left\langle\begin{array}{c} \text{(have a peeled banana)} \Rightarrow \text{eat a banana} \\ \text{(have a banana)} \Rightarrow \text{peel a banana} \\ \Rightarrow \text{get a banana} \end{array}\right\rangle \end{array}\right. \qquad (2)
$$

Rather than encoding a temporal ordering, a competence encodes a *prioritization*. Priority increases in the direction of the vertical arrow on the left. Under this plan, if the monkey is handed a peeled banana, she'll eat it. If she has a whole banana, she'll peel it. Otherwise, she'll try to get a banana. When the goal has been achieved, or if none of the elements can fire, the competence terminates.

This sort of structure could lead to looping, for example if another, larger monkey kept taking the banana away after our agent had peeled it but before she could eat it. To allow termination in this circumstance, competence elements not only have priorities and preconditions, but also (optional) retry limits. When an element has reached its retry limit, it will no longer fire.

5.3   Drive Collections: Some Things Need to Be Checked at All Times

Finally, there must be a way to arbitrate between plan elements or goals. There must be a way to determine the current focus of action-selection attention – to deal with context changes (whether environmental or internal) which require changing between plans, rather than within them. Some hybrid architectures consider this problem the domain of 'deliberation' or 'introspection' – the highest level of a three-layered architecture. But BOD treats this problem as continuous with the general problem of action selection, both in terms of constraints, such as the need for reactiveness, and of solution.

BOD uses a third element type, the *drive collection* for this kind of attention. A drive collection is very similar to a competence. However, it is designed never to terminate – there is no goal condition. The drive collection is the root of the BOD plan hierarchy, and the only element that executes on every program cycle (from hundreds to thousands of times a second, depending on the implementation). The highest priority drive-collection element that triggers passes activation to whatever competence, sequence or action primitive it is currently attending to. Or, if the agent has just been initialized or the element's last attendee has terminated, the element sets its attention to the apex of its plan hierarchy.

$$
life \Rightarrow \left\langle\left\langle\begin{array}{c} \text{(something looming)} \Rightarrow \text{avoid} \\ \text{(something loud)} \Rightarrow \text{attend to threat} \\ \text{(hungry)} \Rightarrow \text{forage} \\ \Rightarrow \text{lounge around} \end{array}\right\rangle\right\rangle \qquad (3)
$$

Drive-collection elements have another feature not shown in the above diagram: they also allow for scheduling, so that a high priority element does not necessarily monopolize program cycles. This increases the parallelism and reactivity of BOD agents.

For working, non-toy examples of plans for complete BOD agents, see [7, 8, 9]; for comparisons to other systems, see [7, 8]. This chapter emphasizes instead the engineering of BOD agents. The next sections return to the question of revising BOD specifications.

6 Revising BOD Specifications

A critical part of the BOD methodology is the set of rules for revising the specifications. The fundamental design principle is *when in doubt, favor simplicity*. A primitive is preferred to an action sequence, a sequence to a competence. Similarly, control state is preferred to learned state, specialized learning to general purpose learning or planning. Given this bias, heuristics are then used to indicate when a simple element should be exchanged for a more complex one.

A guiding principle in all software engineering is to reduce redundancy. If a particular plan or behavior can be reused, it should be. As in OOD, if only part of a plan or a primitive action can be used, then a change in decomposition is called for. In the case of the action primitive, the primitive should be decomposed into two or more primitives, and the original action replaced by a plan element, probably an action pattern. The new plan element should have the same name and functionality as the original action. This allows established plans to continue operating with only minimal change.

If a sequence sometimes needs to contain a cycle, or often does not need some of its elements to fire, then it is really a competence, not an action pattern. If a competence is actually deterministic, if it nearly always actually executes a fixed path through its elements, then it should be simplified into a sequence.

Competences are really the basic level of operation for reactive plans, and learning to write and debug them may take time. Here are two indications provided by competences that the specification of an agent needs to be redesigned:

- *Complex Triggers:* reactive plan elements should not require long or complex triggers. Perception should be handled at the behavior level; it should be a skill. Thus a large number of triggers may indicate the requirement for a new behavior or a new method on an existing behavior to appropriately categorize the context for firing the competence elements. Whether a new behavior or simply a new method is called for is determined by whether or not more state is needed to make that categorization: new state generally implies a new behavior.
- *Too Many Elements:* Competences usually need no more than 5 to 7 elements, they may contain fewer. Sometimes competences get cluttered (and triggers complicated) because they actually contain two different solutions to the same problem. In this case, the competence should be split. If the two paths lead all the way to the goal, then the competence is really two siblings which should be discriminated between at the level of the current competence's parent. If the dual pathway is only for part of the competence, then the original competence should contain two children.

Effectively every step of the competence but the highest priority one is a subgoal. If there is more than one way to achieve that subgoal, trying to express both of them in the same competence can split attention resources and lead to dithering or 'trigger-flipping' (where two plan elements serve only to activate each other's precondition). The purpose of a competence is to focus attention on *one* solution at a time.

The heart of the BOD strategy is rapid prototyping. If one approach is too much trouble or is giving debugging problems, try another. It is important to remember that programmers' experiences are the key selective pressures in BOD for keeping the agent simple. BOD provides at least two paths to simplicity and clarity: modularity and hierarchical reactive plans. Using cyclic development and some trial and error, the programmer should determine which path is best for a particular problem [5, 30]. This is also why modularity and maintainability are key to BOD: programmers are to be encouraged to change the architecture of an agent when they find a better solution. Such changes should be easy to make. Further, they should be transparent, or at least easy to follow and understand, if another programmer encounters them.

Further heuristics, particularly with respect to prioritization and scheduling in drive collections can be found elsewhere [8]. In the next section, I will instead give an example of the most fundamental revision, trading off complexity in plans for that in behaviors.

7   Trading Off Control and Learning

To demonstrate the differences between representational styles, let's think about an insect robot with two 'feelers' (bump sensors), but no other way of sensing its environment.

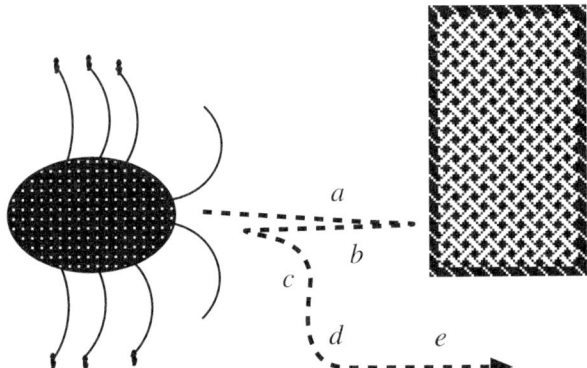

Fig. 1. An insect-like robot with no long-range sensors (e.g. eyes) needs to use its feelers to find its way around a box

7.1 Control State Only

This plan is in the notation introduced earlier, except that words that reference parts of control state rather than primitives are in bold face. Assume that the 'walk' primitives take some time (say 5 seconds) and move the insect a couple of centimeters on the diagram. Also, assume turning traces an arc rather than happening in place. This is about the simplest program that can be written using entirely control state:

$$\textbf{walk} \Rightarrow \left| \left\langle \begin{array}{l} \text{(left-feeler-hit)} \Rightarrow \textbf{avoid-obstacle-left} \\ \text{(right-feeler-hit)} \Rightarrow \textbf{avoid-obstacle-right} \\ \qquad\qquad \Rightarrow \text{walk-straight} \end{array} \right\rangle \right. \tag{4}$$

$$\textbf{avoid-obstacle-left} \Rightarrow \langle \text{walk backwards} \rightarrow \text{walk right} \rightarrow \text{walk left} \rangle \tag{5}$$

$$\textbf{avoid-obstacle-right} \Rightarrow \langle \text{walk backwards} \rightarrow \text{walk left} \rightarrow \text{walk right} \rangle \tag{6}$$

7.2 Deictic State as Well

If we are willing to include a behavior with just one bit of variable state in it, then we can simplify the control state for the program. In the behavior *deictic-avoid*, the bit hit-left? serves as a deictic variable the-side-I-just-hit-on. **Avoid-hit** and **compensate-avoid** (primitives *deictic-avoid* supports) turn in the appropriate direction by accessing this variable. This allows a reduction in redundancy in the plan, including the elimination of one of the action patterns.

$$\xleftarrow{\quad\textbf{avoid-hit, feeler-hit,}\quad}_{\textbf{compensate-avoid}} \boxed{\begin{array}{c} \textit{deictic-avoid} \\ \text{hit-left?} \end{array}} \xleftarrow{\quad \text{feeler info}}$$

$$\textbf{walk} \Rightarrow \left| \left\langle \begin{array}{l} \text{(feeler-hit)} \Rightarrow \textbf{avoid-obstacle} \\ \qquad\quad \Rightarrow \text{walk-straight} \end{array} \right\rangle \right. \tag{7}$$

$$\textbf{avoid-obstacle} \Rightarrow \langle \text{walk backwards} \rightarrow \text{avoid hit} \rightarrow \text{compensate avoid} \rangle \tag{8}$$

7.3 Specialized Instead of Deictic State

Instead of using a simple reference, we could also use a more complicated representation, say an allocentric representation of where the obstacle is relative to the bug, that is updated automatically as the bug moves and forgotten as the bug moves away from the location of the impact. Since this strategy requires the state to be updated continuously as the bug moves, walking must be a method (**find-way**) on this behavior.

$$\xleftarrow{\quad\textbf{back-up, find-way}\quad}_{\textbf{store-obstacle}} \boxed{\begin{array}{c} \textit{specialized-avoid} \\ \text{local-map} \end{array}} \xleftarrow{\quad \text{feeler info}}$$

$$\textbf{walk} \Rightarrow \left| \left\langle \begin{array}{l} \text{(feeler-hit)} \Rightarrow \textbf{react-to-bump} \\ \qquad\quad \Rightarrow \text{find-way} \end{array} \right\rangle \right. \tag{9}$$

$$\textbf{react-to-bump} \Rightarrow \langle \text{store-obstacle} \rightarrow \text{walk backwards} \rangle \qquad (10)$$

If this is really the only navigation ability our bug has, then the vast increase in complexity of the behavior *specialized-avoid* does not justify the savings in control state. On the other hand, if our bug already has some kind of allocentric representation, then it might be sensible to piggy-back the feeler information on top of it. For example, if the bug has a vector created by a multi-faceted eye representing approximate distance to visible obstacles, but has bumped into something hard to see (like a window), it might be parsimonious to store the bump information in the vision vector, providing that updating the information with the bug's own motion isn't too much trouble. Insects actually seem able to do something like this [21], and I've done it with a robot [9].

8 Documenting a BOD Agent

As I said at the end of Sect. 6, maintainability and clarity of design are key to the BOD development process. This is best achieved through self-documenting code. "Self-documenting" is something of a misnomer, because of course the process takes discipline. The primary argument for incorporating documentation into functioning code is that this is the only way to ensure that the documentation will never get out of synchronization with the rest of the software project. The primary argument against this strategy is that code is never really that easy to read, and will never be concisely summarized. BOD at least somewhat overcomes this problem by having two types of summary built into the agent's software architecture. The reactive plans summarize the aims and objectives of the agent, and the plan-primitive / behavior interface documents at a high level the expressed actions of the various behaviors. Further information, such as documentation on the adaptive state used by the agent, can be found in the code of the behavior modules.

8.1 Guidelines for Maintaining Self-Documentation

The following are guidelines to the discipline of making sure BOD agents are self-documented:

Document the plan / behavior interface in one program file. As explained earlier, the primitives of the POSH reactive plans must be defined in terms of methods on the behavior objects. For each behavior library, there should be one code file that creates this interface. In my implementations of POSH action selection, each primitive must be wrapped in an object which is either an *act* or a *sense*. The code executed when that object is triggered is usually only one or two lines long, typically a method call on some behavior object. I cluster the primitives by the behaviors that support them, and use program comments to make the divisions between behaviors clear.

This is the main documentation for the specification – it is the only file likely to have both current *and intended* specifications listed. This is where I list the names of behaviors and primitives determined during decomposition, even before they have been implemented. Intended reactive plans are usually written as scripts (see below.)

Each behavior should have its own program file. Every behavior will be well commented automatically if it is really implemented as an object. One can easily see the state and representations in the class definition. Even in languages that don't require methods to be defined in the class declaration, it is only good style to include all the methods of a class in the same source file with the class definition.

Keep and comment reactive plan scripts. This is the suggestion that requires the most discipline, but having a documented history of the development of an agent can be critical to understanding some of its nuances. Documenting plan scripts effectively documents the history of the agent's development – it is easy to determine when behaviors were added or modified by seeing what primitives are present in a script. Those who can't remember history are doomed to repeat old mistakes. Keeping a complete set of working scripts documenting stages of the agents development also provides a test suite, useful when major changes are made to behavior libraries.

Script comments should contain:

– Its name, to flag if it has been copied and changed without updating the comment.
– What script(s) it was derived from. Most scripts are improvements of older working scripts, though some are shortened versions of a script that needs to be debugged.
– The date it was created.
– The date it started working, if significantly different. Since writing scripts is part of the specification process, some scripts will be ambitious plans for the future rather than working code.
– The date and reasons it was abandoned, if it was abandoned.
– Possibly, dates and explanations of any changes. Normally, changes shouldn't happen in a script once it works (or is abandoned) – they should be made in new scripts, and the old ones kept for a record.

8.2   BOD and Code Reuse

One aspect of OOD that I have not yet explicitly touched on is the issue of code reuse. Code reuse is of course one of the main purposes of both modularity and good documentation. We can think of CCA code as breaking into three categories:

1. *Code useful to all agents.* For BOD, this code is primarily embedded in the POSH planning system. POSH reactive plans are the one set of patterns I believe to be generally useful to modular agents, but not to modular programs which do not have agency.
2. *Code useful to most agents on a particular platform or domain.* By the nature of the BOD methodology, most of this code will wind up in the library of behaviors. However, because this is not the main criteria for determining whether code is in a plan or a behavior, it is possible that some aspects of the reactive plans will also be shared between all agents inhabiting a single platform. Nevertheless, behavior modules can and should be treated exactly as a code library.

3. *Code useful only to a single agent.* Because much of what determines one individual from another is their goals and priorities, the reactive plans will generally succinctly specify individual differences. Though again, it is possible that some behaviors in the behavior library will also only be expressed / referenced by a single agent. Much of what is unique to an agent will not be its program code, but the state within its behavior modules while it is running – this is what encodes its experience and memories.

9   Conclusions

Excellent software engineering is key to developing complete, complex agents. The advances of the last fifteen years in CCA due to the reactive and behavior-based movements come primarily from two engineering-related sources:

1. the trade-off of slow or unreliable on-line processes of search and learning for the one-time cost of development, and
2. the use of modularity.

Of course, the techniques of learning and planning cannot and should not be abandoned: some things can only be determined by an agent at run time. However, constraining learning with specialized representations and constraining planning searches to likely solution spaces greatly increases the probability that an agent can reliably perform successfully. Providing these representations and solution spaces is the job of software engineers, and as such should exploit the progress made in the art of software engineering.

In this chapter I have described the recent state-of-the-art in CCA design, and then described an improvement, the Behavior-Oriented Design methodology. This methodology brings CCA development closer to conventional software engineering, particularly OOD, but with the addition of a collection of organizational idioms and development heuristics that are uniquely important to developing AI.

References

[fip] FIPA-OS: A component-based toolkit enabling rapid development of fipa compliant agents. http://fipa-os.sourceforge.net.

[1] Arkin, R. C. (1998). *Behavior-Based Robotics*. MIT Press, Cambridge, MA.

[2] Ballin, D. (2000). Special issue: Intelligent virtual agents. *Virtual Reality*, 5(2).

[3] Bertolini, D., Busetta, P., Molani, A., Nori, M., and Perini, A. (2002). Designing peer-to-peer applications: an agent-oriented approach. In this volume.

[4] Blumberg, B. M. (1996). *Old Tricks, New Dogs: Ethology and Interactive Creatures*. PhD thesis, MIT. Media Laboratory, Learning and Common Sense Section.

[5] Boehm, B. W. (1986). A spiral model of software development and enhancement. *ACM SIGSOFT Software Engineering Notes*, 11(4):22–32.

[6] Brooks, R. A. (1986). A robust layered control system for a mobile robot. *IEEE Journal of Robotics and Automation*, RA-2:14–23.

[7] Bryson, J. J. (2000). Hierarchy and sequence vs. full parallelism in reactive action selection architectures. In *From Animals to Animats 6 (SAB00)*, pages 147–156, Cambridge, MA. MIT Press.

[8] Bryson, J. J. (2001). *Intelligence by Design: Principles of Modularity and Coordination for Engineering Complex Adaptive Agents*. PhD thesis, MIT, Department of EECS, Cambridge, MA. AI Technical Report 2001-003.

[9] Bryson, J. J. and McGonigle, B. (1998). Agent architecture as object oriented design. In Singh, M. P., Rao, A. S., and Wooldridge, M. J., editors, *The Fourth International Workshop on Agent Theories, Architectures, and Languages (ATAL97)*, pages 15–30. Springer-Verlag.

[10] Bryson, J. J. and Stein, L. A. (2001a). Architectures and idioms: Making progress in agent design. In Castelfranchi, C. and Lespérance, Y., editors, *The Seventh International Workshop on Agent Theories, Architectures, and Languages (ATAL2000)*. Springer.

[11] Bryson, J. J. and Stein, L. A. (2001b). Modularity and design in reactive intelligence. In *Proceedings of the 17th International Joint Conference on Artificial Intelligence*, pages 1115–1120, Seattle. Morgan Kaufmann.

[12] Chapman, D. (1987). Planning for conjunctive goals. *Artificial Intelligence*, 32:333–378.

[13] Chomsky, N. (1980). Rules and representations. *Brain and Behavioral Sciences*, 3:1–61.

[14] Dean, T. and Boddy, M. (1988). An analysis of time-dependent planning. In *Proceedings of the Seventh National Conference on Artificial Intelligence (AAAI-88)*, pages 49–54, Saint Paul, Minnesota, USA. AAAI Press/MIT Press.

[15] Dempster, A. P., Laird, N. M., and Rubin, D. B. (1977). Maximum likelihood from incomplete data via the EM algorithm. *Journal of the Royal Statistical Society series B*, 39:1–38.

[16] Firby, J. (1987). An investigation into reactive planning in complex domains. In *Proceedings of the National Conference on Artificial Intelligence (AAAI)*, pages 202–207.

[17] Fodor, J. A. (1983). *The Modularity of Mind*. Bradford Books. MIT Press, Cambridge, MA.

[18] Gat, E. (1991). *Reliable Goal-Directed Reactive Control of Autonomous Mobile Robots*. PhD thesis, Virginia Polytechnic Institute and State University.

[19] Gat, E. (1998). Three-layer architectures. In Kortenkamp, D., Bonasso, R. P., and Murphy, R., editors, *Artificial Intelligence and Mobile Robots: Case Studies of Successful Robot Systems*, pages 195–210. MIT Press, Cambridge, MA.

[20] Georgeff, M. P. and Lansky, A. L. (1987). Reactive reasoning and planning. In *Proceedings of the Sixth National Conference on Artificial Intelligence (AAAI-87)*, pages 677–682, Seattle, WA.

[21] Hartmann, G. and Wehner, R. (1995). The ant's path integration system: A neural architecture. *Bilogical Cybernetics*, 73:483–497.

[22] Hexmoor, H., Horswill, I., and Kortenkamp, D. (1997). Special issue: Software architectures for hardware agents. *Journal of Experimental & Theoretical Artificial Intelligence*, 9(2/3).

[23] Humphrys, M. (1997). *Action Selection methods using Reinforcement Learning*. PhD thesis, University of Cambridge.

[24] Konolige, K. and Myers, K. (1998). The Saphira architecture for autonomous mobile robots. In Kortenkamp, D., Bonasso, R. P., and Murphy, R., editors, *Artificial Intelligence and Mobile Robots: Case Studies of Successful Robot Systems*, chapter 9, pages 211–242. MIT Press, Cambridge, MA.

[25] Kortenkamp, D., Bonasso, R. P., and Murphy, R., editors (1998). *Artificial Intelligence and Mobile Robots: Case Studies of Successful Robot Systems*. MIT Press, Cambridge, MA.

[26] Larman, C. (2001). *Applying UML and Patterns: An Introduction to Object-Oriented Analysis and Design and the Unified Process*. Prentice Hall, 2nd edition.

[27] Maes, P. (1990). Situated agents can have goals. In Maes, P., editor, *Designing Autonomous Agents : Theory and Practice from Biology to Engineering and back*, pages 49–70. MIT Press, Cambridge, MA.

[28] Malcolm, C. and Smithers, T. (1990). Symbol grounding via a hybrid architecture in an autonomous assembly system. In Maes, P., editor, *Designing Autonomous Agents: Theory and Practice from Biology to Engineering and Back*, pages 123–144. MIT Press, Cambridge, MA.

[29] Minsky, M. (1985). *The Society of Mind*. Simon and Schuster Inc., New York, NY.

[30] Parnas, D. L. and Clements, P. C. (1986). A rational design process: How and why to fake it. *IEEE Transactions on Software Engineering*, SE-12(2):251–7.

[31] Parnas, D. L., Clements, P. C., and Weiss, D. M. (1985). The modular structure of complex systems. *IEEE Transactions on Software Engineering*, SE-11(3):259–266.

[32] Pauls, J. (2001). Pigs and people. *in preperation*.

[33] Perkins, S. (1998). *Incremental Acquisition of Complex Visual Behaviour using Genetic Programming and Shaping*. PhD thesis, University of Edinburgh. Department of Artificial Intelligence.

[34] Sengers, P. (1998). Do the thing right: An architecture for action expression. In Sycara, K. P. and Wooldridge, M., editors, *Proceedings of the Second International Conference on Autonomous Agents*, pages 24–31. ACM Press.

[35] Sierra, C., de Màntaras, R. L., and Busquets, D. (2001). Multiagent bidding mechanisms for robot qualitative navigation. In Castelfranchi, C. and Lespérance, Y., editors, *The Seventh International Workshop on Agent Theories, Architectures, and Languages (ATAL2000)*. Springer.

[36] Tyrrell, T. (1993). *Computational Mechanisms for Action Selection*. PhD thesis, University of Edinburgh. Centre for Cognitive Science.

[37] van Breemen, A. (2002). Integrating agents in software applications. In this volume.

Engineering JADE Agents with the Gaia Methodology

Pavlos Moraïtis [1,2], Eleftheria Petraki [2], and Nikolaos I. Spanoudakis [2]

[1]Dept. of Computer Science
University of Cyprus
P.O. Box 20537, CY-1678 Nicosia, Cyprus
moraitis@ucy.ac.cy

[2]Singular Software,
26th October 43, 54626, Thessaloniki, Greece
{epetraki,nspan}@si.gr

Abstract. Agent Oriented Software Engineering (AOSE) is one of the fields of the agent domain with a continuous growing interest. The reason is that the possibility to easily specify and implement agent-based systems is of a great importance for the recognition of the add-value of the agent technology in many application fields. In this paper we present an attempt towards this direction, by proposing a kind of roadmap of how one can combine the Gaia methodology for agent-oriented analysis and design and JADE, a FIPA compliant agent development framework, for an easier analysis, design and implementation of multi-agent systems. Our objective is realized through the presentation of the analysis, design and implementation phases, of a limited version of a system we currently develop in the context of the IST IMAGE project.

1   Introduction

During the last few years, there has been a growth of interest in the potential of agent technology in the context of software engineering. This has led to the proposal of several development environments to build agent systems (see for example Zeus [3], AgentBuilder [12], AgentTool [6], RETSINA [13], etc), software frameworks to develop agent applications in compliance with the FIPA specifications (see for example FIPA-OS [8], JADE [2], etc). These development environments and software frameworks demanded that system analysis and design methodologies, languages and procedures would support them. As a consequence, many of these were proposed along with a methodology (e.g. Zeus [4], AgentTool [14]) while in parallel have been proposed some promising agent-oriented software development methodologies, as Gaia [15], AUML [1], Tropos [9] and MASE [14]. Also, the Aspect Oriented Programming [11] can be used as a methodology for design and implementation of agent role models. However, despite the possibilities provided by these methodologies, we believe that a further progress must be made, so that agent-based technologies realize their full potential, concerning the full covering of the software life cycle and the proposal of standards to support agent interoperability.

R. Kowalczyk et al. (Eds.): Agent Technology Workshops 2002, LNAI 2592, pp. 77–91, 2003.
© Springer-Verlag Berlin Heidelberg 2003

In this paper we present an attempt to use Gaia in order to engineer a multi-agent system (MAS) that is to be implemented with the JADE framework. The only pretension we have with this paper is to share our experience to conceive and develop a MAS, by combining Gaia and JADE, in the context of the IST IMAGE project, with people who are interested in the development of real life agent-based systems. The Gaia methodology can be applied in a high level design. There is no given way to go from a Gaia model to a system design model. System implementation is still done through object-oriented techniques. Thus, the aim of this paper is to describe a kind of roadmap for implementing a Gaia model using the JADE framework. Towards this end, we provide some additional modeling techniques and make some slight modifications to the Gaia original specification, without obviously altering its philosophy and concepts.

This paper is organized in the following way. In Sects. 2 and 3 we briefly present the Gaia methodology and JADE framework. In Sect. 4 we provide a sample Gaia model. In Sect. 5 we provide a methodology for converting the Gaia model to a JADE implementation. Moreover, we propose some models useful for the detailed design phase. Finally, we discuss on AOSE.

2   Gaia Overview

The Gaia methodology is an attempt to define a complete and general methodology that it is specifically tailored to the analysis and design of MASs. Gaia is a general methodology that supports both the levels of the individual agent structure and the agent society in the MAS development process. MASs, according to Gaia, are viewed as being composed of a number of autonomous interactive agents that live in an organized society in which each agent plays one or more specific roles. Gaia defines the structure of a MAS in terms of a role model. The model identifies the roles that agents have to play within the MAS and the interaction protocols between the different roles.

The objective of the Gaia analysis process is the identification of the roles and the modeling of interactions between the roles found. Roles consist of four attributes: responsibilities, permissions, activities and protocols. Responsibilities are the key attribute related to a role since they determine the functionality. Responsibilities are of two types: liveness properties – the role has to add something good to the system, and safety properties – the role must prevent and disallow that something bad happens to the system. Liveness describes the tasks that an agent must fulfill given certain environmental conditions and safety ensures that an acceptable state of affairs is maintained during the execution cycle. In order to realize responsibilities, a role has a set of permissions. Permissions represent what the role is allowed to do and in particular, which information resources it is allowed to access. The activities are tasks that an agent performs without interacting with other agents. Finally, protocols are the specific patterns of interaction, e.g. a seller role can support different auction protocols. Gaia has formal operators and templates for representing roles and their attributes and also it has schemas that can be used for the representation of interactions between the various roles in a system.

The operators that can be used for liveness expressions-formulas along with their interpretations are presented in Table 1. Note that in liveness formulas activities are written underlined.

Table 1. Gaia operators for liveness formulas

Operator	Interpretation
x . y	x followed by y
x | y	x or y occurs
x*	x occurs 0 or more times
x+	x occurs 1 or more times
x$^{\omega}$	x occurs infinitely often
[x]	x is optional
x || y	x and y interleaved

In the Gaia design process the first step is to map roles into agent types and to create the right number of agent instances of each type. An agent type can be an aggregation of one or more agent roles. The second step is to determine the services model needed to fulfill a role in one or several agents. A service can be viewed as a function of the agent and can be derived from the list of protocols, activities, responsibilities and the liveness properties of a role. Finally, the last step is to create the acquaintance model for the representation of communication between the different agents. The acquaintance model does not define the actual messages that are exchanged between the agents it is rather a simple graph that represents the communication pathways between the different agent types.

3   JADE Overview

JADE is a software development framework fully implemented in JAVA language aiming at the development of multi-agent systems and applications that comply with FIPA standards for intelligent agents. JADE provides standard agent technologies and offers to the developer a number of features in order to simplify the development process:

- Distributed agent platform. The agent platform can be distributed on several hosts, each one of them executes one Java Virtual Machine.
- FIPA-Compliant agent platform, which includes the Agent Management System the Directory Facilitator and the Agent Communication Channel.
- Efficient transport of ACL messages between agents.

All agent communication is performed through message passing and the FIPA ACL is the language that is used to represent the messages. Each agent is equipped with an incoming message box and message polling can be blocking or non-blocking with an optional timeout. Moreover, JADE provides methods for message filtering. The developer can apply advanced filters on the various fields of the incoming message such as sender, performative or ontology.

FIPA specifies a set of standard interaction protocols such as FIPA-request, FIPA-query, etc. that can be used as standard templates to build agent conversations. For every conversation among agents, JADE distinguishes the role of the agent that starts the conversation (initiator) and the role of the agent that engages in a conversation started by another agent (responder). According to the structure of these protocols, the initiator sends a message and the responder can subsequently reply by sending a not-understood or a refuse message indicating the inability to achieve the rational effect of the communicative act, or an agree message indicating the agreement to perform the communicative act. When the responder performs the action he must send an inform message. A failure message indicates that the action was not successful. JADE provides ready-made behaviour classes for both roles, following most of the FIPA specified interaction protocols. Because the FIPA interaction protocols share the same structure, JADE provides the AchieveREInitiator/Responder classes, a single homogeneous implementation of interaction protocols such as these mentioned above. Both classes provide methods for handling all possible protocol states.

In JADE, agent tasks or agent intentions are implemented through the use of be-haviours. Behaviours are logical execution threads that can be composed in various ways to achieve complex execution patterns and can be initialized, suspended and spawned at any given time. The agent core keeps a task list that contains the active behaviours. JADE uses one thread per agent instead of one thread per behaviour to limit the number of threads running in the agent platform. A scheduler, hidden to the developer, carries out a round robin policy among all behaviours available in the queue. The behaviour can release the execution control with the use of blocking mechanisms, or it can permanently remove itself from the queue in run time. Each behaviour performs its designated operation be executing the core method action().

Behaviour is the root class of the behaviour hierarchy that defines several core methods and sets the basis for behaviour scheduling as it allows state transitions (starting, blocking and restarting). The children of this base class are SimpleBehaviour and CompositeBehaviour. The classes that descend from SimpleBehaviour represent atomic simple tasks that can be executed a number of times specified by the developer. Classes descending from CompositeBehaviour support the handling of multiple behaviours according to a policy. The actual agent tasks that are executed through this behaviour are not defined in the behaviour itself, but inside its children behaviours. The FSMBehaviour class, which executes its children behaviours according to a Finite State Machine (FSM) of behaviours, belongs in this branch of hierarchy. Each child represents the activity to be performed within a state of the FSM, with the transitions between the states defined by the developer. Because each state is itself a behaviour it is possible to embed state machines. The FSMBehaviour class has the responsibility of maintaining the transitions between states and selects the next state for execution. Some of the children of an FSMBehaviour can be registered as final states. The FSMBehaviour terminates after the completion of one of these children.

4   A Gaia Model

In order to better understand our proposal on how GAIA and JADE can be combined to conceive and implement a multi-agent system (MAS) we will present a limited version of the system that is currently being implemented in the framework of the IST IMAGE project. We will show how this system can be analyzed, designed and implemented. For this system we have defined the following requirements:

- A user can request a route from one place to another. He can select among a variety of routes that are produced by the Geographical Information System (GIS).

- The MAS maintains a user profile so that it can filter the routes produced by the GIS and send to the user those that most suit him. The profiling will be based on criteria regarding the preferred transport type (private car, public transport, bicycle, on foot) and the preferred transport characteristics (shortest route, fastest route, cheapest route, etc).

- The system keeps track on selected user routes aiming:
  - To receive traffic events (closed roads) and check whether they affect the user's route (if that is the case then inform the user).
  - To adapt the service to user habits and needs.

In the following sections this MAS will be analyzed, designed and implemented.

4.1   The Analysis Phase

The analysis phase has led to the identification of four roles: one role, called Event-sHandler, that handles traffic events, one role called TravelGuide that wraps the GIS, one role, called PersonalAssistant, that serves the user and, finally, a social type role, called SocialType, that should be taken by all agents. A Gaia roles model for our system is presented in Table 2.

Table 2. The Gaia roles model

Role: EventsHandler
Description: It acts like a monitor. Whenever a new traffic event is detected it forwards it to all personal assistants.
Protocols and Activities: <u>CheckForNewEvents</u>, InformForNewEvents.
Permissions: read on-line traffic database, read acquaintances data structure.
Responsibilities:
    Liveness:
    EVENTSHANDLER = (PushEvents)$^\omega$
    PUSHEVENTS = <u>CheckForNewEvents</u>. InformForNewEvents
    Safety: A successful connection with the on-line traffic database is established.

Role: TravelGuide
Description: It wraps a Geographical Information System (GIS). It can query the GIS for routes, from one point to another.
Protocols and Activities: RegisterDF, QueryGIS, RequestRoutes, RespondRoutes.
Permissions: read GIS.
Responsibilities:
> **Liveness**:
> TRAVELGUIDE = RegisterDF. (FindRoutes)$^{\omega}$
> FINDROUTES = RequestRoutes. QueryGIS. RespondRoutes
> **Safety**: A successful connection with the GIS is established.

Role: PersonalAssistant
Description: It acts on behalf of a profiled user. Whenever the user wants to go somewhere it gets the available routes and determines which routes best match the user's profile. These routes are presented to the user. Moreover, it can adapt (i.e. using learning capabilities) to a user's habits by learning from user selections. Finally, it receives information on traffic events, it checks whether such events affect its user's route and in such a case it informs the user.
Protocols and Activities: InitUserProfile, UserRequest, InferUserNeeds, PresentRoutes, LearnByUserSelection, CheckApplicability, PresentEvent, RequestRoutes, RespondRoutes, InformForNewEvents.
Permissions: create, read, update user profile data structure, read acquaintances data structure.
Responsibilities:
> **Liveness**:
> PERSONALASSISTANT = InitUserProfile. ((ServeUser)$^{\omega}$ || (ReceiveNewEvents)$^{\omega}$)
> RECEIVENEWEVENTS = InformForNewEvents. CheckApplicability. [PresentEvent]
> SERVEUSER = UserRequest. RequestRoutes. RespondRoutes. [InferUserNeeds]. PresentRoutes. LearnByUserSelection
> **Safety**: true

Role: SocialType
Description: It requests agents that perform specific services from the DF. It also gets acquainted with specific agents.
Protocols and Activities: RegisterDF, QueryDF, SaveNewAcquaintance, IntroductsNewAgents.
Permissions: create, read, update acquaintances data structure.
Responsibilities:
> **Liveness**:
> SOCIALTYPE = GetAcquainted. (MeetSomeone)$^{\omega}$
> GETACQUAINTED = RegisterDF. QueryDF. [IntroductNewAgent]
> MEETSOMEONE = IntroductNewAgent. SaveNewAcquaintance
> **Safety**: true

Here it must be noted that another role is involved in the MAS operation. It is the Directory Facilitator (DF) FIPA role that is supported by JADE. However, this role concerns the operational level of the MAS and not the application itself, that's why a Gaia representation is not supplied for this role. Moreover, interactions with it are not presented as protocols, as they are defined in Gaia methodology, but as activities. Indeed, the activities RegisterDF (denoting the registration to the DF) and QueryDF (querying for agents of specific types or that have registered specific services activities) are DF services provided directly by JADE framework, provided not as a result of interaction between agents, but as method invocations.

The Gaia interaction model denotes which action returns from a request along with the roles that can initiate a request and the corresponding responders. Table 3 holds the necessary information for our model.

Table 3. Gaia interactions model

Protocol	Introduct-NewAgent	InformFor-NewEvents	RequestRoutes
Initia-tor(s)	SocialType	EventsHandler	PersonalAssistant
Re-ceiver(s)	SocialType	PersonalAssistant	TravelGuide
Respond-ing Action	-	-	RespondRoutes
Purpose/ Parame-ters	Introduce an agent to other agents. Possible content is the services and name associated with the initiator agent.	Inform an assistant that a new traffic event has occurred.	The assistant agent requests a set of routes from one place to another. The response includes different routes with different characteristics (shortest, fastest, cheapest) and for different transportation (private car, public transport, on foot).

4.2   The Design Phase

During this phase the Agent model is achieved, along with the services and acquaintance models.

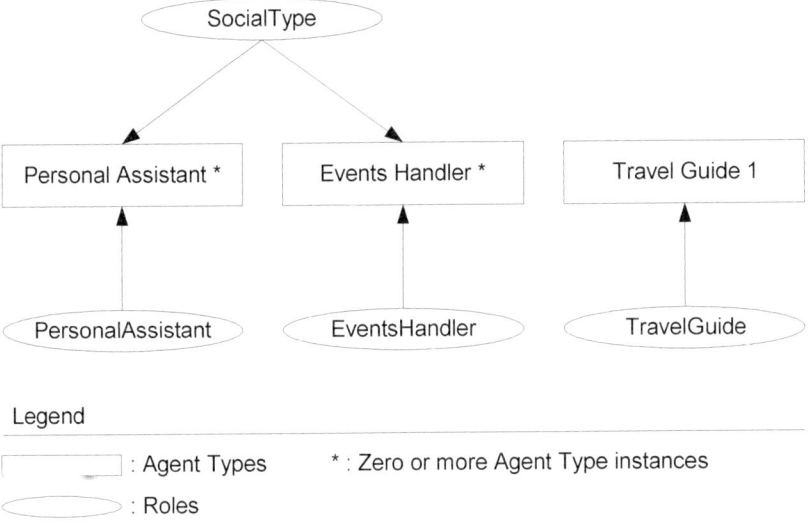

Fig. 1. Gaia agent model

The Agent model creates agent types by aggregating roles. Each emerging agent type can be represented as a role that combines all the aggregated roles attributes (activities, protocols, responsibilities and permissions). The agents model for our system will include three agent types: the personal assistant agent type, who fulfills the PersonalAssistant and SocialType roles, the events handler agent type, who fulfills the EventsHandler and SocialType roles and the travel guide agent type, who fulfills the TravelGuide role.

There will be one travel guide agent, as many personal assistants as the users of the system and zero or more events handlers. The Agent model is presented graphically in Fig. 1.

The services model for our system is presented in Table 4.

Table 4. Gaia services model

Service	Obtain route	Get notified on a relevant to the user's route traffic event
Inputs	Origin, destination	-
Outputs	A set of routes	The description of the event
Pre-condition	A personalized assistant agent is instantiated and associated with the user	A personalized assistant agent is instantiated and associated with the user. The user has selected a route to somewhere. A traffic event that is relevant to the user's route has happened
Post-condition	User selects a route	-

Finally we define the acquaintances model. For this model we propose a slight modification compared to the original definition presented in Gaia.

Table 5. Gaia acquaintances model

	Personal Assistant	Travel Guide	Events Handler
Personal Assistant		I, A	I
Travel Guide	I		
Events Handler	I, A		
Legend:			
I: Interacts (read "I" occurrences in rows, e.g. the **personal assistant** agent type interacts with **travel guide** and **events handler** agent types).			
A: Is acquainted, has the agent type in his **acquaintances** data structure (read "A" occurrences in rows, e.g. the **personal assistant** agent type knows **travel guide** agent types).			

We believe that this modification takes better into account the idea that an agent can interact with another agent (e.g. just responding to a request) without having necessarily any knowledge (information) about him. Therefore an analyst needs not only to specify which agent interacts with which, but which agent is acquainted with whom (i.e. knows whom) also. So, the **personal assistants** are acquainted and interact with **travel guides** and just interact with **events handlers**. **Events handlers** are acquainted and interact with **personal assistants**, while **travel guides** are not aware of the other agents however they interact with (service requests of) **personal assistants**. The above scheme is illustrated in Table 5.

At this point the abstract design of the system is complete, since the limit of Gaia has been reached. More effort must be done in order to obtain a good design though. At the end of the design process the system must be ready for implementation.

5   Developing JADE Agents from a Gaia Model

When moving from the Gaia model to an implementation using the JADE framework we have to make some assumptions and definitions.

Let's consider the liveness part of each role as its behaviour (usually having the same name with the role) in correspondence with the JADE terminology. Thus a simple or a complex behaviour represents each role. This behaviour is considered as the top-level behaviour of the role. Each behaviour may contain other behaviours, as in the JADE behaviours model. Let the contained behaviours be called lower level behaviours. The SocialType role of our system, for instance, has the SocialType top behaviour. This behaviour has two lower level behaviours, GetAcquainted and MeetSomeone.

The ω and ‖ operators on Gaia liveness formulas now have the following meaning. The ω means that a lower level behaviour is added by the behavior that contains it in the Gaia liveness formula and is only removed from the agent's scheduler when the behavior that added it, is removed itself. If such behaviours are more than one, they are connected with the ‖ symbol which denotes that they execute "concurrently". Concurrency in JADE agent behaviours is simulated. As noted before, only one thread executes per agent and behaviour actions are scheduled in a round robin policy.

5.1   Detailed Design

Many important design issues have yet to be covered when trying to implement a Gaia model with the JADE framework. Some of them are: a) ACL Messages (ontologies, protocols, content), b) Data structures, c) Algorithms and software components.

The ACL messages should be defined with respect to the FIPA ACL Message Structure Specification [7] and the JADE ACL Message class fields. The ACL Messages RequestRoutes and RespondRoutes are presented in Fig. 2. It is obvious that FAILURE and REFUSE ACL messages should be defined for global use in cases that, an action failed, is not supported or is denied.

ACL Message: RequestRoutes	ACL Message: RespondRoutes
Sender: Personal Assistant Agent	Sender: Travel Guide Agent
Receiver: Travel Guide Agent	Receiver: Personal Assistant Agent
FIPA performative: REQUEST	FIPA performative: INFORM
Protocol: RequestRoutes	Protocol: RequestRoutes
Language: SL	Language: SL
Ontology: ImageOntology	Ontology: ImageOntology
Content: Ontology action: RequestRoutes	Content: Ontology concept: Routes

Fig. 2. ACL messages definition

Finally, the designer of the MAS will easily implement agents whose internal structures and methods are pre-defined. To that end, the data structures and the AI tools that are to be used should be defined in this stage. For our system the following structures and methods should be clarified at this point:

- The acquaintances structure: It will contain information about other agents. Some of this information might include names, types (or services) and addresses (as will see in the following in JADE address and name are quite the same). The SocialType role maintains this structure (see the permissions field of the role definition in Table 2).
- The user profile structure: What information will be known about the user, how is it organized. Such questions must be answered at this point. The PersonalAssistant role maintains this structure (see the permissions field of the role definition in Table 2).

- The route structure: What is a route, what attributes are associated with a route. This structure is needed by both the TravelGuide and PersonalAssistant roles, the former instantiates such objects by information that it gets from the GIS (QueryGIS activity), while the latter filters the route structure objects according to the user profile (InferUserNeeds activity).
- The traffic event structure: What is a traffic event, what attributes are associated with it, how is it associated with a route. Both the EventsHandler and PersonalAssistant roles use this structure. The former instantiates such objects by information that it gets from external sources (CheckForNewEvents activity), while the latter checks whether a traffic event structure object is in a user's active route (CheckApplicability activity).
- The learning method: What will be learned about the user, how, where is it going to be stored, which machine learning algorithm will be used (different goals can indicate different algorithms). A learning method will be used by the LearnByUserSelection activity of the PersonalAssistant role.
- The components and technologies that will enable communication with external systems. Such systems are the on-line traffic database and the GIS. If the GIS services are available as web services then a suitable SOAP (Simple Object Access Protocol) client must be developed along with an XML parser that will translate the SOAP message content to an ontology concept or a JAVA object.

5.2   The JADE Implementation

At this point the MAS designer should have the full plan on how to implement the system. In our case the framework is JADE and the purpose of this paragraph is precisely to explain how the Gaia model is translated to a JADE implementation.

The procedure is quite straightforward. All Gaia liveness formulas are translated to JADE behaviours. Activities and protocols can be translated to JADE behaviours, to action methods (which will be part of finite state machine - FSM like behaviours) or to simple methods of behaviours. The JADE behaviours that can be useful for our model are the SimpleBehaviour, FSMBehaviour, AchieveREResponder and AchieveREInitiator.

The behaviours that start their execution when a message arrives, can receive this message either at the beginning of the action method (simple behaviours) or by spawning an additional behaviour whose purpose is the continuous polling of the message box (complex behaviours). For behaviours that start by a message from the Graphical User Interface (GUI), a GUI event receiver method should be implemented on the agent that starts the corresponding behaviour. Finally, those behaviours that start by querying a data source, or by a calculation, should be explicitly added by their upper level behaviour. For example, the SocialType role adds both the GetAcquained and the MeetSomeone behaviours. The difference is that GetAcquained will set itself as finished after executing once, while the MeetSomeone will continue executing forever – or until the agent is "killed".

The safety properties of the Gaia roles model must be taken into account when designing the JADE behaviours. Some behaviours of the role, in order to execute properly, require the safety conditions to be true. Towards that end, one at least behaviour is responsible for monitoring each safety condition of a role. Whenever a safety condition is found to be false, the functionality of the behaviours that depend on this safety condition is suspended and the monitoring behaviour initializes a procedure that will reestablish the validity of safety conditions. This procedure, for instance, can be the addition to the agent scheduler of a specific behaviour that will address the task of restoring the validity of safety conditions. In general, this procedure depends on the nature of the implemented system and the safety conditions. When the safety conditions are restored, the suspended functionalities are reactivated.

In our case, the safety requirement of the TravelGuide role is the establishment of communication with the GIS. The FindRoutes behaviour is responsible for monitoring the validity of this safety requirement. Whenever a connection fails to be established the FindRoutes behaviour sends to the agent GUI an event that results in a connection failure message, while responding to the personal assistant agent with a FAILURE ACL message. The system administrator must act in order to restore the GIS communication.

All behaviours of the lowest level are implemented first:

- PushEvents: A SimpleBehaviour that queries a database and if it gets a new event prepares an ACL message and sends it to all personal assistant agents.

- FindRoutes: It is a SimpleBehaviour that waits until it receives a specific ACL message, queries the GIS and sends back to the original sender a responding ACL message.

- ReceiveNewEvents: A SimpleBehaviour that waits until it receives a specific ACL message verifies if it is of interest for the specific user and sends an appropriate event to the GUI.

- ServeUser: a complex behaviour more like an FSMBehaviour with three states. At the first state it gets the user request (it is added to the agent scheduler as a consequence of that request) and sends an ACL message to the travel guide agent. Then it waits for its response. Alternatively, after getting the user request it could add an AchieveREInitiator behaviour. When it gets the response (second state) it infers on which routes should be forwarded to the user, forwards them and terminates its execution. If the user selects a route through the GUI, the GUI event catcher method of the agent starts this behaviour, but sets it immediately at the third state, which employs the learning algorithm in order to gain knowledge from the user action.

- GetAcquainted: This is a SimpleBehaviour that registers the agent to the DF, gets all needed agents from the DF and finally sends appropriate IntroductNewAgent ACL messages to all agents whom this agent wants to notify about his appearance. After the execution of these tasks the behaviour removes itself.

- MeetSomeone: a SimpleBehaviour that waits until it receives a specific ACL message then updates its acquaintance data structure with a new contact and the services that the new contact provides.

A good architecture paradigm contains no functionality at the top-level behaviour, instead, the agents tasks are embedded in lower level behaviors. Thus, the top-level behaviours that represent the actions performed in the setup phase of the agent are:

- EventsHandlerAgent: initialize the Acquaintances data structure, add the PushEvents, GetAcquainted and MeetSomeone behaviours.

- TravelGuideAgent: register to the DF and add the FindRoutes behaviour.
- PersonalAssistantAgent: initialize the Acquaintances data structure, get the initial user profile, add the GetAcquainted, MeetSomeone, ServeUser and ReceiveNewEvents behaviours.

Summarizing, the following steps should be followed in order to easily translate a Gaia model to a JADE implementation:

1. Define all the ACL messages by using the Gaia protocols and interactions models.
2. Design the needed data structures and software modules that are going to be used by the agents by using the Gaia roles and agents models.
3. Decide on the implementation of the safety conditions of each role.
4. Define the JADE behaviours. Start by implementing those of the lowest levels, using the various Behaviour class antecedents provided by JADE. The Gaia model that is useful in this phase is the roles model. Behaviours that are activated on the receipt of a specific message type must either add a receiver behaviour, or receive a message (with the appropriate message filtering template) at the start of their action. Gaia activities that execute one after another (sequence of actions that require no interaction between agents) with no interleaving protocols can be aggregated in one activity (behaviour method or action). However, for reusability, clarity and programming tasks allocation reasons, we believe that a developer could opt to implement them as separate methods (or actions in an FSM like behaviour).
5. Keep in mind that Gaia roles translated to JADE behaviours are reusable pieces of code. In our system, the same code of the behaviours GetAcquainted and MeetSomeone will be used both for the personal assistant and events handler agents.
6. At the setup method of the Agent class invoke all methods (Gaia activities) that are executed once at the beginning of the top behaviour (e.g. RegisterDF). Initialize all agent data structures. Add all behaviours of the lower level in the agent scheduler.

6   Discussion

In this paper we have presented the analysis, design and implementation phases of a limited version of a system developed in the context of the IST IMAGE project. As we already have said before, the only pretension we have with this paper is to share our experience on how one can combine the Gaia methodology and the JADE development environment in order to implement a real multi-agent system. Gaia methodology is an easy to use agent-orient software development methodology that however presently, covers only the phases of analysis and design. On the other hand JADE is a FIPA specifications compliant agent development environment that gives several facilities for an easy and fast implementation. Our aim was to reveal the mapping that may exists between the basic concepts proposed by Gaia for agents specification and agents interactions and those provided by JADE for agents implementation, and therefore to propose a kind of roadmap for agents developers. Presently we have introduced a slight modification for the Gaia acquaintances model and our future work, through our main work on IMAGE project, will be to examine if there could be proposed some modifications in both, Gaia and JADE, that would help to make more efficient their combination.

Other works in this volume address also the modeling of agent systems (see [5, 10]) and the transition from models to implementation [5]. More precisely in [5], authors use UML–like diagrams in order to model multi-agent systems. They also provide a tool that produces JADE or FIPA-OS implementation assistance. Their major difference with Gaia is that they first model agents and then their roles-behaviours. Besides this, they also define patterns in system, agent, behaviour and action levels. They use patterns as components in multi-agent system design and implementation, as Gaia uses services (agent interactions), agents and behaviours. Finally, in [10] the author extends UML in order to address the problem of multi-agent system modeling.

Acknowledgements. We gratefully acknowledge the Information Society Technologies (IST) Programme and specifically the Research and Technological Development (RTD) "Intelligent Mobility Agent for Complex Geographic Environments" (IMAGE, IST-2000-30047) project for contributing in the funding of our work.

References

1. Agent UML: http://www.auml.org/
2. Bellifemine, F., Caire, G., Trucco, T., Rimassa, G.: Jade Programmer's Guide. JADE 2.5 (2002) http://sharon.cselt.it/projects/jade/
3. Collis, J. and Ndumu, D.: Zeus Technical Manual. Intelligent Systems Research Group, BT Labs. British Telecommunications. (1999)
4. Collis, J. and Ndumu, D.: Zeus Methodology Documentation Part I: The Role Modelling Guide. Intelligent Systems Research Group, BT labs. British Telecommunications (1999)
5. Cossentino, M., Burrafato, P., Lombardo, S., Sabatucci, L.: Introducing Pattern Reuse in the Design of Multi-Agent Systems. In this volume
6. DeLoach S. and Wood, M.: Developing Multiagent Systems with agentTool. In: Castelfranchi, C., Lesperance Y. (Eds.): Intelligent Agents VII. Agent Theories Architectures and Languages, 7th International Workshop (ATAL 2000, Boston, MA, USA, July 7-9, 2000),. Lecture Notes in Computer Science. Vol. 1986, Springer Verlag, Berlin (2001)
7. FIPA specification XC00061E: FIPA ACL Message Structure Specification (2000) http://www.fipa.org
8. FIPA-OS: A component-based toolkit enabling rapid development of FIPA compliant agents: http://fipa-os.sourceforge.net/
9. Giunchiglia, F., Mylopoulos, J., Perini, A.: The Tropos Software Development Methodology: Processes, Models and Diagrams, in AAMAS02
10. Huget, M. P.: Agent UML Class Diagrams Revisited. In this volume
11. Kendall, E.A.: Role Model Designs and Implementations with Aspect Oriented Programming. Proceedings of the 1999 Conference on Object- Oriented Programming Systems, Languages, and Applications (OOPSLA'99)
12. Reticular Systems Inc: AgentBuilder An Integrated Toolkit for Constructing Intelligent Software Agents. Revision 1.3. (1999) http://www.agentbuilder.com
13. Sycara, K., Paolucci, M., van Velsen, M. and Giampapa, J.: The RETSINA MAS Infrastructure. Accepted by the Journal of Autonomous Agents and Multi-agent Systems (JAAMS)

14. Wood, M.F. and DeLoach, S.A.: An Overview of the Multiagent Systems Engineering Methodology. AOSE-2000, The First International Workshop on Agent-Oriented Software Engineering. Limerick, Ireland (2000)
15. Wooldridge, M., Jennings, N.R., Kinny, D.: The Gaia Methodology for Agent-Oriented Analysis and Design. Journal of Autonomous Agents and Multi-Agent Systems Vol. 3. No. 3 (2000) 285-312

Designing Peer-to-Peer Applications: An Agent-Oriented Approach

D. Bertolini[1], P. Busetta[1], A. Molani[2], M. Nori[1], and A. Perini[1]

[1] ITC-Irst, Via Sommarive, 18, I-38050 Trento-Povo, Italy
{bertolini,busetta,nori,perini}@irst.itc.it – phone +39 0461 314 330
[2] University of Trento, via Sommarive 14, I-38050 Trento-Povo, Italy
molani@science.unitn.it

Abstract. This paper focuses on design issues to be faced when developing knowledge management (KM) applications based on the integration of peer-to-peer and multi-agent technologies. The reasons for using these technologies rest on the requirements posed by the specific KM paradigm that has been adopted, which emphasizes aspects such as *autonomy* and *distribution* of knowledge sources. We adopt an agent-oriented approach that extends *Tropos*, a software engineering methodology introduced in earlier papers. We present a characterization of peer-to-peer in terms of a general architectural pattern, a set of design guidelines for peer-to-peer applications, and a framework that integrates multi-agent and peer-to-peer concepts and technologies.

1 Introduction

Emerging application areas such as *distributed knowledge management* (DKM) [2] call for open as well as robust and secure software systems. Several technological solutions – sometimes combined – have been proposed to address some of these non-functional requirements, including e-services, application service provision, multi-agent systems (MAS), and peer-to-peer architectures (P2P).

A question naturally arises: how can we compare these technologies and evaluate which one is better given an application's requirements, or decide if they can be integrated in our application? For instance, some P2P concepts and technologies – such as *peer* and *peer group* – are well suited to capture and support certain aspects, such as the dynamism of the social organizations using a distributed system; typical MAS concepts – such as *agent* and *multi-agent systems* – are better suited to capture reasoning and collaboration among distributed components; e-services and ASPs promise to be widely available and to become important technological players in the near future, in spite of their current clumsiness in supporting real openness and robustness.

We present here some results of our work on a development framework for distributed knowledge management applications, which attempt to give some answers to the question raised above. In particular, we focus on the problem-solving process underlying the design of DKM applications. Objective of this process is to specify how to implement the system's functional requirements while respecting

R. Kowalczyk et al. (Eds.): Agent Technology Workshops 2002, LNAI 2592, pp. 92–106, 2003.

the constraints imposed by the non-functional requirements (NFR). We adopt an agent-oriented approach by extending a software engineering methodology, called *Tropos*, presented in earlier papers [7].

The main contributions of this work are three. The first is a characterization of peer-to-peer in terms of an architectural pattern, obtained by applying the *Tropos* methodology. The second is a set of guidelines for a system designer on when and how to apply a P2P approach; these guidelines are targeted at a specific class of applications (DKM) but can be easily generalized to others. The third result is an architectural framework that integrates P2P and multi-agent systems, basically proposing P2P as a way of deploying an open MAS and providing some supporting services, most notably discovery.

This paper is structured as follows. Section 2 briefly presents the distributed knowledge management approach that motivated this research, and other work relevant to this paper. Section 3 introduces some basic Tropos concepts and notations, which are used, in Sect. 4, to discuss some typical peer-to-peer architectures and to identify their common patterns. In Sect. 5 we focus on JXTA, the P2P framework we are currently using. Section 6 presents the guidelines for designing the architecture of a DKM system applying both P2P and MAS concepts. Section 7 describes a scenario taken from the Health Care domain, and shows how to apply the guidelines in practice. Finally, conclusions and future works are presented in Sect. 8.

2   Background and Related Work

Common knowledge management (KM) systems support the collection and categorization of knowledge with respect to a single, shared perspective, and its redistribution to its users by a variety of means. In many instances, this leads to the construction of one or a few repositories of documents, organized around a single ontology or other meta-structures. Users are given tools to search, browse, or receive documents as soon as they become available, varying from simple Web interfaces to sophisticated agents running on the users' desktops.

However, common wisdom is that building the shared perspective at the core of a KM systems is an expensive task, sometimes impossible to achieve when users have substantially different goals and background. To tackle this issue, our research group is investigating a novel approach called *distributed knowledge management* (DKM) [2]. The idea at the basis of DKM is supporting the integration of autonomously managed KM systems, without forcing the creation of centralized repositories, indexes, or shared ontologies. This integration involves the deployment of a set of complex protocols and algorithms for information retrieval, natural language processing, machine learning, deductive and inductive reasoning, and so on, sometimes requiring the direct participation of human users. Initial publications include [10,3].

To build such systems, a clear understanding of how available technologies can be combined to deal with functional and non-functional requirements becomes a critical issue. Software engineering provides the designer with a set of principles aimed at decomposing the system design problem into subproblems and at favoring the re-use of previous solutions. We are considering these issues from an agent-oriented perspective, by extending the *Tropos* software enginee-ring methodology.

The extensions to Tropos for the support of DKM are part of the development of a larger framework, whose aim is to provide both a methodology and techno-logical support to analysts, designers, and developers. As part of this framework, we are investigating both multi-agent and peer-to-peer technologies.

Different lines of research are relevant to the work presented here, ranging from peer-to-peer and agents applications, to agent oriented software engineering in general [17]. For instance, two recent works have attempted to blend peer-to-peer and agents into a single framework: Anthill [1] and InfoQuilt [13]. Both see peer-to-peer as a paradigm for networks of autonomous systems that may join or leave at any time. Relevant to our work are also studies on peer-to-peer systems that have been conducted using i^* [18] (the modeling language at the basis of Tropos), devoted to the characterization of the base technology with respect to critical issues such as security and neighborhood analysis [9,14]

3   An Overview of Tropos

The *Tropos* methodology [7] adopts an agent oriented approach to software de-velopment starting from the very early stage of requirement specifications, when the environment and the system-to-be are analyzed, down to system design and implementation. The methodology identifies a number of phases (*early requi-rements, late requirements, architectural design, detailed design*, and *implemen-tation*) and has been applied to several case studies [8]; an example is in this volume [4].

The core process of the methodology consists in performing conceptual mo-deling using a visual language that provides intentional and social concepts such as actor, goal, belief, plan and dependency. An *actor* models an entity that has strategic goals and intentionality, such as a physical *agent*, a *role* or a *position*. A *role* is an abstract characterization of the behavior of an actor within some specialized context, while a *position* represents a set of roles, typically covered by one agent. The notion of actor in Tropos is a generalization of the classical AI notion of software agent. *Goals* represent the strategic interests of actors. A *dependency* between two actors indicates that an actor depends on another in or-der to achieve a goal, execute a plan, or exploit a resource. Tropos distinguishes between hard and soft goals, the latter having no clear-cut definition and/or cri-teria as to whether they are satisfied. Softgoals are useful for modeling software qualities [5], such as security, performance and maintainability. A Tropos model is represented as a set of diagrams. In particular, actor and dependency models are graphically represented as *actor diagrams* in which actors are depicted as circles, their goals as ovals. The network of dependency relationships among ac-tors are depicted as two arrowed lines connected by a graphical symbol varying according to the dependum: a goal, a plan or a resource. Actor goals and plans can be analyzed from the point of view of the individual actor using three basic reasoning techniques: *means-end analysis, contribution analysis,* and *AND/OR decomposition.* During this analysis, new actor dependencies can be identified.

Tropos is currently being extended with concepts suitable to model some specific notions – such as distributed knowledge, autonomy and coordination – that are peculiar to DKM, in order to support an early requirement model of a DKM domain [11].

In this paper, we focus on a later phase in software development: given a requirement analysis of a DKM problem, we provide guidelines for defining one or more suitable system architectures. According to *Tropos,* these types of issues are typically faced during the *architectural design* phase.

4   The Peer-to-Peer Virtual Community Pattern

In this section, we use Tropos to analyze which functional and non-functional requirements are successfully satisfied by peer-to-peer systems [12]. In particular, we consider two well known systems, Gnutella and Napster, both supporting file sharing (in particular MP3 sound files).

The logical architecture of *Gnutella*[1] is described in the *Tropos* actor diagram depicted in Fig. 1. Both actors model a basic role, the Servent (Gnutella's terminology for peer). Each servent has the goals of discovering other servents, looking for MP3 files, and providing its own MP3s, and depends on all other servents for achieving them. In other words, the goals of a servent are achieved only by means of a virtual community of peers, each one playing the role of servent. In the diagram, this is represented as goal dependencies between two generic servents. The dependencies are symmetric for every servent; for simplicity, we have shown them going only in one direction. The discovery goal enables the community to be dynamic, since peers can join and leave the community at any time without having to register their presence anywhere. There is no centralized service local to any one peer; conversely, all the peers provide (and take advantage of) the same sets of services.

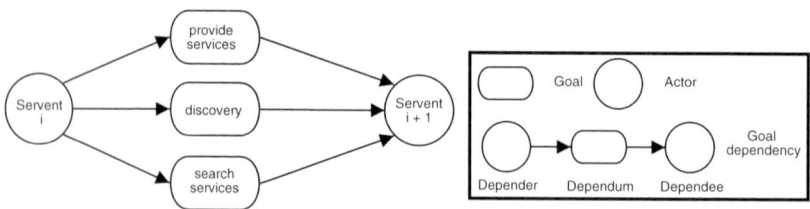

Fig. 1. Gnutella: architectural design, according to the Tropos methodology

---
[1] http://www.gnutella.com

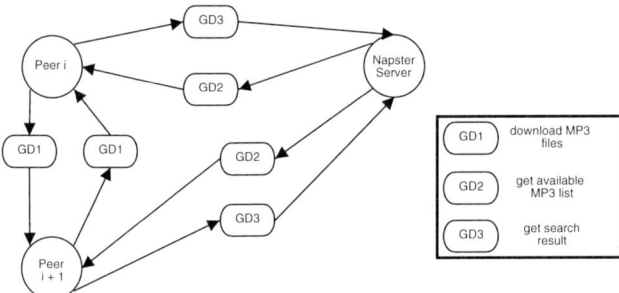

Fig. 2. Napster: architectural design

In *Napster*[2], each client, represented by Peer actors in Fig. 2, has similar goals to those of Servents in Gnutella (search and download files). However, in order to obtain a search result, the client must contact a central server (Napster Server). This server maintains a list of all the active clients and a list of all the shared MP3 files. The server depends on the clients for building these lists. This situation is represented by a set of goal dependencies in Fig. 2: the generic Peer depends on Napster Server for getting a search result (goal dependency GD3) and Napster Server depends on Peer to get the lists of available MP3 files (goal dependency GD2). Analogously to Gnutella, two generic peers depend on each other for downloading MP3 files (goal dependency GD1). So, the individual goal is obtained by means of a community of peers (Napster clients) that coordinate with an actor playing a distinct role in the community, i.e. the server.

We abstract these two architectures, and others not discussed here, in the basic model depicted in Fig. 3. The Individual actor models somebody who has at least one of two goals: accessing a resource (or, equivalently, using a service); and, letting others access her own resources (or use her own services). An individual has a set of constraints and preferences (shown as softgoals in the diagram) which drive her behavior when achieving these goals. In particular, an important requirement is being autonomous, i.e. being able to control what to access or to make available to the external world, and when.

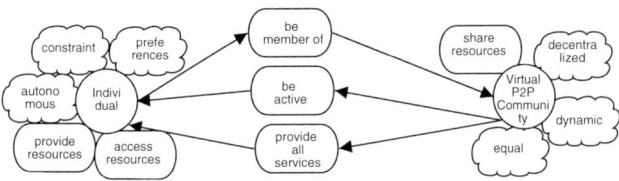

Fig. 3. The peer-to-peer virtual community pattern

---

[2] http://opennap.sourceforge.net

The actor Virtual P2P community has the main goal of letting its members cooperate for sharing a specific type of resource or service. Three additional requirements (shown as softgoals) have been identified: being highly dynamic – available resources and services provided by members can be added or removed at any time –; being decentralized, i.e. the community is able to achieve its main goal independently of any specific member or component; and finally, being composed of peers on an equal basis, that is, every member is compelled to provide (at least potentially) resources or services and has the right to access the services of others.

The dependency be member of captures the fact that the individual's goals can be satisfied by joining the community. The act of joining implies the acceptance of the community's main goal and rules, highlighted above. Conversely, a community can achieve its goal only if it is active, and this can be accomplished only by having individuals as members. Finally, the dependency provide all services models the rule that in a peer-to-peer community all members are equal, i.e. provide the same set of services.

Going back to the systems discussed previously, Gnutella is a "perfect" P2P system, since it satisfies all non-functional requirements of the community highlighted in Fig. 3. Similarly, it may be argued that Napster is not a real P2P system, since the community depends on a centralized service.

5   JXTA P2P: Logical Architecture Elements

We focus now on *JXTA*[3], a set of open, generalized peer-to-peer protocols that allow devices to communicate and collaborate through a connecting network.

From our perspective, the fundamental concepts of JXTA are three: *peer*, *peer group* and *service*.

A *peer* is "any device that runs some or all the Project JXTA protocols". The complex layering of the JXTA protocols and the ability for a peer to simultaneously participate to more than one peer group (described below) imply that a peer is – at least conceptually – a multi-threaded program.

A *peer group* is a collection of peers that have agreed upon a common set of rules to publish, share and access "codats" (shorthand for code, data, applications), and communicate among themselves. Each peer group can establish its own membership policy, from open (anybody can join) to highly secure and protected (join only if you have sufficient credential). Thus, a peer group is used to support structuring (based on social organizations or other criteria) on top of the basic peer-to-peer network. As mentioned above, a peer may be part of many groups simultaneously.

A peer provides to each of its group a set of *services*, which are advertised to the other members of the group. In other words, groups in JXTA support a model for service publishing which is alternative to the traditional, centralized directory services model. Each group specifies a set of services that have to be

---

[3] A P2P open source effort started in April 2001. http://www.jxta.org/ and http://spec.jxta.org/v1.0/docbook/JXTAProtocol.html

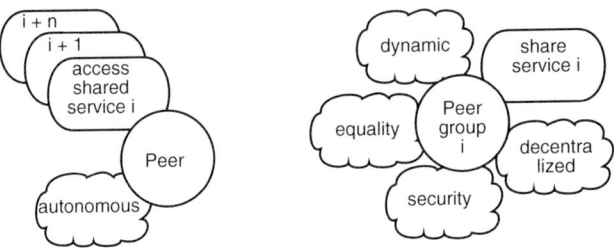

Fig. 4. JXTA: architectural design overview

provided by its members; they may include supporting basic JXTA protocols (e.g., discovery) as well as user-written applications, such as file sharing.

From a high-level perspective, these concepts can be modeled in terms of roles in the architectural design diagram depicted in Fig. 4, which extends the pattern depicted in Fig. 3. A Peer has n goals access shared service i, one for each type of service it needs, and the requirement (specified as a softgoal) of being autonomous. A generic actor PeerGroup has the goal of sharing a service of a specific type i and a set of requirements: to support the decentralized distribution of services, to allow for dynamic membership of the group, to support security management, and to require that all members provide equal services. JXTA tackles these requirements using a number of mechanisms, represented as goal decompositions in Fig. 5. The publishing and discovery mechanisms, together with a message-based communication infrastructure and peer monitoring services, support decentralization and dynamism. security is supported by a membership policy (which authenticates any peer applying to a peer group) and an access protocol (for authorization control).

In summary, it can be said that the creation of a PeerGroup follows from the need to define a set of peers that are able to communicate and interact with a

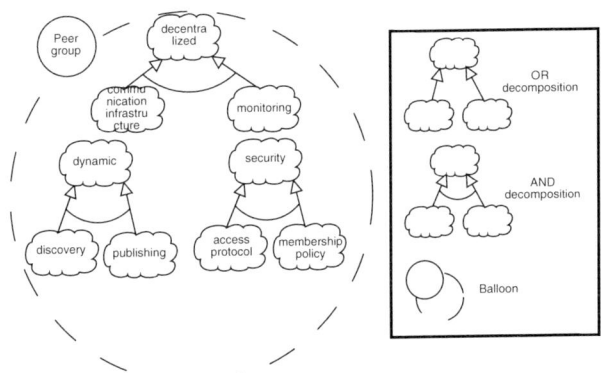

Fig. 5. JXTA: PeerGroup architectural design

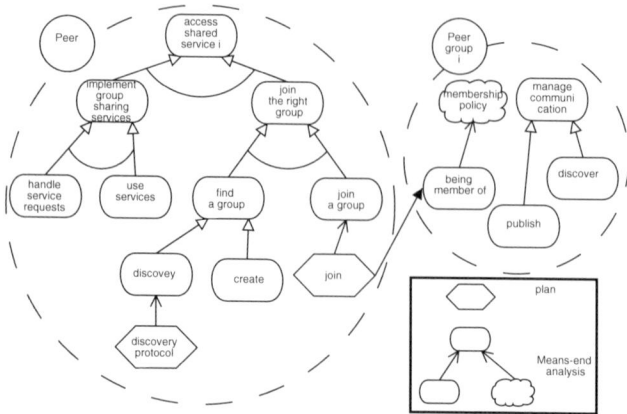

Fig. 6. JXTA: Peer architectural design

common set of capabilities and protocols, and that are aggregated for security reasons (i.e., no extraneous peer can intervene amongst them).

Figure 6 shows that a Peer's main goal (access shared service *i*) can be decomposed into two subgoals: join the right group and implement group sharing services. The first means that the peer must find a group that it is authorized to join and that provides the services it needs. This goal can be further decomposed into join a group and find a group. The latter can be satisfied by using the JXTA discovery service; if no adequate group is found, a new one can be created. Joining a group that has been discovered or created depends on its membership policy. Once a group has been joined, a Peer must implement all the services required by that group, implementing both the client side (use services) and the server side (handle service requests). It is important to stress that a Peer is autonomous in deciding how to provide a service (only protocols are predefined), and that a Peer can join different PeerGroups in order to achieve different goals.

6   Architectural Design of Communities in a DKM System

In this section, we sketch the guidelines for designing the architecture of a DKM system, focusing on the support for virtual communities. Input to the process described here is, in Tropos terms, the output of a late requirements phase (Sect. 3); that is, a domain analysis which included the system-to-be. The late requirements phase led to the identification of the stakeholders, including knowledge and service sources and of their users; how this analysis is performed is outside the scope of this paper. The guidelines consist of the following steps:

1. apply the P2P virtual community pattern;
2. design the JXTA peer group implementing the virtual community;
3. design the agents that implement a peer service.

Each step is briefly described below, while in the next we section present their practical application.

Applying the P2P Virtual Community Pattern. At this step, the designer has to answer to a basic question: can a Tropos late requirement model of a DKM problem be refined as a virtual P2P community model? The importance of reusing known patterns during architectural design is stressed in [6].

By definition, actors in a DKM scenario, as well as individuals in the community pattern, want to keep their autonomy. A designer, then, should look for the following conditions to be realized (they are illustrated in Fig. 7 where two P2P virtual communities can be identified):

- there is a subset C of the actors that have pair-wise dependencies (e.g., actor A1 depends on A7 for goal α and A7 depends on A1 for β) in such a way that any actor is dependent on many, if not most, other actors in C. In other words, each actor is at the center of a sort of "star" network of reciprocal dependencies with many others;
- these dependencies can be generalized to one or a few types only. In particular, they are of type "access" and "provide" a service or a resource.

If these conditions are satisfied, then the designer can apply the P2P virtual community pattern to generate a model for a community supporting C. The main goals of this virtual community are suitable abstractions of the generalized dependencies linking the actors in C. The goals of an individual are also generated from each generalized dependency, and are two: achieve the objective of the dependency, and, conversely, satisfy it. A validation step is necessary, and consists of two main activities:

- verifying that the goals of the actors in C can be effectively satisfied by adopting the goals of the individual in the community model; that is, verify that the P2P virtual community helps in achieving the actors' goals. This may be performed as a means-end analysis, which decomposes those hard

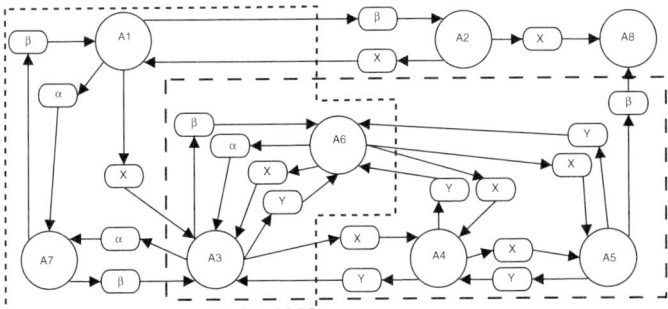

Fig. 7. Identifying P2P virtual communities. The *first community* (dashed line) is composed by actors involved in pair dependencies for *goals X and Y*, the *second* (dotted lines) by actors involved in pair dependencies for *goals α and β*

goals into plans that, eventually, are satisfied by adopting the individual's goals;
- verifying that the general soft goals of a P2P virtual community – dynamism, equality, and decentralization – are at least partially achieved by the community being designed.

Designing a JXTA Peer Group. Once that one or more P2P virtual communities have been identified, their supporting infrastructure can be designed. Assuming that JXTA is adopted as the basic communication technology, our major – and natural – choice is then to associate a JXTA peer to each individual, and a JXTA peer group to each P2P virtual community. Thus, a social actor (no matter if it is a single person or a group) will have, as its supporting system, a peer participating to as many peer groups as communities that have been identified during the requirement analysis of its DKM problems.

Agents as JXTA Services. The final step is to design how individuals implement their goals with respect to each of the communities they form part of. From a JXTA perspective, it is necessary to specify which application services have to be provided by a peer, and how they are published on the network. JXTA, however, leaves the designer total freedom concerning their communication protocols and internal implementation. JXTA provides its own communication mechanisms (unreliable message queues called "pipes" and an XML based encoding format), but the application may decide to use something different, since JXTA supports publishing and discovery of communication end-points of any type (i.e., the address to which to send messages in order to obtain a service from a certain peer) as long as they can be represented in XML. Moreover, JXTA allows the publishing of additional service-specific information – also encoded as XML documents – along with their communication end-points; this gives a nice opportunity for targeted discoveries and filtering of potential peers.

The services to be provided by the peer associated to an individual are easily identified from the individual's goals with respect to the community. Once the peer protocols and publishing information have been defined, the design of the actors participating to the community can be performed in parallel. The Tropos methodology naturally leads to the design of multi-agent systems; thus, it is most likely that the implementation of services are agents themselves, interacting both with agents internal to their own actor and with other agents implementing other actors of the same community.

7   A Case Study

We show a practical application of the architectural guidelines given above to a DKM problem, taken from the Health Care (HC) domain. Health care is interesting both from a multi-agent perspective – as pointed out in [16,19], where an approach is proposed to construct an information system centered on the individual user (customer-patient), instead of on the providers of information and

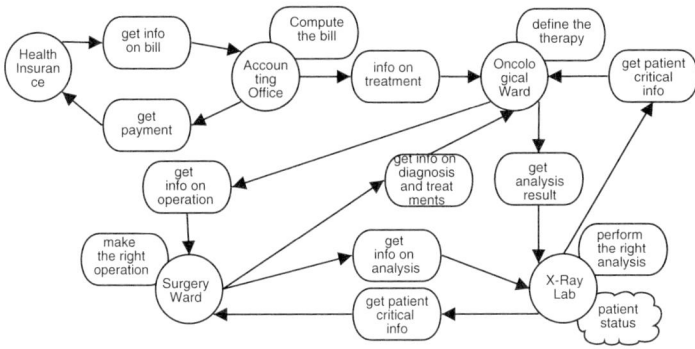

Fig. 8. The hospital wards scenario: an actor diagram

services – and also from a knowledge management perspective [15]. Our scenario consists of the hospital wards involved in a specific patient treatment along with other related stakeholders. In our example, a patient with cancer may be admitted and pharmacologically treated by the oncological ward, operated by the surgical ward, and sent to the X-ray lab for analysis. All these stages may happen at different times, and possibly for different reasons. Each ward has its own local knowledge and a specific way of organizing and managing activities, processes, and information, according to its own primary objective. We assume that the output of a deep requirement analysis led to the model depicted in Fig. 8, where the actors represent the knowledge management systems of the entities involved in the scenario. The main goal of each actor has been delegated to it by its local users (not shown in the diagram). For instance, the Oncological Ward has the goal define the therapy. In order to achieve this goal, the Oncological Ward has to collect information on the surgery that the patient was subjected to and on possible post operation events and treatments. This is represented by the goal dependency get info on operation between the actor Oncological Ward and the actor Surgery Ward. Analogously, the Oncological Ward needs information also on the MRI or x-ray analysis performed by the X-Ray Lab, represented by the goal dependency get analysis result. Conversely, the X-Ray Lab depends on the Oncological Ward to get critical information about the patient. We represent additional requirements as softgoals; for example, X-Ray Lab's perform the right analysis goal is influenced by the patient status (e.g., a patient could be pregnant).

To realize their mission, hospital wards need to cooperate, as highlighted by the goal dependencies among their KM systems. The result of this collaboration is a *distributed and virtual* clinical patient report. Privacy policy is a sensitive issue in this context. Following the first step of the guidelines presented in Sect. 6, we analyze the model depicted in Fig. 8 and apply the virtual P2P community pattern (Fig. 9). The result is a model where the community has the goal share patient info, and each individual has the generic goals of finding patient record information (access available patient info) and making available patient record information (make available patient info). It can be easily seen that these generic

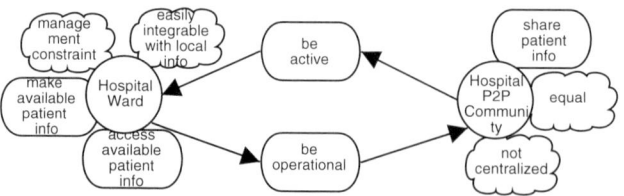

Fig. 9. Applying the P2P virtual community pattern

goals do actually contribute to achieving the wards' and labs' original goals. The community is not centralized, since all members keep their own information (as noted above, the members of this group have a strong interests in maintaining their autonomy), and all of them are on an equal basis because they need both to access and to provide information. The generic goal dependency, **be member of**, between an individual and a P2P community described in Fig. 3 has been specialized as **be operational**. Dynamism of the community is probably minimal in this setting, since the number of wards and labs tend to be stable over time; however, new actors may join over time, for instance new wards, or mobile emergency teams. One of the advantages of designing a virtual P2P community is that, *after* the system's deployment, new actors can enter the community as long as their goals can be reformulated as the individual's goals. Figure 10 represents the output of the second step of our guidelines: a model for the peer group supporting the community identified above, according to the model of a JXTA peer group depicted in Fig. 4. A custom **security** policy is specified, while

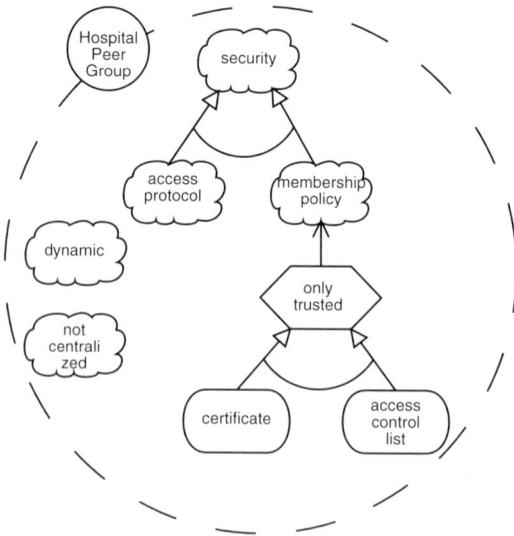

Fig. 10. The HC PeerGroup.

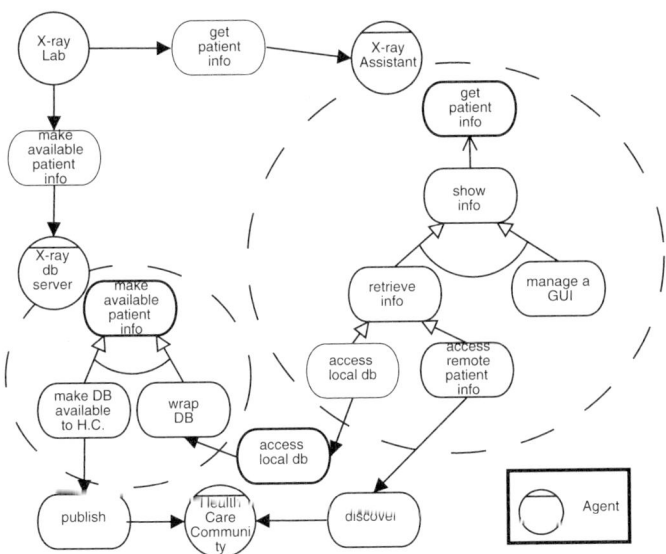

Fig. 11. The X-Ray Lab Peer and the PeerGroup

all other other goals are adequately supported by the standard JXTA protocols and do not need to be analyzed any further. The membership policy allows only trusted peers to join the community; this control could be performed by using certificates and digital signatures, maintaining a list of trusted or revoked certificates, and granting particular credentials to every peer. Each time a peer uses a service, the access protocol checks the credentials of the peer in order to verify if it is authorized to access that particular service (not shown in the diagram). The final result of this second step is the definition of an environment – the peer group – that supports the joining, leaving and dynamic discovery of members, while at the same time providing a known level of reliability in terms of security and tolerance to failures of single components.

We finally apply the third step of our guidelines. The service provided by the peers of our community is an interface to their local databases containing patient information. We do not elaborate here on published information and protocols, which can be inspired by works on information integration systems, while we present a plausible architecture for a specific actor, the X-Ray Lab. Figure 11 contains only what is relevant to this discussion. The actor has a goal of its own (perform the right analysis), which, after a decomposition (not shown here), includes the sub-goal get patient info. This is delegated to a specific agent, the X-Ray Assistant. After further decomposition, the goal is achieved by accessing the community (access remote patient info, which is a reformulation of the generic individual's access available patient info of Fig. 9).

Participating to the community implies that the X-Ray Lab has to satisfy the individual's make available patient info goal. This is delegated to a second agent, the X-ray db server, which operates both for the community and for local agents (see the access local db goal delegated by the X-Ray Assistant).

8   Conclusions and Future Work

This paper sketches initial results of a study aimed at defining a framework for building DKM applications, covering all development phases. The process is based on *Tropos*, an agent oriented software engineering methodology. We focused on the architectural design phase, when the application requirements are analyzed in the light of known architectural patterns, and defined some guidelines driving the development of DKM systems as a combination of a specific peer-to-peer infrastructure, JXTA, and multi-agent technologies. We illustrated these guidelines with reference to a scenario taken from the health care domain.

Our long term objective is to complete the framework, working both on the conceptual aspects required to model a DKM application and on the technology, currently being tested in a real-world application.

Acknowledgments. The work presented in the paper is realized inside the *EDAMOK* project (http://edamok.itc.it/), funded by the Provincia Autonoma di Trento (P.A.T.). We thank John Mylopoulos and Eric Yu for their precious suggestions, and Mark Carman for his review.

References

1. O. Babaoglu, H. Meling, and A. Montresor. Anthill: A framework for the development of agent-based peer-to-peer systems. Technical Report UBLCS-2001-09, University of Bologna, Italy, 2001. http://www.cs.unibo.it/projects/anthill/.
2. M. Bonifacio, P. Bouquet, and P. Traverso. Enabling Distributed Knowledge Management: Managerial and technological implication. *Novatica and Informatik/Informatique*, 3(1), 2002.
3. P. Bouquet, A. Donà, L. Serafini, and S. Zanobini. Contextualized local ontologies specification via ctxml. In *MeaN-02 – AAAI workshop on Meaning Negotiation*, Edmonton, Alberta, Canada, 2002.
4. P. Bresciani and F. Sannicolò. Requirements analysis in tropos: a self referencing example. In *this volume*, 2002.
5. L. K. Chung, B. A. Nixon, E. Yu, and J. Mylopoulos. *Non Functional Requirements in Software Engineering*. Kluwer Publishing, 2000.
6. M. Cossentino, P. Burrafato, S. Lombardo, and L. Sabatucci. Introducing pattern reuse in the design of multi-agent systems. In *this volume*, 2002.
7. F. Giunchiglia, A. Perini, and J. Mylopoulus. The Tropos Software Development Methodology: Processes, Models and Diagrams. In C. Castelfranchi and W.L. Johnson, editors, *Proceedings of the first international joint conference on autonomous agents and multiagent systems (AAMAS02)*, pages 63–74, Bologna, Italy, July 2002. ACM press.

8. F. Giunchiglia, A. Perini, and F. Sannicolò. Knowledge level software engineering. In J.-J.C. Meyer and M. Tambe, editors, *Intelligent Agents VIII*, LNCS 2333, pages 6–20. Springer-Verlag, Seattle, WA, USA, Proceedings of the eighth International Workshop on Agent Theories, Architectures, and Languages (ATAL) edition, August 2001. Also IRST Technical Report 0112-22, Istituto Trentino di Cultura, Trento, Italy.

9. L. Liu, E. Yu, and J. Mylopoulos. Analyzing security requirements as relationships among strategic actors. In *Proc. of the 2nd Symposium on Requirements Engineering for Information Security (SREIS'02)*, Raleigh, North Carolina, October 2002.

10. B. Magnini, L. Serafini, and M. Speranza. Linguistic based matching of local ontologies. In *MeaN-02 – AAAI workshop on Meaning Negotiation*, Edmonton, Alberta, Canada, 2002.

11. A. Molani, P. Bresciani, A. Perini, and E. Yu. Intentional analysis for knowledge management. Technical report, ITC-IRST, 2002.

12. Andy Oram, editor. *Peer-to-Peer Harnessing the Power of Disruptive Technologies.* O'Reilly Associates, 2001.

13. S. Patel and A. Sheth. Planning and optimizing semantic information requests using domain modeling and resource characteristics. In *Proc. of the International Conference on Cooperative Information Systems (CoopIS 2001)*, Trento, Italy, September 2001. http://lsdis.cs.uga.edu/proj/iq/iq_pub.html.

14. L. Penserini, L. Liu, J. Mylopoulos, M. Panti, and L.Spalazzi. Modeling and evaluating cooperation strategies in p2p agent systems. In *Proceedings of the Agents and Peer-to-Peer Computing, AAMAS02*, 2002.

15. M. Stefanelli. The socio-organizational age of artificial intelligence in medicine. *AI in Medicine Journal*, 23(1):25–48, 2002.

16. P. Szolovits, J. Doyle, and W.J. Long. Guardian angel: Patient-centered health information systems. Technical Report TR-604, MIT/LCS, 1994.

17. M. Wooldridge, P. Ciancarini, and G. Weiss, editors. *Proc. of the 2nd Int. Workshop on Agent-Oriented Software Engineering (AOSE-2001)*, Montreal, CA, May 2001.

18. E. Yu. *Modelling Strategic Relationships for Process Reengineering*. PhD thesis, University of Toronto, Department of Computer Science, 1995.

19. E. Yu and L.M. Cysneiros. Agent-oriented methodologies – towards a challange exemplar. Technical report, Faculty of Information System, University of Toronto, 2002.

Introducing Pattern Reuse in the Design of Multi-agent Systems

Massimo Cossentino[1], Piermarco Burrafato[2], Saverio Lombardo[2], and
Luca Sabatucci[2]

[1] ICAR/CNR – Istituto di Calcolo e Reti ad Alte Prestazioni/Consiglio Nazionale
delle Ricerche, c/o CUC, Viale delle Scienze, 90128 Palermo, Italy
cossentino@pa.icar.cnr.it
http://www.csai.unipa.it/cossentino
[2] DINFO – Dipartimento di Ingegneria Informatica, Universita' degli Studi di
Palermo Viale delle Scienze, 90128 Palermo, Italy

Abstract. In the last years, multi-agent systems (MAS) have proved
more and more successful. The need of a quality software engineering
approach to their design arises together with the need of new methodological ways to address important issues such as ontology representation,
security concerns and production costs. The introduction of an extensive pattern reuse practice can be determinant in cutting down the time
and cost of developing these systems. Patterns can be extremely succes-
ful with MAS (even more than with object-oriented systems) because
the great encapsulation of agents allows an easier identification and dis-
position of reusable parts. In this paper we discuss our approach to the
pattern reuse that is a phase of a more comprehensive approach to agent-
oriented software design.

1 Introduction

In the last years, multi-agent systems (MAS) have proved successful in more
and more complex duties; as an example, e-commerce applications are growing
up quickly, they are leaving the research field and the first experiences of industrial applications are appearing. These applicative contexts require high-level
qualities of design as well as secure, affordable and well-performing implementation architectures. In our research we focus on the design process of multi-agent
systems considering that this activity implies not only modelling an agent in
place of an object but also capturing the ontology of its domain, representing
its interaction with other agents, and providing it with the ability of performing
intelligent behaviours. Several scientific works that address this topic can be fo-
und in literature; it is possible to note that they come from different research
fields: some come from Artificial Intelligence (Gaia [12,27], BOD [28]) others
from Software Engineering (MaSE [11], Tropos [24,26]) but there are also methodologies coming directly from Robotics (Cassiopeia [13]). They give different
emphasis to the different aspects of the process (for example the design of goals,

R. Kowalczyk et al. (Eds.): Agent Technology Workshops 2002, LNAI 2592, pp. 107–120, 2003.

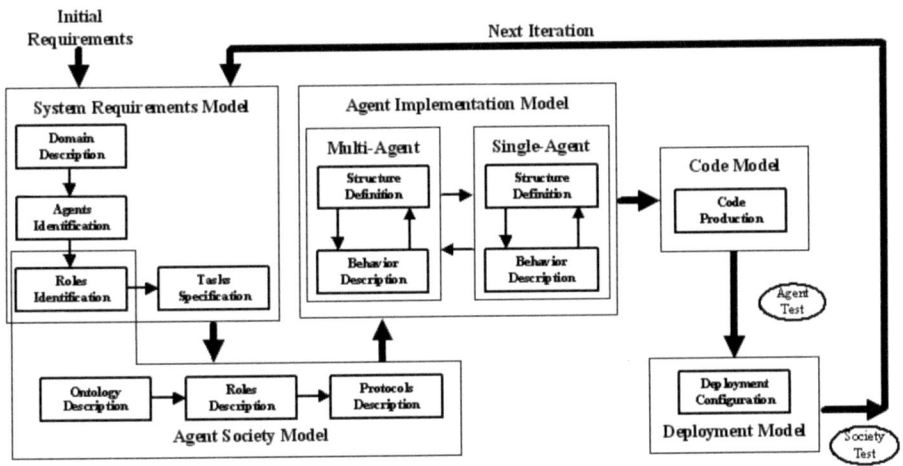

Fig. 1. The models and phases of the PASSI methodology

communications, roles) but almost all of them deal with the same basic elements although in a different way or using different notations/languages.

We can consider that the process of designing a MAS is not very different from other software design processes, if we look at the process inputs and outputs. In order to increase the results, we think that an important role can be played by an analysis of the inputs and of the activities to be performed as well as by the automation of as many steps of the process as possible (or similarly by providing a strong automatic support to the designer). In pursuing these objectives we developed a design methodology (PASSI, "Process for Agent Societies Specification and Implementation" [5]) specifically conceived to be supported by a CASE tool that automatically compiles some models that are part of the process, using the inputs provided by the designer.

PASSI is a step-by-step requirement-to-code method for developing multi-agent software that integrates design models and philosophies from both object-oriented software engineering and MAS using UML notation. It has evolved from a long period of theory construction [16,1,2] and experiments in the development of embedded robotics applications [3,4,15,17] and now is the design process used in a more comprehensive approach to robotics that encompasses a flexible vision architecture and an extensive modelling of environmental knowledge and ontology [6]. Moreover, it also proved successful in designing information systems [19].

The design process is composed of five models (see Fig. 1): the System Requirements Model is an anthropomorphic model of the system requirements in terms of agency and purpose; the Agent Society Model is a model of the structure of the agents involved in the solution, of their social interactions and dependencies; the Agent Implementation Model is a model of the solution architecture in terms of classes and methods; the Code Model is a model of the solution at the code level and the Deployment Model is a model of the distribution of the parts of the system (agents) across hardware processing units, and their movements across the different available platforms.

In PASSI great importance has the reuse of existing patterns. We define a pattern as a representation and implementation of some kind of (a part of) the system behaviour. Therefore each pattern in our approach is composed of a model of (dynamic) behaviour, a model of the structure of the involved elements, and the implementation code.

During a PASSI design process, the designers will use a Rational Rose add-in that we have specifically produced. In this procedure they move gradually from the problem domain (described in the System Requirements Model and Agent Society Model) towards the solution domain (mainly represented by the Agent Implementation Model) and, passing through the coding phase, to the dissemination of the agent in their world.

While they face the problem domain they need to determine the functionalities required for the system, identify the agents (assigning the previously identified functionalities to them) and their roles, represent the ontology of the domain, and describe the agents' communication. We have not introduced an explicit model of the goals of the system because several contributes can already be found in literature ([7,8,9,10]), and can be used in order to perform this activity.

In the solution domain the designer essentially produces some representations of the structure of the agents and of their dynamic behavior. From this specification, the designer (or more likely the programmer), after having chosen the implementation architecture, produces the code and deploys it as described in the deployment model.

It is in this progress of activities, mainly looking at the work performed in the solution domain, that we identified the most useful structure for our patterns: one structural representation of the agent (a class diagram), one dynamical representation of the behavior expressed by the agent(s) (an activity diagram) and the corresponding code.

We propose a classification of our patterns in four different categories: the action pattern (a functionality expressed by an agent – for example a specific task – usually it represents only a portion of the agent), the behavior pattern (again a portion of an agent but it addresses a more complex functionality, often performed by the agent using more than one of its tasks), a component pattern (a complete agent capable of performing some kind of behaviors), a service pattern (composed by at least two component patterns where the involved agents interact in order to actuate some kind of cooperative behavior).

Now we are working on the Agent Factory Project funded within the Agent-cities initiative [22], and our goal is to implement a service for the network community, that is composed of a pattern-based agent production process and a repository that will contain the patterns that we will identify and many others that will be introduced by other members of the community.

In order to support the localization of our patterns in both of the two different most diffused FIPA [18] platforms (FIPA-OS [14] and JADE [23]) we are planning to represent the models and the code of each pattern using XML. In the case of the models we will use the diffused XMI representation of the UML diagrams while for the code we will introduce a meta-representation of the agent using XML. Then applying to it an XSLT transformation we will instantiate the code localized for the selected platforms. Obviously this approach is possible because FIPA-OS and JADE are based on the Java language and share a similar structure, while the solution could become more difficult without these favourable conditions.

The remaining part of the paper is organized as follows: Sect. 2 gives a quick overview of the PASSI methodology; Sect. 3 presents the Agent Factory Project; Sect. 4 provides a discussion on patterns; meta-language representation of agents is discussed in Sect. 5, and some conclusions are presented in Sect. 6.

2   The PASSI Methodology

PASSI [5] is composed of five models that address different design concerns and twelve steps in the process of building a model.

In PASSI we use UML as the modelling language because it is widely accepted both in the academic and industrial worlds. Its extension mechanisms (constraints, tagged values and stereotypes) facilitate the customized representation of agent-oriented designs without requiring a completely new language.

The models and phases of PASSI are (see Fig. 1):

1. **System Requirements Model**. An anthropomorphic model of the system requirements in terms of agency and purpose. Developing this model involves four steps: Domain Description (D.D.): A functional description of the system using conventional use-case diagrams. Agent Identification (A.Id.): Separation of responsibility concerns into agents, represented as stereotyped UML packages. Role Identification (R.Id.): Use of sequence diagrams to explore each agent's responsibilities through role-specific scenarios. Task Specification (T.Sp.): Specification through activity diagrams of the capabilities of each agent.

2. **Agent Society Model**. A model of the social interactions and dependencies among the agents involved in the solution. Developing this model involves three steps in addition to part of the previous model: Role Identification (R.Id.). See the System Requirements Model. Ontology Description (O.D.): Use of class diagrams and OCL constraints to describe the knowledge ascribed to individual agents and the pragmatics of their interactions. Role Description (R.D.): Use of class diagrams to show distinct roles played by agents, the tasks involved that the roles involve, communication capabilities and inter-agent dependencies. Protocol Description (P.D.): Use of sequence diagrams to specify the grammar of each pragmatic communication protocol in terms of speech-act performatives like in the AUML approach [25,29].

3. **Agent Implementation Model**. A model of the solution architecture in terms of classes and methods, the development of which involves the following steps: Agent Structure Definition (A.S.D.): Use of conventional class

diagrams to describe the structure of solution agent classes. Agent Behaviour Description (A.B.D.): Use of activity diagrams or state charts to describe the behaviour of individual agents.

4. **Code Model.** A model of the solution at the code level requiring the following steps to produce: Code Reuse Library (C.R.): A library of class and activity diagrams with associated reusable code. Code Completion Baseline (C.C.): Source code of the target system.

5. **Deployment Model.** A model of the distribution of the parts of the system across hardware processing units, and their migration between processing units. It involves one step: Deployment Configuration (D.C.): Use of deployment diagrams to describe the allocation of agents to the available processing units and any constraints on migration and mobility.

Testing: The testing activity has been subdivided into two different steps: the (single) agent test is devoted to verifying its behaviour with regards to the original requirements of the system solved by the specific agent. During the society test, the validation of the correct interaction of the agents is performed, in order to verify that they concur in solving problems that need cooperation.

3   The Support of the CASE Tool in Designing with PASSI

In this section we describe the CASE tool we have developed for designing multi-agent systems following the PASSI methodology.

Our work starts from the consideration that most commercial CASE tools are only object-oriented. Besides, the design of a MAS is often very difficult for unskilled users. We believe that the support of an agent-oriented CASE tool can simplify the MAS designer's work, increase the reuse of code (through a database of agents/tasks patterns), and permit the automatic production of a considerable part of the code. Moreover, our tool helps untrained users to follow a proper software engineering approach.

We have realized our tool by building an add-in for the commercial UML-based CASE tool Rational Rose. It enables the user to follow the PASSI's process of analysis and design, providing a set of functionalities that are specific for each phase of the process by means of sub- and pop-up menus that appear after having selected some UML elements (classes, use cases and so on). The tool also allows the designers to perform check operations, which are based on correctness of single diagrams and consistency between related steps and models. The main functionalities of our tool are as follows:

Automatic Cmpilation of Diagrams: Our Rose add-in allows us to save analysis and design time by totally or partially drawing some diagrams in an automatic way. This enables designers to go through the twelve steps of PASSI in a very fast and easy way. For example, the Agent Identification is totally drawn by the tool once the user has chosen what functionality to insert into an agent and a consistency check is positively performed. This simple identification of an agent triggers a series of automatic operations: a) a Task Specification diagram is assigned to the created agent; b) the model of the agent skeleton is depicted in the Multi- and Single-Agent Structure Definition (SASD); c) a Single-Agent Behaviour Description (SABD) diagram is assigned to the new agent; and so forth. Other diagrams, such as Communication Ontology Description, Roles Description and Single-Agent Structure Definition are partially drawn as pieces of new information are gradually inserted into the PASSI's models.

Automatic Support to the Execution of Recurrent Operations: Apart full and partial assembling of diagrams, the tool enable developers to also modify the model at any point and obtain the automatic update of all of the diagrams that depend on the modification.

Project Consistency: In general, the tool permits a check of the model. When it is invoked, it verifies the entire model correctness and consistency between the diverse diagrams yielded till that point. Furthermore, a check operation is automatically run whenever the user completes any phase. The check will inform the user about any error or inconsistency.

Automatic Compilation of Reports and Design Documents: The add-in can produce a report of the entire model in a Microsoft Word format. Together with the diagrams, the document will contain textual descriptions and some tables summarizing the agent tasks, roles, communication, ontology, etc.

Access to a Database of Patterns: In order to increase the productivity and lower the developing time, the tool includes a repository of pattern whose functionalities are described in the next sections.

Generation of Code and Reverse Engineering: The Rose add-in can generate code from the diagrams of the implementation model. The code that is produced is actually an agent skeleton written in Java, including tasks subclasses. Furthermore, our add-in enables reverse engineering by the creation of a Single-Agent Structure Definition diagram in the Rose model from a source Java file. During this operation, the tool will refresh all the related diagrams. More consistent parts of code will be automatically produced with the introduction of the patterns.

4   The Agent Factory Project

The Agent Factory Project is an Agentcities.NET [22] deployment project. Our commitment is to deploy a service that provides Agentcities' designers with a tool for either building new agents and/or upgrading their existing ones. The agents are meant to be FIPA compliant that means they need to belong to platforms such as Jade or FIPA-OS. The key issue of the project is to produce a repository of agents' patterns (see next section) that will be described as pieces of model and code.

We intend to provide the Agentcities Network with a web-based application that enables the MAS designers to easily build their own agents and upgrade them. As for the building operation, the users will be able to select either the

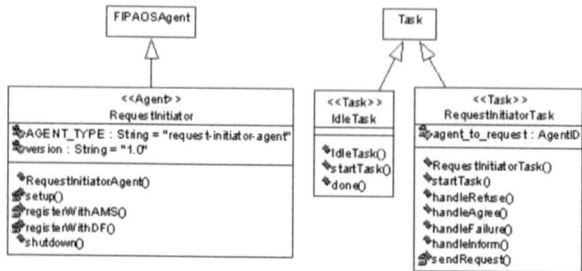

Fig. 2. Static structure of an agent and its tasks

agent platform for deployment – mainly FIPA-OS and JADE – and the functionalities they want to introduce from the repository. As for the upgrading operation, the users will be able to input their UML models of an agent and add new capabilities from the repository so as to get back the upgraded models. For both operations, the application will eventually provide some validation mechanisms for designers' inputs.

Agent Factory is thought to accelerate analysis and design phases of multi-agent systems by easy reuse of patterns to be identified and plugged into the project's models and/or the code. Its two main features – creation of new agents and upgrading of existing ones – will allow multi-agent systems designers to speed up prototyping.

As most Agentcities' members are concerned with the world of agents, and most of their teams are involved in the development of multi-agent systems and platforms, we believe that the service we are going to develop will be a valid support for their work, allowing them to obtain relevant benefits in terms of productivity.

Among the other things, our work will also focus on the possibility of giving a contribution to standardization activities such as those of FIPA. Building up a large repository of patterns may render us important information about their possible generalization. This could make us address feasible ways of standardization for patterns of agents, and explore the opportunity of proposing a related specification for FIPA.

With regard to the solution strategy, we imagine a pattern as a couple of diagrams: a structural and a behavioral diagram (see next section). Users will be able to input their original agents' models in the XMI format, which will then be transformed into XML representations of the agents (see Sect. 5). This will provide an easy way to instantiate the target code from such a structured representation of data.

5 Patterns

Regarding the patterns of agents, many works have been proposed (among the others, [20,21]); as already discussed, our concept of pattern addresses an entity

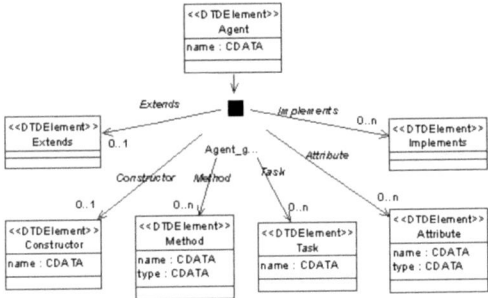

Fig. 3. The DTD related to the agent structure

composed of portions of design and the corresponding implementation code. We look at an agent as a composition of a base agent class and a set of task classes – this is the *structure*. The behavior expressed by the agents using their structural elements can be detailed in a dynamic diagram such as an activity/state chart diagram – this is the *behavior*.

From the structural point of view, we consider four classes of patterns. They are described as follows:

- *Action pattern.* A functionality of the system; it may be a method of either an agent class or a task class.
- *Behavior pattern.* A specific behavior of an agent; we can look at it as a collection of actions; it usually represents a task.
- *Component pattern.* An agent pattern; it encompasses the entire structure of an agent together with its tasks.
- *Service pattern.* A collaboration between two or more agents; it is an aggregation of components.

As we know, some elementary pieces of behavior are largely used. For example, if two agents communicate using one of the FIPA standard protocols, the parts of code devoted to dealing with communication can be reused. On the one hand, we can consider tasks as encapsulating behaviours that can be put in patterns (patterns of behaviour). On the other hand, if we consider an agent as an entity capable of pursuing specific goals, carrying out some operations (e.g. communication, moving across platforms, getting some hardware resources), we see that we can also identify patterns of agents. Furthermore, we may put together two or more patterns of agents to obtain a pattern of service. Thus, we can access and use single patterns or a composition of them.

It is now important to highlight that we can identify some specific patterns for some specific fields of application. For example, as for robotics, patterns of tasks may be useful to reuse common behaviours like planning and obstacle avoidance. Hence, it turns out that some application domain classification needs to be done. Looking at the functionality of the patterns, we can consider four categories:

- *Mobility.* These patterns describe the possibility for an agent to move from a platform to another, maintaining its knowledge.
- *Communication.* They represent the solution to the problem of making two agents communicate by a communication protocol.
- *Elaboration.* They are used to deal with the agent's functionality devoted to perform some kind of elaboration on relevant amounts of data.
- *Access to hardware resources.* They deal with information retrieval and manipulation of source data streams coming from hardware devices, such as cameras, sensors, etc.

The Repository of Patterns will be structured as a database that grants easy update and retrieval of patterns' models. As stated above, we are going to represent the latter as XML files. Each pattern record in the database will have a field containing a link to the related XML file. The repository will provide some mechanisms to ensure that coherence rules are met.

6 Meta-language Representation of Agents

Although patterns represent a good technique for generalization, we however need to notice that they may depend on the particular target programming language. This is also true if we talk of multi-agent systems, as different agent platforms and frameworks exist.

In our work we have adopted the FIPA-OS and the Jade frameworks. As both of them present agent structures as classes containing attributes, methods and inner-classes, we have thought to adopt a hierarchical meta-language to describe the agent and his properties. The hierarchy's root level entity is the agent, which contains inherent properties such as attributes, methods and tasks (the inner classes). We have chosen XML as our meta-language as it is oriented to tree data structure representations. It proved very useful for managing agents and tasks. As a matter of fact, this allowed us to easily manipulate their structure and add, edit or delete agents' properties very easily. This is the key point in the application of a pattern to an existing agent in terms of skills and behaviors. We believe that the use of a meta-language can give us a straight way to create and maintain the Repository of Patterns mentioned above. Moreover, agents' source code can be automatically generated from meta-language representation without any manual support.

The choice of XML is also valid in the context of the Agent Factory Project. As a matter of fact, because of the XML easy portability, agents' patterns will be shared in a web server, so that designers of the Agentcities community will be able to access and update them. In what follows we describe the main issues of our meta-language.

Language Definition: In order to build an XML agent representation, we retrieve information coming from diagrams such as the SASD and MABD of the PASSI design methodology. In the Single-Agent Structure Definition diagram (see Fig. 2) an agent is represented by a class, which inherits from a super class

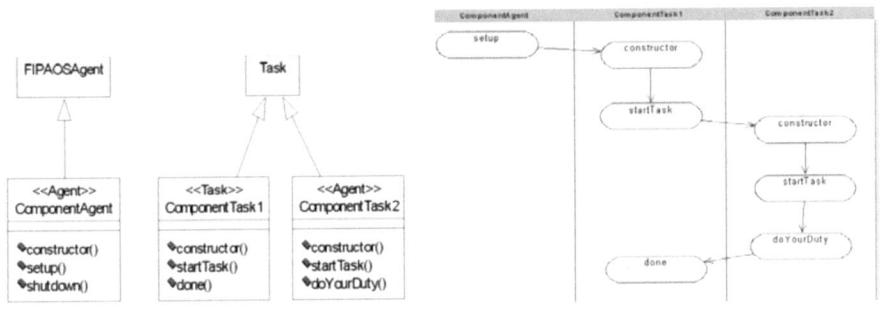

(a) The structure of the pattern   (b) The behavior of the pattern

Fig. 4. A very simple pattern composed of the agent base class and two behavior classes

called either FIPAOSAgent or Agent depending on the agent platform selected. Attributes and methods of this class correspond respectively to data structure and functions that the agent owns. In the XML file an agent is described inside an Agent tag. The DTD fraction that describes the Agent tag is shown in Fig. 3. Agent properties such as attributes or tasks are represented as inner elements of the structure. Each of them contains other sub-elements that describe their properties. In the same manner, a Task tag has got sub-elements to specify its characteristics. For example the Parent tag describes parent-child relationships between tasks. This specifies that a task (called parent) instantiates and executes another task (child) from its methods as it has been specified in the Multi-Agent Behaviour Description (MABD) diagram. In FIPA-OS this information could be used to automatically add the related done method to the parent task; this method will be called from the task manager when the child task terminates its duty.

In Fig. 4(a) we can see the structural representation (a SASD diagram) of a toy-pattern of FIPA-OS agent that can be useful to understand our approach. It is composed of the base agent class and two task classes. The behaviour of the agent is described in Fig. 4(b): the setup method of the agent is invoked from its constructor. It calls the first task that performs some kind of operation and then starts the second task. At the end of the second task the done method is invoked in the parent task. The consequent XML description of this agent is shown in Fig. 5.

Patterns and Constraints: When a pattern is applied to a project it modifies the context in which it is placed, that is: it introduces new functionality into the system. These additions need to satisfy some constraints. For example in FIPA-OS, when we insert a communication task pattern into an existing agent, the Listener Task should have a handleX method to catch performative acts of a particular type. This relationship between the pattern and existing elements could be expressed with a constraint. A constraint is a rule composed of two

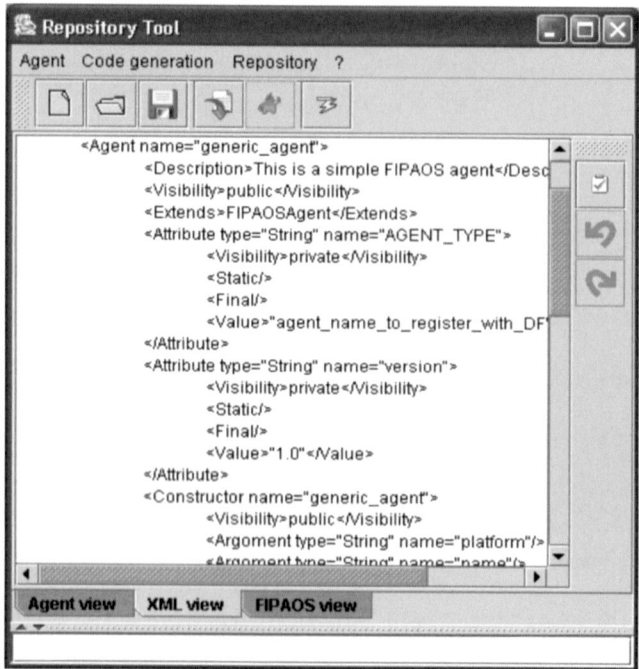

Fig. 5. The XML representation of the pattern described in Fig. 4

elements: a target and a content. The target specifies what agent/task will be influenced from the rule. The content expresses the changes to be applied when the pattern is inserted into the project; it could be an aggregation of attributes, constructors or methods.

Code Generation: As briefly mentioned before, XSLT application grants to export an agent described with our meta-language into a specific programming language. This is possible because the PASSI's Agent Structure Definition intrinsically represents an implementation viewpoint. As a matter of fact, UML classes correspond to Java classes, and UML attributes and methods correspond to the Java classes' attributes and methods. This important characteristic of the agent structure allows us to look at the source code as one of the possible views of an agent: we could imagine agent representation as an intermediate layer between agent design and agent development. The use of XSLT enables code generation for both FIPA-OS and Jade frameworks by only changing the transformation sheet. Although using FIPA-OS and Jade implies different design processes, because of different mechanisms (e.g., message handling or task execution control), the same meta-language could be used to represent agents independently from the used platform.

In Fig. 6 we can see the JAVA code of the toy-pattern presented in Fig. 4 obtained applying the FIPA-OS transformation sheet.

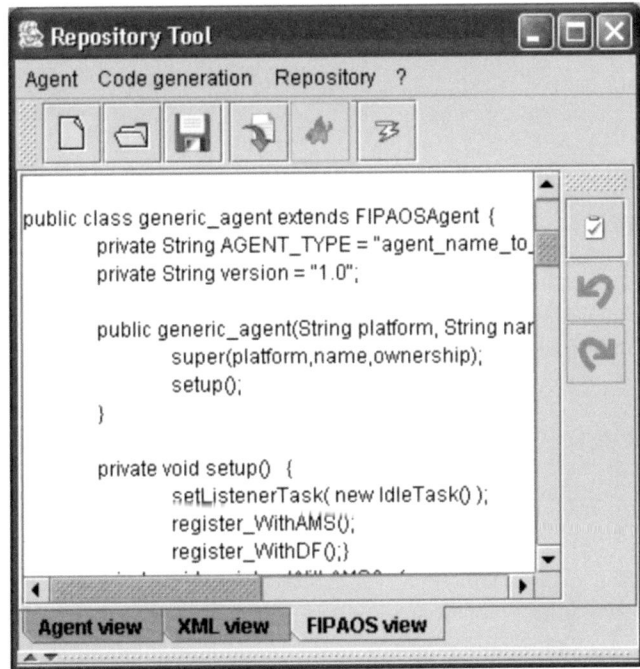

Fig. 6. (Part of) The JAVA code obtained from the XML agent description of Fig. 5

7   Conclusions

Our conviction is that pattern reuse is a very challenging and interesting issue in multi-agent systems as it has been in object-oriented ones. However we are aware that the problems arising from this subject are quite delicate and risky. Nonetheless, we believe, thanks to previous experience made in fields such as robotics, that we can succeed in creating a very useful service for the Agentcities community.

Taking advantage of other projects we are at present working on, we think it is feasible to create a repository that could contain patterns coming from diverse fields of research and application – among the others, image processing and robotics. We are also confident that the contribute of Agentcities members will be precious in order to quickly broaden our database to include more and more useful elements.

Acknowledgements. This research was partially supported by grants from Engineering Ingegneria Informatica S.p.A, Rome (Italy) and the Agentcities.NET initiative [22].

References

1. Chella, A., Cossentino, M., Lo Faso, U.: Designing agent-based systems with UML. Proc. of ISRA'2000 Conference. Monterrey, Mexico (2000)
2. Chella, A., Cossentino, M., Infantino, I., Pirrone, R.: An agent based design process for cognitive architectures in robotics. Proc. of Workshop on Objects and Agents, WOA'01. Modena, Italy (2001)
3. Chella, A., Cossentino, M., Tomasino, G.: An environment description language for multirobot simulations. Proc. of ISR 2001. Seoul, Korea (2001)
4. Chella, A., Cossentino, M., Pirrone, R., Ruisi, A.: Modeling Ontologies for Robotic Environments. Proc. of the Fourteenth International Conference on Software Engineering and Knowledge Engineering (SEKE'02). Ischia, Italy (2002)
5. Cossentino, M., Potts, C.: A CASE tool supported methodology for the design of multi-agent systems. Proc. of the 2002 International Conference on Software Engineering Research and Practice (SERP'02). Las Vegas, NV (2002)
6. Infantino, I., Cossentino, M., Chella, A.: An Agent Based Multilevel Architecture for robotics vision systems. Proc. of the 2002 International Conference on Artificial Intelligence (IC-AI'02). Las Vegas, NV (2002)
7. Antón, A.I., Potts, C.: The Use of Goals to Surface Requirements for Evolving Systems. Proc. of International Conference on Software Engineering (ICSE '98). Kyoto, Japan (1998)
8. van Lamsweerde, A., Darimont, R., Massonet, P.: Goal-Directed Elaboration of Requirements for a Meeting Scheduler: Problems and Lessons Learnt. Proc. 2nd International Symposium on Requirements Engineering (RE'95). York, UK (1995)
9. Potts, C.: ScenIC: A Strategy for Inquiry-Driven Requirements Determination. Proc. of IEEE Fourth International Symposium on Requirements Engineering (RE'99). Limerick, Ireland (1999)
10. Yu, E., Liu, L.: Modelling Trust in the i* Strategic Actors Framework. Proc. of the 3rd Workshop on Deception, Fraud and Trust in Agent Societies at Agents 2000. Barcelona, Catalonia, Spain (2000)
11. DeLoach, S.A., Wood, M.F., Sparkman, C.H.: Multiagent Systems Engineering. International Journal on Software Engineering and Knowledge Engineering 11, 3, 231–258
12. Wooldridge, M., Jennings, N.R., Kinny, D.: The Gaia Methodology for Agent-Oriented Analysis and Design. Journal of Autonomous Agents and Multi-Agent Systems 3,3 (2000) 285–312
13. Collinot, A., Drogoul, A.: Using the Cassiopeia Method to Design a Soccer Robot Team. Applied Articial Intelligence (AAI) Journal, 12, 2—3 (1998), 127–147
14. Poslad S., Buckle P., Hadingham R.: The FIPA-OS Agent Platform: Open Source for Open Standards. Proc. of the 5th International Conference and Exhibition on the Practical Application of Intelligent Agents and Multi-Agents. Manchester, UK (2000) 355–368
15. Chella, A., Cossentino, M., Pirrone, R.: Multi-Agent Distributed Architecture for a Museum Guide Robot. Proc. of the GLR worshop at the 2001 AI*IA conference. Bari, Italy (2001)
16. Chella, A., Cossentino, M., Lo Faso, U.: Applying UML use case diagrams to agents representation. Proc. of AI*IA 2000 Conference. Milan, Italy (2000)
17. Chella, A., Cossentino, M., Infantino, I., Pirrone, R.: A vision agent in a distributed architecture for mobile robotics in Proc. Of Worskshop "Intelligenza Artificiale, Visione e Pattern Recognition" in the VII Conf. Of AI*IA. Bari, Italy (2001)

18. O'Brien, P., Nicol, R.: FIPA – Towards a Standard for Software Agents. BT Technology Journal 16,3(1998),51–59
19. Burrafato, P., Cossentino, M.: Designing a multi-agent solution for a bookstore with the PASSI methodology. Fourth International Bi-Conference Workshop on Agent-Oriented Information Systems (AOIS-2002). May 2002, Toronto, Ontario, Canada at CAiSE'02
20. Kendall, E.A., Krishna, P.V.M., Pathak C. V., Suresh, C. B.: Patterns of intelligent and mobile agents. Proc. of the Second International Conference on Autonomous Agents. Minneapolis (1998) 92–99
21. Aridor, Y., and Lange, D.B.: Agent Design Patterns: Elements of Agent Application Design. Proc. of the Second International Conference on Autonomous Agents. Minneapolis (1998) 108–115
22. Agentcities.NET: http://www.agentcities.net
23. Bellifemine, F., Poggi, A., Rimassa, G.: JADE – A FIPA2000 Compliant Agent Development Environment. In Proc. Agents Fifth International Conference on Autonomous Agents (Agents 2001). Montreal, Canada (2001) 216–217
24. Castro, J , Kolp, M., Mylopoulos, J.: Towards Requirements-Driven Information Systems Engineering: The Tropos Project. To appear in Information Systems, Elsevier, Amsterdam, The Netherlands (2002)
25. Odell, J., Van Dyke Parunak, H., Bauer, B.: Extending UML for Agents. AOIS Workshop at AAAI 2000. Austin, Texas (2000).
26. Bresciani, F. Sannicolo. Requirements Analysis in Tropos: A Self Referencing Example. In this volume.
27. P. Moraitis, E. Petreki, N. Spanoudakis. Engineering Jade Agents with the GAIA Methodology. In this volume.
28. J. Bryson. The Behavior-Oriented Design of Modular Agent Intelligence. In this volume.
29. M.-P.Huget. Agent UML Class Diagrams Revisited. In this volume

Specifying Reuse Concerns in Agent System Design Using a Role Algebra

Anthony Karageorgos[1], Simon Thompson[2], and Nikolay Mehandjiev[1]

[1] Dept. of Computation, UMIST, Manchester M60 1QD, UK
{karageorgos,nikolay}@computer.org
[2] Intelligent Systems Research, BTexact Technologies, Ipswich, UK
simon.2.thompson@bt.com

Abstract. During the design of an agent system many decisions will be taken that determine the structure of the system for reasons that are clear to the designer and customers at the time. However, when later teams approach the system it may not be obvious why particular decisions have been taken. This problem is particularly acute in the case of designers attempting to integrate complex "intelligent" services from many different service providers. In this paper a mechanism for recording these decisions is described and grouping functionality into Roles which can then be combined using the recorded design knowledge is subsequent development episodes. We illustrate how design decisions can be captured, discuss the semantics of the constructs we introduce and how these abstractions can then be used as the basis of reuse in an extension of the Zeus agent toolkit.

1   Introduction and Motivation

Multi-agent system architectures can be viewed as organised societies of individual computational entities e.g. [6,15,19], and the problem of designing a multi-agent system can be viewed as designing an *agent organisation.* The criteria affecting an agent organisation design decision are numerous and highly dependent on factors that may change dynamically. Therefore, there is no standard best organisation for all circumstances [14,15]. As a result, agent organisation design rules are often left vague and informal, and their application is mainly based on the creativity and the intuition of the human designer. This can be a serious drawback when designing large and complex real-world agent systems. Therefore, many authors argue that social and organisational abstractions should be considered as first class design constructs and that the agent system designer should reason at a high abstraction level, e.g. [4,7,9,12,20].

The Zeus tool [11] was released as open source software in 1999, and used in a number of projects internally at BT both before and after release. It became clear that a number of factors limited the usefulness of the tool for real world application development. These problems were especially acute when the tool was used by commercial

R. Kowalczyk et al. (Eds.): Agent Technology Workshops 2002, LNAI 2592, pp. 121-136, 2003.

developers, who had no background in agent development or AI, but prodigious skills with respect to e-business and distributed systems development. One of the key issues that was repeatedly raised by real world developers was the lack of a reuse model for agent code that they had developed. After some analysis the reasons for these problems became clear to us.

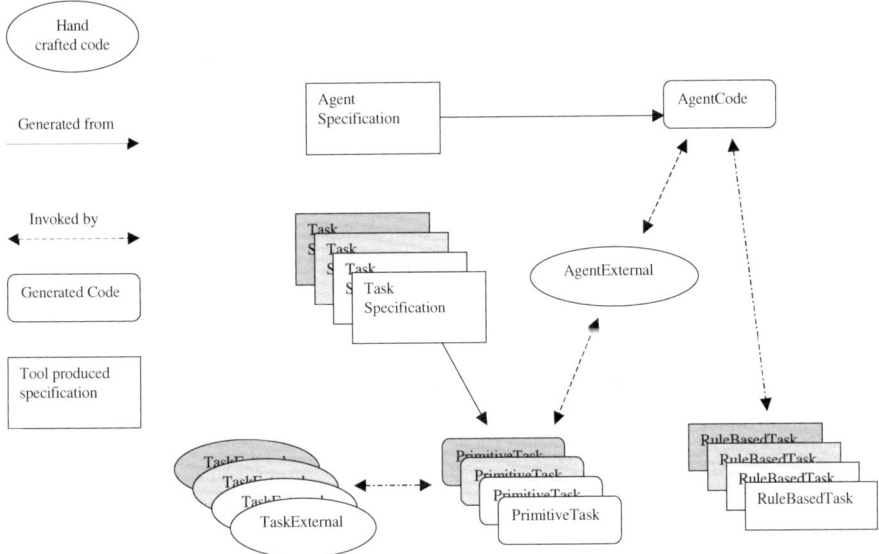

Fig. 1. Modules that define a Zeus agent

The Zeus system, requires software engineers to produce an "Agent External" (which is called when the agent is run in order to set-up things like database connections and user interfaces), and a set of tasks that the agent uses. The tasks are either "primitive" or "rule-based". Primitive tasks are executed by the planner and rule-based tasks are executed by an expert-system style inference engine. The agent code itself is automatically generated from a specification loaded into a tool. In later versions of Zeus the concept of a "task external" was introduced to allow code to be cleanly added to the task model of the agent, but this exacerbated the proliferation of components that made up the agent model. This problem is illustrated in Fig. 1, which shows the relationships between 7 different components that must be defined to create an agent. All of these components are required to support a BDI agent model, and the components use knowledge representations that are appropriate for the use of each (for example production rules for rule based tasks and Java code for user interfaces).

But because BDI agents like Zeus are so complex to define, the design imperative of modularity appears to actually be harmful to developers that are attempting to understand the system and reuse components of it. In Fig. 1 we use shading to indicate four modules which are part of the agent developed using modular design. Logic for these modules may reside in any of the components of the agent, and may also be entangled in the code for the agent external. Zeus developers have no way of indicating what module a component or piece of code is a part of, and because of this, reusing the various subsystems from project to project becomes virtually impossible. It rapidly became clear to us that the standard abstractions for development were deficient when it came to dealing with this problem and that we needed to be able to encode the relationships between components in some way so that they could be separated. We also identified a requirement that the non-functional concerns of the developer in the context of this particular system could be separated from the functional concerns common to any use case for the component.

2   Designing Agent Organisations

The need to develop agent systems of realistic complexity in a reliable and systematic manner has spawned a number of methodologies for agent-oriented analysis and design such as MESSAGE [5], GAIA [19] and SODA [12]. All these methodologies use a number of analysis and design sub-models, each emphasising a particular analysis or design aspect. These can be further improved in the following ways:

- A more systematic way to construct large agent system design models from the analysis models. The steps involved in transforming analysis to design models are not specified to a detail that enables at least some degree of automation by a software tool [16].
- By considering non-functional requirements on design time. The aim should be to optimise the agent organisation before it has been deployed. To achieve this, we need to identify some means for considering non-functional requirements before actually deploying the multi-agent system. This hypothesis is along the lines of similar works [13,14] where the behaviour of a multi-agent system is modelled and studied before actual system deployment. Non-functional requirements in a system can be elicited by a number of standard techniques, for example by using requirement templates [22].
- By reusing organisational settings. This view regarding reuse of organisational settings has been inspired by the concepts introduced in [21]. It is believed that work can be further extended by classifying known organisational patterns, and by providing some rigorous means for selecting them in a particular design context.

3   Our Approach

Figure 2 shows an overview of our approach. Our objective is to develop a technology that supports software engineers in describing the design concerns that have motivated choices about models of implementation in particular systems, and to place this knowledge in a repository. The repository would then be used by subsequent engineers who wished to reuse subsystems or to modify and rebuild the legacy system itself.

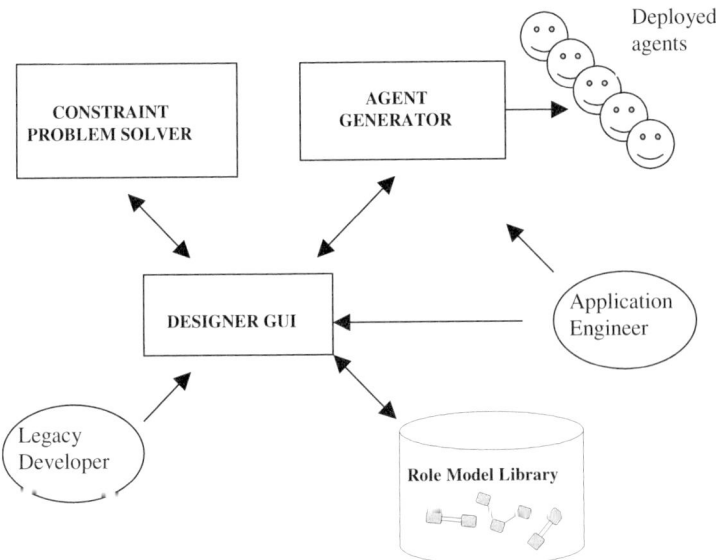

Fig. 2. Conceptual view of proposed system

We believe that further value can be added to the system by providing advice to the engineer using constraint satisfaction technology that can provide possible solutions to compositional design problems. Finally the resulting designs can be used to generate template systems linked to libraries of domain specific implementation code.

The rest of this paper describes the role modeling abstractions that we have developed in the context of an example derived from a real development project. We then describe how we have implemented a prototype that demonstrates that this knowledge can be used to support design reuse.

3.1   Role-Based Design

Existing role-based approaches to multi-agent system design stress the need to identify and characterise relations between roles [1,9]. However, only a small number of them, e.g. [9], investigate the influence of role relations on the design. This is partly due to lack of formal foundations of role relationships. Therefore, in this work we first identified role relations that would affect multi-agent system design and then we formalised them in an algebraic specification model. Role identification was based on organisational principles and in particular on *role theory* [3].

Role theory emphasises various relations between roles. For example, an examiner cannot also be a candidate at the same time and therefore appointing these roles to a person at the same time results to inconsistency. Role relations can be complex. For example, a university staff member who is also a private consultant may have conflicting interests. In this case, appointing these roles to the same person is possible but it would require appropriate mechanisms to resolve the conflicting behaviour.

3.2   Role Characteristics

Following [9], a role is defined as associated with a *position* and a set of *characteristics*. Each characteristic includes a set of *attributes*. Countable attributes may further take a range of values. More specifically, a role is capable of carrying out certain *tasks* and can have various *responsibilities* or *goals* that aims to achieve. Roles normally need to interact with other roles, which are their *collaborators*. Interaction takes place by exchanging messages according to *interaction protocols*. A collection of interacting roles representing collective behaviour constitutes a *role model*.

Roles can be used to create specialised roles by a process called role *specialisation* or *refinement* [1,9]. Specialised roles represent additional behaviour on top of the original role behaviour in a manner similar to inheritance in object-oriented systems.

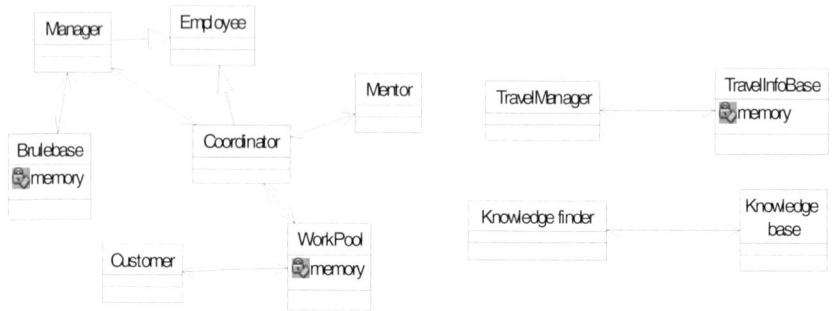

Fig. 3. Role models for the telephone repair service teams case study

For roles to pragmatically represent behaviour in an application domain, they need to model issues relevant to non-functional requirements in that domain. Therefore, the above role definition is extended to include *performance variables*. Performance variables are parameters whose value defines the run-time behaviour represented by a role. For example, if the behaviour a role represents requires using some resource like storage space, the resource capacity can be modelled by a performance variable. Performance variables can also be defined at an agent level. In that case, their value is a function of the function of the respective performance variables of all roles the agent is capable of playing. This allows us to apply design heuristics by imposing constraints on the values of the agent performance variables that must be observed when allocating roles to agents. This is illustrated in the example given below.

4   Example: Supporting Mobile Work Teams

For this example we consider a case study concerning telephone repair service teams. The aim is to build an agent system that would assist field engineers to carry out their work. Among the issues involved in such a system are those of travel management, teamwork coordination, and knowledge management [17,18].

Travel management is about supporting mobile workers for moving between repair task locations. It involves finding the position of each worker, obtaining relevant travel information, planning the route to the next repair task location and allocating travel resources as required. Teamwork coordination is about allocating and coordinating the execution of repair tasks in a decentralised manner taking into account the personal preferences and working practices of the mobile workers. Work knowledge management concerns storage and dissemination of expertise about work-related tasks.

4.1   Role Identification

To model this system in terms of roles, we need to first identify the roles involved in the case study. According to [10] a way to identify roles in an application domain is to start from identifying use cases, associating each use case with a goal, creating a goal hierarchy from the use case hierarchy and coalescing semantically relevant goals in roles. For the purpose of our example we consider the following three use cases: *Teamwork coordination*, *Travel management* and *Work Knowledge Management*.

We can use the methodology from [10], the identify the following roles (Fig. 3):

1. *Employee*: This role describes generic behaviour of the members of the customer service teams. An example of this type of behaviour is accessing common team resources including work practice announcements and business news.
2. *Coordinator:* The Coordinator role describes the behaviour required to coordinate the work of a field engineer. This includes bidding for and obtaining repair work tasks from a work pool, negotiating with other workers and the team manager as required and scheduling and rescheduling work task execution.
3. *Manager:* This role models the team manager. This includes confirming task allocation, monitoring work and ensuring that business rules are followed.
4. *Mentor:* This role provides assistance to field engineers for non-technical issues.
5. *WorkPool:* The *WorkPool* role maintains a pool of telephone repair requests. Customers interact with this role to place requests and engineers interact with this role to select tasks to undertake.
6. *Customer:* The *Customer* role models the behaviour of a customer. In involves placing telephone repair requests, receiving relevant information and arranging appointments with field engineers.
7. *Brulebase:* This role maintains a database of business rules. It interacts with manager providing information about the current work policy of the business.
8. *TravelManager:* The *TravelManager* role provides travel information to the field engineer including current location, traffic information and optimal route to next telephone repair task.
9. *TravelinfoBase:* This role store travel information from various travel resources i.e. GPS and traffic databases.
10. *Knowledgefinder:* This role searches for experts and obtains assistance regarding complex work tasks.
11. *Knowledgebase:* The *Knowledgebase* role maintains and manages a database of expertise about telephone repair tasks.

4.2   Specifying Design Constraints

Having identified the roles in our case study, we can proceed to define some constrains regarding how they can be combined in agents. We call these *compositional
constraints*, and use a simple *Role Constraint Language (RCL)* to specify them. The
compositional constraints relevant to our case study are shown on Fig. 4. *RCL* is simple declarative constraint language we introduced to represent design constraints on
agent and role characteristics. Any non-obvious use of *RCL* in Fig. 4 is described below together with the relevant design constraints. *RCL* itself is described in detail in
[8].

```
/* ROLE DEFINITIONS */                 /* ROLE CONSTRAINTS */

Role employee, coordinator, mentor,    in(employee, coordinator);
     customer, travelmanager,          in(employee, manager);
     knowledgefinder;
                                       not(customer, employee);
Role workpool, brulebase, workerassistant,    not(customer, travelinfobase);
     travelinfobase, knowledgebase {   not(customer, knowledgebase);
     int memory;                       not(customer, travelmanager);
}                                      not(mentor, manager);
     workpool.memory = 1;              not(manager, coordinator);
     brulebase.memory = 1;
     travelinfobase.memory = 2;        and(mentor, employee);
     knowledgebase.memory = 2;
     workerassistant.memory = 1;       merge(coordinator, travelmanager,
                                       knowledgefinder,
Role manager {                         workerassistant);
     collaborators = {Coordinator,
                      Brulebase};      /* GENERAL CONSTRAINTS */
     protocols = {contracting};
}                                      Constraint Y {
                                           forall a:Agent {
                                           a.memory <= 2
                                           }
                                       }
```

Fig. 4. Compositional constraints for telephone repair service teams roles

Roles in *RCL* are specified in a manner similar to programming languages. Roles
that directly manipulate databases require access to some storage space. This is modelled by the performance variable *memory*. The memory requirements of each role are
different. For example, *Travelinfobase* and *Knowledgebase* require twice as much
memory as *Workpool* and *Brulebase*.

Part of the definition of the characteristics of the *Manager* role is shown in more
detail in Fig. 4. The collaborators of the *Manager* role are the *Coordinator* and *Brulebase* roles and its interaction protocol is the Contract Net. The Employee role is contained in both *Manager* and *Coordinator* roles. Furthermore, a *Manager* cannot coexist with *Mentor* or *Coordinator* and for security purposes a *Customer* cannot coexist
with *Employee*, *Travelinfobase* or *Knowledgebase*. In order for an agent to be *Mentor*
it must also be an *Employee*.

When an agent plays all three *Coordinator*, *TravelManager* and *KnowledgeFinder* roles then overheads occur in synchronising results from these three different activities. This is modelled as a merging of the *Coordinator*, *Travelmanager* and *Knowledgefinder* resulting to the *WorkerAssistant* role. The *WorkerAssistant* role requires some storage space to store intermediate synchronisation results.

An example of non-functional requirements is the limit to the memory each agent could occupy. In this case study, agents supporting field engineers should be able to operate in PDAs with limited amount of memory. This is modelled as a general design constraint on the performance variable *memory*.

4.3   From Roles to Agents

Having defined our example roles, we need to aggregate them into agents. We define this task as a search problem: we must search the space of all possible system designs in order to find a design that doesn't violate any of the constraints that are defined in the role relationships and satisfies the context that the system is to be used in.

A possible agent organisation satisfying the above design constraints is shown in Fig. 5. The process of allocating roles to agents can be extremely complex because it is necessary to provide designers with options for the number of types of agents that must be produced and to allow them to "hardpin" roles to particular types. However, a simple algorithm illustrates how this process works in general.

1. Allocate all roles to a single agent
2. Make a list of constraints that are violated in the current design
3. Find the role with the most violated constraints
4. If the role can be moved to any agent in the design without violating any constraints, move it to that agent and add to that agent any roles that are *required* by the newly added role.
5. Else create a new agent and move the role to that agent and add any *requires* roles to that agent
6. Repeat 2 to 5 until the list produced at 2 is empty
7. Create the plays list by applying all *mergeswith* relations for each agent

Algorithm 1. Simple role allocation procedure

This algorithm assumes that we prefer agent systems with many roles allocated to few agents. Other algorithms can also be constructed easily, which, for example, prefer designs with many *mergeswith* roles grouped in particular agents. It should be noted that the above procedure and many others that we have considered is not guaranteed to terminate and that the efficiency and scalability of design algorithms of this sort are the topic of ongoing research and are beyond the scope of this paper.

5   A Role Algebra for Agent System Design

In order to execute the role allocation algorithm described above, we need to have a formally specified relationships and constraints between roles. We have used role theory [3] and case studies of human activity systems, e.g. [17], to identify six basic role relations, which we have used in the example so far. In this section, we define a formal model of role relations, referred by the term *role algebra*. Using relations from the role algebra, constraints driving the assignment of roles to agents can be specified and hence the agent organisation design process can be partially automated as described in the previous section.

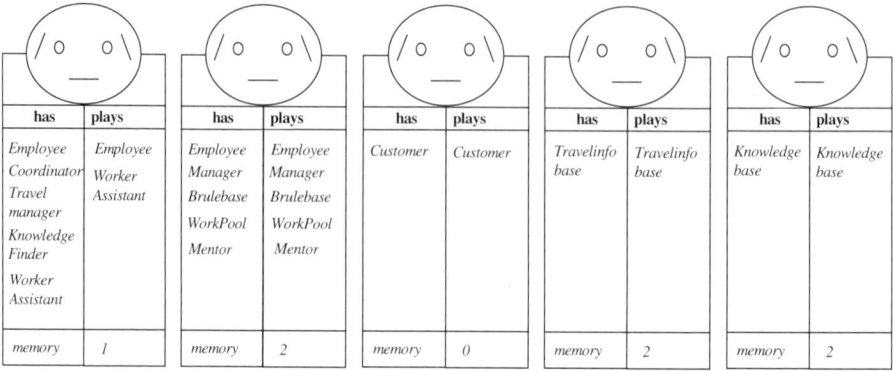

has	plays	has	plays	has	plays	has	plays	has	plays
Employee Coordinator Travel manager Knowledge Finder Worker Assistant	Employee Worker Assistant	Employee Manager Brulebase WorkPool Mentor	Employee Manager Brulebase WorkPool Mentor	Customer	Customer	Travelinfo base	Travelinfo base	Knowledge base	Knowledge base
memory	1	memory	2	memory	0	memory	2	memory	2

Fig. 5. Agent types for the telephone repair service teams case study

5.1   Relations of the Role Algebra

Let R be a set of roles. For any $r_1, r_2 \in R$, the following binary relationships may hold:

1   **Equals (eq)** — This means that r_1 and r_2 describe exactly the same behaviour. For example, the terms *Employee* and *Member of staff* can be used to refer to people employed in an organisation. When two roles are equal, an agent playing the first one also plays the other one at the same time. The relation *Equals* $\subseteq$ $R{\times}R$ is an equivalence relation since it is reflexive, symmetric and transitive:
$$\forall\, r : R\ (r \text{ eq } r)$$
$$\forall\, (r_1, r_2) : R{\times}R\ (r_1 \text{ eq } r_2 \Rightarrow r_2 \text{ eq } r_1)$$
$$\forall\, (r_1, r_2, r_3) : R{\times}R{\times}R\ ((r_1 \text{ eq } r_2) \wedge (r_2 \text{ eq } r_3) \Rightarrow (r_1 \text{ eq } r_3))$$

2   **Excludes (not)** — This means that r_1 and r_2 cannot be assigned to the same agent simultaneously. For example, in a conference reviewing agent system, an agent should not be playing the roles of paper author and paper reviewer at the same time. In the mobile working example above we have specified that a customer *excludes* employee not(customer, employee). Furthermore, a role cannot exclude itself — if it would then no agent would ever play it. Therefore, the relation *Excludes* $\subseteq$ $R{\times}R$ is anti-reflexive and symmetric:
$$\forall\, r : R\ (\neg(r \text{ not } r))$$
$$\forall\, (r_1, r_2) : R{\times}R\ (r_1 \text{ not } r_2 \Rightarrow r_2 \text{ not } r_1)$$

3   **Contains (in)** — This means that a role is a sub-case/specialisation of another role. Therefore, the behaviour the first role represents completely includes the behaviour of the second role. For example, a role representing *Manager* behaviour completely contains the behaviour of the *Employee* role. When two roles such that the first contains the second are composed, the resulting role contains the characteristics of the first role only. Therefore, the relation *Contains* $\subseteq R \times R$ is reflexive and transitive:

$\forall\ r : R\ (r \text{ in } r)$

$\forall\ (r_1, r_2, r_3) : R \times R \times R\ ((r_1 \text{ in } r_2) \wedge (r_2 \text{ in } r_3) \Rightarrow (r_1 \text{ in } r_3))$

4   **Requires (and)** — The *Requires* relation can be used to describe that when an agent is assigned a particular role, then it must also be assigned some other specific role as well. This is particularly applicable in cases where agents need to conform to general rules or play organisational roles. For example, in a university application context, in order for an agent to be a *Library_Borrower* it must be a *University_Member* as well. Although the behaviour of a *Library_Borrower* could be modelled as part of the behaviour of a *University_Member*, this would not be convenient since this behaviour could not be reused in other application domains where being a *Library_Borrower* is possible for everyone. In the mobile working example we specify that a mentor *requires* employee - that in order to act as a mentor you must also be an employee. Each role requires itself. Intuitively, the roles that some role *r* requires are also required by all other roles that require *r*. Therefore, the relation *Requires* $\subseteq R \times R$ is reflexive, and transitive:

$\forall\ r : R\ (r \text{ and } r)$

$\forall\ (r_1, r_2, r_3) : R \times R \times R\ ((r_1 \text{ and } r_2) \wedge (r_2 \text{ and } r_3) \Rightarrow (r_1 \text{ and } r_3))$

5   **Addswith (add)** — The *Addswith* relation can be used to express that the behaviours two roles represent do not interfere in any way. For example, the *Customer* and the *Football_Player* roles describe non-excluding and non-overlapping behaviours. Hence, these roles can be assigned to the same agent without any problems. The relation *Addswith* $\subseteq R \times R$ is reflexive and symmetric:

$\forall\ r : R\ (r \text{ add } r)$

$\forall\ (r_1, r_2) : R \times R\ ((r_1 \text{ add } r_2) \Rightarrow (r_2 \text{ add } r_1))$

6   **Mergeswith (merge)** — This relation can be used to express that the behaviours of two roles overlap to some extend or that different behaviour occurs when two roles are put together. For example, a *Student* can also be a *Staff_Member*. This refers to cases when PhD students start teaching before they complete their PhD or they register for another degree (e.g. an MBA) after their graduation. Although members of staff, these persons cannot access certain information (e.g. future exam papers) due to their student status. Also, their salaries are different. In the mobile working example the roles of Travelmanager, Coordinator and Knowledgefinder merge to Workerassistant because otherwise a synchronisation overhead would be incurred. In cases like this, although the two roles can be assigned to the same agent, the characteristics of the composed role are not exactly the characteristics of the two individual roles put together. This relation is used to achieve role synergy. The relation *Mergeswith* $\subseteq R \times R$ is symmetric:

$\forall\ (r_1, r_2) : R \times R\ (r_1 \text{ merge } r_2 \Rightarrow r_2 \text{ merge } r_1)$

5.2 Semantics of Role Relations

We describe the semantics of role relations by using a two-sorted algebra (Fig. 6) to represent agent organisation. The algebra includes two sorts, *A* representing agents and *R* representing roles.

AGENT ORGANISATION

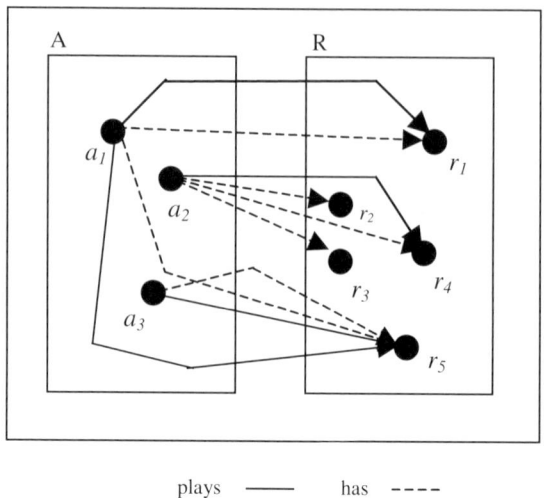

plays ———   has – – – –

Fig. 6. Semantics of role relations

Let *Has*: $A \rightarrow R$ be a relation mapping agents to roles. The term "has" means that a role has been allocated to an agent by a role allocation procedure or tool. It is possible for an agent to have roles that do not contribute to defining the agent behaviour. For example, this happens when roles merge with other roles. For each $a \in A$, let *a.has* be the set of roles that the agent *a* maps to in the relation *Has*. In other words, *a.has* denotes the relational image of the singleton $\{a\} \subseteq A$ in the relation *Has*.

Let *Plays*: $A \rightarrow R$ be a relation mapping agents to roles again. The term "plays" means that that the behaviour a role represents is actively demonstrated by the agent, for example the role does not merge with other roles that are also played by the agent. For each $a \in A$, let *a.plays* denote the set of roles that the agent *a* maps to in the relation *Plays*. In other words, *a.plays* denotes the relational image of the singleton $\{a\} \subseteq A$ to the relation *Plays*.

The meaning of the relations between roles can now be described as follows:

Equals — An agent has and plays equal roles at the same time.

$\forall\ a\ :\ A,\ (r_1,\ r_2)\ :\ R{\times}R\ \cdot\ (r_1\ eq\ r_2 \Leftrightarrow ((r_1 \in a.has \Leftrightarrow r_2 \in a.has) \wedge (r_1 \in a.plays \Leftrightarrow r_2 \in a.plays)))$

Excludes — Excluded roles cannot be assigned to the same agent.

$\forall\ a : A,\ (r_1,\ r_2) : R{\times}R\ \cdot\ (r_1\ not\ r_2 \Leftrightarrow \neg(r_1 \in a.has \wedge r_2 \in a.has))$

Contains — Contained roles must be assigned and played by the same agent as their containers.

$$\forall\, a : A,\, (r_1, r_2) : R{\times}R \cdot (r_1 \text{ in } r_2 \Leftrightarrow ((r_2 \in a.has \Rightarrow r_1 \in a.has) \wedge (r_2 \in a.plays$$
$$\Rightarrow r_1 \in a.plays)\,))$$

Requires — Required roles must be played by the same agent as the roles that require them.

$$\forall\, a : A,\, (r_1, r_2) : R{\times}R \cdot (r_1 \text{ and } r_2 \Leftrightarrow (r_1 \in a.plays \Rightarrow r_2 \in a.plays))$$

MergesWith — When two roles merge, only the unique role that results from their merge is played by an agent.

$$\forall\, a : A,\, (r_1, r_2) : R{\times}R \cdot (r_1 \text{ merge } r_2 \Leftrightarrow \exists 1\; r_3 : R \cdot((r_1 \in a.has \wedge r_2 \in a.has) \Rightarrow$$
$$(r_1 \notin a.plays \wedge r_2 \notin a.plays \wedge r_3 \in a.has)))$$

For example, let us assume that roles r_2 and r_3 merge resulting to role r_4. Based on the above semantic definition, if an agent has r_2 and r_3 then it must also have r_4 and it must not play r_2 and r_3 (the agent may or may not play r_4 depending on the relations of r_4 with the other roles the agent has). The example of a Mergeswith relation between roles r_2, r_3, and r_4, where r_4 is played by the agent, is depicted in Fig. 1.

AddsWith — There is no constraint in having or playing roles that add together.

$$\forall\, a : A,\, (r_1, r_2) : R{\times}R \cdot (r_1 \text{ add } r_2 \Leftrightarrow (r_1 \in a.has \Rightarrow ((r_2 \in a.has \vee r_2 \notin a.has) \wedge$$
$$(r_2 \in a.plays \vee r_2 \notin a.plays))))$$

Using the above semantic axioms, it is trivial to verify that the properties of role relations that we have introduced hold.

Relations between more than two roles can be defined in a similar manner. In that case, a predicate notation is more convenient to represent role relations. For example, when three roles r_1, r_2, and r_3 merge to r_4 this can be noted by $merge(r_1, r_2, r_3, r_4)$. In this paper, we will not provide any formal definitions for relations among roles with arity greater than two.

Role relations, as defined in the above algebra, constrain the way that roles can be allocated to agents. Therefore, the agent organisation design problem is transformed to a constraint satisfaction problem that must be solved for roles to be allocated to agents. The problem can be constrained further by including constraints based on general design heuristics. These constraints are expressed on the performance variables of the agents. For example, the system designer should be able to define the maximum number of roles that an agent could play, or an upper limit to the resource capacity that the roles an agent plays would require. Furthermore, role allocation heuristics could also be specified. For example, roles requiring access to similar resources could be required to be assigned to the same agent.

6   Implementation

We have developed an experimental implementation of the system shown in Fig. 2 that provides support for engineers to describe design concerns in the role algebra described above and then to reuse previous role models in their implementations.

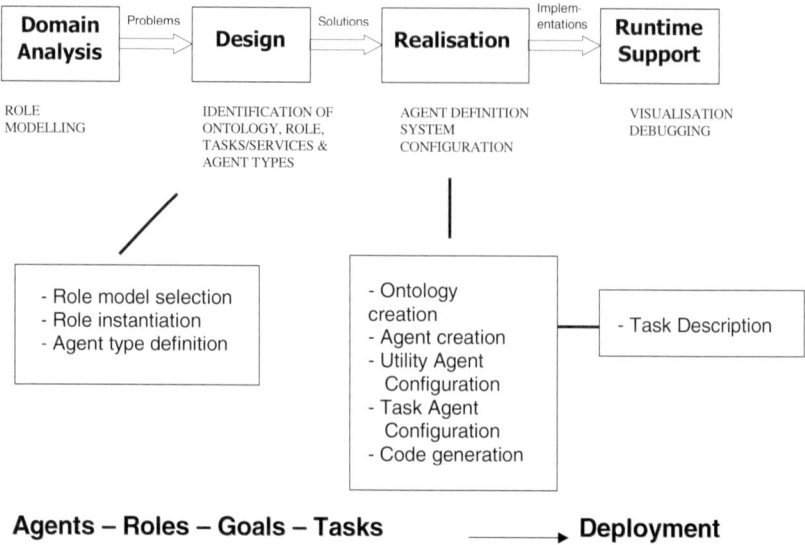

Fig. 7. New agent design process in Zeus

In order to implement the tools shown on Fig. 2 we needed to change the development process assumed by the Zeus toolkit. This change is detailed in Fig. 7, where we show how role selection and the notion of agent type definition and selection is incorporated into the process at the design phase before realisation and configuration and then deployment.

7   Promoting Reuse

In the introduction and motivation section of this paper a discussion of the problems e-commerce developers encountered when they attempted to use Zeus was given, and we emphasized the barrier to adoption that the difficulties in reusing entangled agent application code presents. A description of a system of modularity based on role modeling was then given, along with the mobile working example which shows how this may work in a practical situation. However, we have not addressed how developing agents using role models can promote reuse.

Roles in our model are a set of constraints that are attached to a group of primitive and rule-based tasks and some "external" code which can then be deployed into an agent. In this way the agent concerned is equipped with problem solving knowledge in the form of new plan atoms (primitive tasks) and reactive behaviors (rule based tasks). In the mobile working example we developed a *travelmanager* role, which contains a set of tasks for planning a route using traffic information and current location. This role could be very useful when we later want to develop a Personal Travel Assistant (PTA).

In a PTA we would not have a *employee* role, but we would have a *customer* role, a *airtravelProvider, hotelProvider* and *broker*. The *customer* and *travelmanager* roles are imported from the mobile working role model, and the other roles are developed from scratch. We then specify a set of simple Role Constraints which govern the way these roles combine:

```
not (customer, travelmanager)
not (broker, hotelProvider)
not (broker, travelinfobase)
not (broker, airtravelProvider)
not (broker, customer)
not (hotelProvider, customer)
not (airtravelProvider, customer)
```

When we apply Algorithm 1 to the problem it can be seen that this will lead to a role allocation of

```
A1 :customer
A2: broker, travelmanager
A3: airtravelProvider, hotelProvider
```

We now have a robust and systematic manner to re-use behavioral models (roles and role models) in new applications, and to allocate roles to agents during system design.

8   Conclusions and Further Work

To enable effective design of complex multi-agent systems, we need to support the complexity involved in this process by semi-automating the transformation of analysis into design and reusing previous design knowledge. This is a weakness of the majority of existing approaches to agent organisation design and a focus of the work reported here. In this paper, a simple role algebra enabling automatic allocation of roles to agents has been introduced. This approach enables reuse of organisational design settings by representing them as role models being able to be manipulated considering the proposed role algebra.

In this paper we illustrate our approach with an example from the field of collaborative working, however, we have also considered systems from other fields such as e-commerce. Of course, it is an open question as to how widely the roles and methods that we outline can be applied, however we believe that this is a general programming and compilation method which can be used by developers of many different types of distributed system.

Some issues have not been addressed yet and require further work. For example, agents can play different roles in different contexts and hence the possible contexts should be considered when designing agent organisations. Furthermore, it is planned to use the role algebra to enable allocating and de-allocating roles to agents dynamically on run-time. This will require alterations to the mechanisms for naming and namespace management that are currently used in agent standards and agent systems, and may provide support for an open agent type management system.

More importantly than either of these considerations for further action is the need to demonstrate the value of this approach. We have shown that role modelling is a feasible mechanism for recording design knowledge and reusing it at a later date. We have no evidence that our method is a better way of doing this than any other approach, or indeed that it is better than not doing it at all! Providing answers to the questions posed by development teams before they commit to a technology ("how much will it cost", "what sort of training will we need", "what it the return on investment we can expect", "what impact will this have on our development timescales") will require a substantial investigation to measure the impact of industrialised versions of this technology on a number of real world projects. This investigation will have to study the impact on code reuse, productivity, cost and quality of product, and show that the technology can be used by developers with typical skills and abilities.

Before such a study can be mounted a great deal of further work is necessary. We need to move our implementations beyond the small examples tried so far. We will need to implement tools that are attractive and intuitive for developers and we will need to integrate the technology with widespread *de-facto* standards. Last, but by no means least, we will be required to document and implement a large number of role models to provide our trail subjects with sufficient material to test the validity of the approach.

Acknowledgements. This work has been supported by BT under a grant from the office of the Chief Technologist (No. ML816801/MH354166).

References

[1]   Andersen, E., *Conceptual Modelling of Objects: a role modelling approach*, in *Dept of Computer Science*. 1997, University of Oslo: Oslo.

[2]   Artikis, A. and J. Pitt, *A Formal Model of Open Agent Societies*, in *Proceedings of Autonomous Agents 2001*. 2001: Montreal. p. 192-193.

[3]   Biddle, B.J., *Role Theory: Expectations, Identities and Behaviours*. 1979, London: Academic Press.

[4]   Depke, R., R. Heckel, and J.M. Kuster, *Improving the Agent-oriented Modeling Process by Roles*, in *Proceedings of the fifth international conference on Autonomous Agents*. 2001, ACM Press: Montreal, Canada.

[5]   Evans, R., *MESSAGE: Methodology for Engineering Systems of Software Agents*. 2000, BT Labs: Ipswich.

[6]   Ferber, J. and O. Gutknecht. *A meta-model for the analysis and design of organisations of Multi-Agent systems*. in *International Confernce in Multi-Agent Systems (ICMAS 98)*. 1998. Paris, France: IEEE Press.

[7]   Hilaire, V., et al., *Formal Specification and Prototyping of Multi-Agent Systems*, in *Engineering Societies in the Agents' World ESAW'00 (in ECAI'00)*. 2000: Berlin.

[8]   Karageorgos, A. and N. Mehandjiev, *Specifying Role Constraints in RCL*. 2001, UMIST: Manchester. p. 35.

[9]   Kendall, E.A., *Role models - patterns of agent system analysis and design*. BT Tech. Journal, 1999. **17**(4): p. 46-57.

[10] Kendall, E.A. and L. Zhao. *Capturing and Structuring Goals*. in *Workshop on Use Case Patterns, Object Oriented Programming Systems Languages and Architectures (OOPSLA)*,. 1998.

[11] Nwana, H.S., et al., *Zeus: A toolkit for Building Distributed Multi-Agent Systems*. Applied Artificial Intelligence Journal, 1999. **13**(1): p. 187-203.

[12] Omicini, A. *SODA : Societies and Infrastructures in the Analysis and Design o Agent-based Systems*. in *Workshop on Agent-Oriented Software Engineering*. 2000. Limetick, Ireland.

[13] Parunak, V., J. Sauter, and S. Clark, *Toward the Specification and Design of Industrial Synthetic Ecosystems*, in *Intelligent Agents IV: Agent Theories, Architectures, and Languages*, M.P. Singh, A. Rao, and M.J. Wooldridge, Editors. 1998, Springer Verlag: Berlin. p. 45-59.

[14] Scott, W.R., *Organisations: Rational, Natural and Open Systems*. 1992, New York: Prentice Hall International.

[15] So, Y.-p. and E.H. Durfee, *Designing Organisations for Computational Agents*, in *Simulating Organisations: Computational Models of institutions and groups*, M.J. Prietula, K.M. Carley, and L. Gasser, Editors. 1998, AAAI Press. p. 47 64.

[16] Sparkman, C.H., S.A. DeLoach, and A.L. Self. *Automated Derivation of Complex Agent Architectures from Analysis Specifications*. in *Agent-Oriented Software Engineering (AOSE-2001)*. 2001. Montreal, Canada.

[17] Stark, J., et al., *ACSOSS: a case study applying the MESSAGE analysis method*. 2001, BT Labs: Ipswich.

[18] Thompson, S.G. and B.R. Odgers. *Collaborative Personal Agents for Team Working*. in *Proceedings of the AISB Symposium*. 2000. Birmingham, England.

[19] Wooldridge, M., N.R. Jennings, and D. Kinny, *The Gaia methodology for agent-oriented analysis and design*. International Journal of Autonomous Agents and Multi-Agent Systems, 2000. **3**(3): p. 285-312.

[20] Yu, L. and B.F. Schmid. *A Conceptual Framework for Agent Oriented and Role Based Workflow Modelling*. in *CaiSE Workshop Conference on Agent Oriented Information Systems (AOIS '99)*. 1999. Heidelberg: MIT Press.

[21] Zambonelli, F.; Jennings, N.R.; and Wooldridge, M. J. 2000. Organizational Abstractions for the Analysis and Design of Multi-Agent Systems. In *Proceedings of the 1st Workshop on Agent-Oriented Software Engineering*, Springer-Verlag.

[22] Robertson, J. & Robertson, S. "Volere Requirements Specification Template Edition 8" http://www.guild.demon.co.uk/SpecTemplate8.pdf

Comparison of Some Negotiation Algorithms Using a Tournament-Based Approach

Peter Henderson, Stephen Crouch, Robert John Walters, and Qinglai Ni

Declarative Systems and Software Engineering
Department of Electronics and Computer Science
University of Southampton
Southampton, UK, SO17 1BJ
{ph,stc,rjwl,qn}@ecs.soton.ac.uk

Abstract. This paper provides some results and analysis of several negotiation algorithms. We have used a tournament-based approach to evaluation and applied this within a community of Buyers and Sellers in a simulated car hire scenario. An automated negotiation environment has been developed and the various negotiation algorithms made to compete against each other. In a single tournament, each algorithm was used as both a Buyer-negotiator and a Seller-negotiator. Each negotiating algorithm accommodates the parameters for negotiation as a set of desirable goals, represented as examples of product specifications. It was the task of each negotiating algorithm to get the best deal possible from every one of their opposites (i.e. Buyer versus Seller) in the sense of being close to the examples they were given as goals. One algorithm proved to be superior to the others against which it was made to compete.

1 Introduction

A significant problem in distributed e-commerce applications is the choice of algorithm used to carry out automated negotiation on behalf of a client [3,4,6,7,8,12]. Even very simple algorithms can have behaviour which is acceptable in a restricted scenario but which might be unpredictable in a more liberal environment. In order to gain some confidence in algorithms we were planning to deploy, we decided to establish a simulation environment in which they could be evaluated.

In 1984 Robert Axelrod published The Evolution of Cooperation [1], a book that amongst many other things discussed the results of two tournaments that attempted to find the best automated algorithm at playing the iterated Prisoner's Dilemma, a deceptively simple game with its origins in economic game theory. In this game, competitors are required either to co-operate or defect (i.e. not co-operate) in a series of rounds. Each participant can observe the behaviour of its opponent and choose to collaborate or defect on the next round according to how it feels the opponent may perform. The rewards are highest for a defector whose opponent collaborates. But they are lowest if both defect. The optimal long-run strategy is for both to collaborate, where the rewards are not as high as they are for a lone defector but where they are much higher than if both defect.

R. Kowalczyk et al. (Eds.): Agent Technology Workshops 2002, LNAI 2592, pp. 137-150, 2003.

One of the many interesting aspects of the work was that the algorithm that emerged victorious against all the others was incredibly simple. Tit-For-Tat simply cooperated with its opponent on the first round, and from then on just reciprocated whatever its opponent did on the previous round. Surprisingly, this algorithm accumulated more rewards than the others, although it would never actually win a complete game. It won because it encouraged high scoring games with its opponents, and although it was always either drawn with or beaten, it subsequently attained the highest score overall (see [2]).

Three characteristics formed the basis for its success: it never was the first to defect, it didn't hold grudges, but it was retaliatory. Cooperate with it, and you both do well. Defect against it and it does the same. We wondered if a similar result might hold where participants were engaged in negotiation. We have chosen to build a tournament that pits algorithms against each other, which, although simple, are of the sort that are actually used in commercial scenarios.

At an abstract level, similarities may be identified between the tournament presented in this paper and the tournament conducted by Axelrod. However, there are some very notable differences, since this experiment deals with interaction on a far more detailed, and therefore semantically rich, level.

Firstly, the scoring system has to be more complex. In negotiation, often there is no absolute notion of cooperation and defection as in the Prisoner's Dilemma. For example, what one Seller views as defection by a Buyer is not always what another Seller would view as defection. The existence of this perceptual grey area means a more detailed scoring system to ensure consistency across scores was required.

Secondly, the algorithms in this tournament are each given a set of negotiation goals, in the form of desired product specifications. Interpreting and reasoning about these goals in some way is an issue that has to be dealt with by the algorithms.

Thirdly, in this tournament, all algorithms face each other, including instances of themselves, as Buyers versus Sellers.

Previously we have looked at architectures for e-commerce systems [10,11,12] and been interested in how federations of applications co-operate, particularly when new applications can join the federation at any time. Networks of e-commerce negotiation algorithms have exactly this property and have become a test case for us.

Section 2 describes the nature and context of the experiment. The car hire negotiation scenario is discussed in Sect. 2.1 and the negotiation environment is outlined in Sect. 2.2. Sect. 2.3 details the generic behaviour of the negotiators, and Sect. 2.4 introduces the use of example specifications as goals, and describes the two sets of examples used for the experiments.

Section 3 provides a behavioural description and brief discussion of each of the seven algorithms.

Section 4 presents the results of the experiments and provides some analysis and discussion of these results, which are further discussed in Sect. 5.

2 The Experiment

2.1 The Car Hire Scenario

The chosen scenario for the negotiation tournament was car hire. If we consider a single Buyer and Seller pair in the tournament, a Buyer's objective is to secure the best deal possible for hiring a car, with respect to a given set of car specifications. The Buyer has a set of examples of deals they would accept. Each entry in this set consists of four attribute name and value pairs for the following attributes:

- **Days:** the length of time we wish to hire the car
- **Price:** the price we would like to pay
- **Features:** some linear, quantified grade of features (e.g. air conditioning, electric windows), higher number represents more features
- **Class:** the desired size of the car, higher number represents larger car

A set of examples consists of car specifications, each representing an acceptable outcome of negotiation. The Seller also has a set of examples, representing the cars they wish to hire out, reflecting their stock constraints.

Specifying negotiation criteria as examples provides an abstract yet flexible method of stating a negotiator's desires, although the potential exists for ambiguity between these example criteria. There is not always a clear correlation between these examples, and interpreting them in the context of the negotiation process and using this understanding to guide actions are behavioural tasks of the negotiator [13].

Table 1. A set of examples to be used as Buyer goals

Days	Price	Features	Class
9	250	4	2
6	150	2	2

Consider the examples in Table 1. If these are examples used by a Buyer, then we see that they are after a particular class of car and want about 9 days of hire. They are prepared to compromise on days (and features) but only for a significant saving in cost. If the Buyer using these examples receives an offer that is close to one of these examples, they would be inclined to accept it. If they have to make a counter offer, they will construct one using an algorithm which takes into account offers they have received and which attempts to stay close to these examples. It is algorithms of this sort (for Sellers as well as Buyers) that we wish to evaluate.

2.2 The Negotiation Environment

A negotiation environment was developed within which multiple automated negotiators could compete. Two applications form this environment:

- **Supervisor:** responsible for initiating the environment, including the negotiators. Maintains a list of algorithms, one of which can be adopted by each negotiator

– **Negotiator:** given a set of negotiation examples and environment parameters (round cut-off, opponent negotiator identities) and is responsible for conducting negotiation

The environment allowed for a configurable number of Buyer and Seller negotiators to be instantiated for a single tournament, each with their own algorithm for conducting negotiation, and each with a set of either Buyer or Seller examples. These examples provide a set of acceptable goals for each negotiator. Once the scenario has been initiated, it remains fixed. Therefore, it is not possible to simulate situations where a negotiator switches its behaviour to that of other algorithms during a negotiation. However, this can be achieved by modelling multiple behaviours within a single algorithm. In which case, the algorithm can switch between them when it sees fit.

The negotiation environment was designed and implemented such that the process of inserting a new algorithm into the tournament and allowing it to compete with the others was simple and rapid. This process was as follows:

– **Description:** the algorithm is described in pseudocode
– **Translation & Compilation.** this pseudocode is translated into Visual Basic and compiled into an executable
– **Add to Supervisor List:** the name of the algorithm is added to the Supervisor's list of available algorithms

To a great extent, by careful design of the pseudocode language, the translation to Visual Basic is mechanistic.

In a typical tournament, a Buyer's target is to secure one car from each Seller, whilst the Seller's target is to sell one car to each Buyer. If we adopt a global view of all negotiations, we essentially observe a series of pair-wise negotiations between each possible Buyer/Seller/algorithm permutation, with successful negotiations resulting in the exchange of a car from a Seller to a Buyer. In other words, after a tournament is complete, every Buyer will have negotiated once with every Seller, and vice-versa, and every algorithm in a tournament will be represented as both a Buyer and Seller.

In a typical tournament, although all negotiations between Buyers and Sellers are handled concurrently, the actions of each Buyer do not affect other Buyers, and the same is true for Sellers. This is because each Seller potentially has one car to hire out to each Buyer. However, if we give each Seller less cars than there are Buyers, the actions of a Buyer have possible ramifications for other Buyers, since those which typically take more time in reaching agreement may not secure a car. This makes it possible for us to run a tournament with the added element of competition for resources, where an algorithm's efficiency contributes to success. This has not been done in the experiments reported here.

To measure the success of a negotiator following a tournament, a simple scoring system was devised and applied to each outcome of each negotiation for a negotiator. Only two possible outcomes of negotiation between a Buyer and a Seller exist:

– **Accept:** After a negotiator has received an offer, they can choose to accept it. However, if instead they make a counter-offer, that received offer can no longer be accepted (unless the same offer is made again). For each acceptance a negotiator manages to secure with another, either by accepting an offer themselves, or having one of their offers accepted, a measure of 'distance' between that offer and their given set of examples provides us with a base score. Therefore, accepting an offer (or having it accepted) that perfectly matches one of their examples will get them the best score possible. An acceptance facilitates one car being passed from the Seller to the Buyer as a resource.

– **Quit:** Determined and imposed by the negotiation environment, if a Buyer-Seller pair is still negotiating after a given number of rounds, their negotiations are terminated and they each receive a score of zero. In addition, the Seller does not sell a car, and the Buyer does not receive one. This outcome represents the penalty for not reaching an agreement, and therefore provides an incentive for each algorithm to reach agreement quickly. However, they are not told prior to the tournament how many rounds they are not allowed to exceed. Similar in motivation to the Axelrod tournament, algorithms cannot therefore attempt to do better than their negotiation opponents by using their knowledge of the maximum number of rounds to try to take advantage [5].

It is in the best interests of a negotiator's algorithm to reach agreement with their opposites under any circumstances, and to do so quickly. It is intentional that securing a bad deal quickly and receiving a low score is a better outcome than not securing a deal and receiving a zero score. Another approach would have been to offer algorithms the choice of quitting negotiations themselves instead of making another offer, and many scoring methods could have been employed. However, it was decided that the main objective of the tournament is to ascertain how well each algorithm can negotiate with each other, not how well strategically they can quit negotiations. An algorithm that knows when to quit against another, perhaps to attain the best payoff, does not tell us very much about how effectively it negotiates. However, such an algorithm can be simulated. It would simply repeat its final (rejected) offer until such time as the supervisor intervened.

The algorithm used for calculating the distance between an accepted offer and a set of examples was straightforward. For each attribute in an offer, a minimum and a maximum allowed value are imposed. No penalty is awarded for going outside of these ranges, but any offending attributes are constrained within those ranges. These range values are accessible by an algorithm, and this mechanism therefore provides a sanity check against offers that may inhibit the operation of the system, but more importantly these range values provide a scope for scoring algorithms. The scoring function takes an accepted offer and an example and returns an inverse measure of 'distance' between the two, as a value between 0 and 1. i.e. the higher the score, the closer the offer to the given example.

This function is applied to all examples and the highest score of these represents the negotiator's overall score for that accepted offer. To ensure a more representative spread of results, and to reduce the effect of anomalies, the tournament was executed many times and the results averaged. Following this process, each negotiator is given an average sub-total that represents the negotiator's average score over all executions of the tournament. This sub-total is then multiplied by the factor of resources the negotiator was able to purchase or sell, depending on whether they were a Buyer or a Seller, to determine a final score.

2.3 Negotiation Behaviour

The negotiation process between a Buyer and Seller consists of a series of offers and counter-offers being made until an agreement is reached, with a single offer consisting of Days, Price, Features and Class attribute and value pairs. Of course, during negotiation it may prove impossible for a Buyer to acquire exactly what they want from the Seller, or vice versa, so each negotiator must be able to compromise on certain attributes in order for negotiation to be successful. However, it is obvious that it is not in the best interests of each negotiator to over-compromise, simply because this could mean they secure a deal which does not match with their desired criteria. The manner and degree in which a negotiator deviates from those criteria is dictated by the negotiation strategy they employ.

Figure 1 shows the messages that will flow between two negotiators, one configured as a Buyer and one as a Seller. The supervisor will tell one or other (we will always use the Buyer) to start.

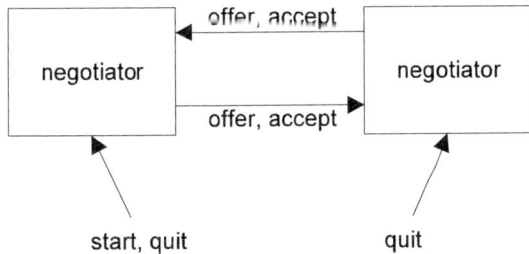

Fig. 1. The negotiation scenario

Offers will alternate according to a predefined behaviour specified in the negotiator.

Pseudocode representation of the behaviour of the negotiator

```
on receive start from supervisor {
    compute initial offer;
    send offer to partner;
}
on receive offer from partner {
    evaluate offer;
    if (offer is acceptable) {
        send accept to partner;
        exit;
    } else {
        compute new offer;
        send offer to partner;
    }
}
on receive accept from partner {exit;}
on receive quit from supervisor {exit;}
```

Eventually either one of the negotiators accepts or the supervisor tells both to quit. Each negotiator is initialised with this behaviour, and is told that it is buying or selling against a set of examples with which it has been furnished. Each algorithm will take a different approach to implementing one or other of the basic actions:

- compute initial offer,
- evaluate offer,
- compute new offer.

Before we go into further detail of the actual algorithms that we use, we need to say a little more about the examples.

2.4 The Examples

In order to understand how and to what extent the examples contribute to the results, two very simple sets were used, each with different qualities. Set one is shown in Table 2 and Table 3.

Table 2. Buyer example set 1

Days	Price	Features	Class
9	200	4	2
8	190	3	2

Table 3. Seller example set 1

Days	Price	Features	Class
7	300	3	2
4	150	2	1

Essentially, for the Buyer and the Seller, each of their examples is roughly consistent with each other, and appears quite rational. Comparing the Seller's examples with the Buyer's examples shows the Buyer would take one less day and a slightly less featured car for $10 less, whilst the Seller would like to hire out a less featured, smaller car for half the price.

Set two is shown in Table 4 and Table 5.

Table 4. Buyer example set 2

Days	Price	Features	Class
10	200	3	4
8	140	2	3

Table 5. Seller example set 2

Days	Price	Features	Class
8	260	2	2
6	170	2	1

These examples were designed to be less consistent and more difficult to reason about. If we consider the Buyer's second example, the Buyer wishes to hire a big, low features car for 8 days at $140, yet would be willing to pay $60 more for two extra days and a slightly better featured, larger car. However, the Seller's second example is $170 for a small car with low features for 6 days. His first however, is a great deal more for only two extra days and a moderately sized car. It should be noted that the first set of examples appears to provide a little more room for 'negotiation manoeu-vring' than the second set. In other words, the Buyer and Seller negotiation criteria are further apart in the first set than the second set. Giving the algorithms a smaller bar-gaining arena gives us the opportunity to observe how well they perform under such tight circumstances.

Although we have run experiments with larger example sets, the results are essen-tially as those reported here. The smaller example sets make clearer what is going on.

3 The Algorithms

Seven algorithms were developed and submitted to compete in the tournament. Each algorithm had to address the following three questions:

- What constitutes an initial offer if the algorithm is a Buyer?
- Under what circumstances is an offer accepted?
- If the most recently received offer is not accepted, how is a counter-offer formu-lated?

In these terms, let us describe the seven algorithms that we compared.

The first two algorithms, Random and JustAccept, were trivial. These algorithms were included purely for comparison with other algorithms. Obviously, any algorithm should always do better than Random or JustAccept, so these two algorithms provide a 'comparison bar' for the lowest level of performance. Interestingly, we discovered that some algorithms were not able to beat either of these trivial choices.

3.1 Random

Random simply produced random offers, by picking a random number between the minimum and maximum ranges for each attribute. Each time an offer was received, random would have a 10% chance of accepting it.

- **Compute Initial Offer:** choose arbitrary values for each attribute within permitted ranges
- **Evaluate Offer:** accept probabilistically (for these experiments, with 10% chance)
- **Compute New Offer:** choose arbitrary values for each attribute within permitted ranges

3.2 JustAccept

JustAccept simply accepted the first offer it received, and if it was a Buyer, its first offer was simply its first example.

- **Compute Initial Offer:** choose first example
- **Evaluate Offer:** always accept
- **Compute New Offer:** never happens

3.3 AgreeRandomAttribute

This algorithm only attempts to negotiate with respect to its first example, which forms its first offer if it is a Buyer. After receiving an offer, it uses its first example as an offer template. Into this offer template it randomly substitutes an attribute value from the opponent's offer. This offer template forms the new offer. It will accept an offer if it only has one attribute different from any one of its examples.

- **Compute Initial Offer:** choose first example
- **Evaluate Offer:** accept if agreement in all but at most one attribute
- **Compute New Offer:** alter one attribute to equal value received from opponent

3.4 AgreeProgressive

AgreeProgressive was a more accommodating, and more sophisticated, version of AgreeRandomAttribute. It utilises a matrix that acts as a mask for merging an example and an offer to form a new offer. The merging process is simple: the best matching example to the last received offer is used as the template, and the matrix decides which attribute values in the template to substitute. In the first four rounds, the algorithm will accept an offer with only one different attribute to one of its examples. Otherwise, on a per-round basis, it cycles through each attribute, substituting the appropriate attribute in the closest matching example for the corresponding attribute in the last received offer. The closest matching example is the one with the least number of different attributes from the last received offer.

In the next 6 rounds, the number of attributes to substitute is increased to two. Every possible permutation of two attributes is attempted. A received offer is accepted if it differs from one of its examples by only two attributes.

In rounds 11 to 14, all possible permutations of three attributes are attempted, and offers are accepted if different from an example by only three attributes. When the algorithm reaches round 15, it will accept whatever offer is sent by the opponent.

- **Compute Initial Offer:** choose first example
- **Evaluate Offer:** accept if agreement in all but at most one (two, three, ...) attributes
- **Compute New Offer:** alter one (two, three, ...) attributes to equal value in example nearest to offer received from opponent

3.5 Tit-For-Tat

This algorithm represents a simple interpretation of Axelrod's Tit-For-Tat in the context of negotiation. It replicates on an attribute-attribute basis the inverse behaviour of the opponent. This behaviour is determined by simply comparing the opponent's most recent offer with the one received before that. e.g. if the opponent (as a Seller) deducts $10 off the price, the Buyer as Tit-For-Tat will add $10. Until Tit-For-Tat has two offers to compare, it initially cooperates by adding 10% onto its previous offer.

- **Compute Initial Offer:** choose first example
- **Evaluate Offer:** accept if offer within a margin of one example
- **Compute New Offer:** for each attribute, reflect opponent's behaviour by moving the same degree in the opposite direction: if opponent closes gap, then close gap. If opponent opens gap, then open gap

3.6 Retreat

This algorithm begins by offering the first example. As negotiation progresses, it then proceeds to 'back away' from this example in the opposite direction of the opponent's last received offer. If the opponent's offer is close to this example, it accepts the offer.

- **Compute Initial Offer:** choose first example
- **Evaluate Offer:** accept if agreement within a margin
- **Compute New Offer:** for each attribute, regardless of whether opponent opens or closes gap, open gap by 10%

3.7 TestAlgorithm

This algorithm employed a numerical method to dictate its offers, and to determine whether to accept an opponent's offer. This is an attempt to emulate the kind of rational algorithm which is often deployed in practice, where some quantitative knowledge of the domain is used to refine its decision making process.

- **Compute Initial Offer:** choose first example
- **Evaluate Offer:** accept if agreement within a margin
- **Compute New Offer:** numerical method of moving within region of disagreement with opponent

4 Results

The results with the first example set are given in Table 6 and Table 7. The results with the second example set are given in Table 8 and Table 9. To ensure fair testing and comparable results, the configuration of the environment remained the same; only the example sets were different for both rounds of experimentation. The results are not entirely as we would have expected. We did not predict that AgreeProgressive would do as well as both Buyer and Seller. Nor did we expect the rational TestAlgorithm to do so badly as a Seller.

Table 6. Buyers using the first example set

Negotiator	Algorithm	Resource Used	Score	Final Score
Buyer4	AgreeProgressive	1	0.937	0.937
Buyer5	Tit-For-Tat	1	0.901	0.901
Buyer1	TestAlgorithm	1	0.898	0.898
Buyer7	JustAccept	1	0.883	0.883
Buyer2	Random	0.986	0.799	0.788
Buyer6	Retreat	0.857	0.786	0.674
Buyer3	AgreeRandomOne	0.7	0.673	0.471

Table 7. Sellers using the first example set

Negotiator	Algorithm	Resource Used	Score	Final Score
Seller4	AgreeProgressive	1	0.948	0.948
Seller6	Retreat	0.993	0.920	0.913
Seller7	JustAccept	1	0.860	0.861
Seller2	Random	0.993	0.781	0.776
Seller3	AgreeRandomOne	0.843	0.819	0.691
Seller5	Tit-For-Tat	0.857	0.794	0.680
Seller1	TestAlgorithm	0.857	0.737	0.631

Table 8. Buyers using the second example set

Negotiator	Algorithm	Resource Used	Score	Final Score
Buyer4	AgreeProgressive	1	0.860	0.860
Buyer7	JustAccept	1	0.823	0.823
Buyer1	TestAlgorithm	1	0.816	0.816
Buyer2	Random	0.993	0.748	0.742
Buyer5	Tit-For-Tat	0.843	0.757	0.638
Buyer6	Retreat	0.571	0.511	0.292
Buyer3	AgreeRandomOne	0.457	0.441	0.202

Table 9. Sellers using the second example set

Negotiator	Algorithm	Resource Used	Score	Final Score
Seller4	AgreeProgressive	1	0.923	0.923
Seller7	JustAccept	1	0.732	0.732
Seller2	Random	0.986	0.692	0.682
Seller3	AgreeRandomOne	0.836	0.795	0.664
Seller6	Retreat	0.757	0.699	0.529
Seller5	Tit-For-Tat	0.714	0.618	0.441
Seller1	TestAlgorithm	0.571	0.472	0.270

Some algorithms never reach agreement, even though doing so entails such a severe penalty. The reasons for this are threefold: firstly, the algorithms do not know how many rounds of negotiation they are allowed. If they did, they could simply accept the last offer made by their opponents before the cut-off. Secondly, each negotiator faces the simple dilemma of whether to accept the other negotiator's latest offer, or to make another offer. Because they have no global view of how negotiations will turn out, they cannot know at any point during negotiations whether the most recent offer received is the best they will ever get. Thirdly, the negotiation behaviour that emerges as a result of the inherent nature of each algorithm, when faced with the other, may guarantee they never reach agreement. Retreat, for example, could never reach agreement with AgreeRandomAttribute if each of their examples were sufficiently far apart.

AgreeProgressive was more successful than we expected. The behaviour reported here was repeated in other experiments, including for larger example sets. Most importantly, it is the only algorithm that will always reach agreement as long as the negotiation cut-off is at least 15 rounds (for 4 attributes). At round 15 it eventually agrees with whatever the opponent is then offering. As long as this is the case, this ensures that the algorithm is never penalised for not reaching agreement early enough. Secondly, the nature of the algorithm means that it gradually alters its negotiation strategy from initially very stubborn (only agreeing to one attribute), to very conciliatory (agreeing with all four attributes, and therefore accepting the offer). Every possible permutation of offer agreement within these two extremes is presented to the opponent, and therefore the likelihood that an offer will be accepted increases with every iteration. Thirdly, because the algorithm is initially very stubborn, this allows the algorithm to take advantage of any concessions that may be made by the opponent in the earlier stages of negotiation, before it begins to compromise on a greater scale. This can be seen with TestAlgorithm. Unlike algorithms such as Retreat and TestAlgorithm, AgreeProgressive does not waste time making 'bluff' offers. It immediately attempts to find a formula for mutual agreement. Tournament cut-off permitting, this will always be the case.

However, if the cut-off is set to less than 15 rounds, AgreeProgressive does not do so well, for a very specific reason. Let us take the results of a cut-off of 12 rounds as an example. Whilst the other algorithms generally maintain their ranking order, AgreeProgressive as a Buyer slips to around fifth in the rankings, whilst the Seller slips to around sixth. The reason for this poor performance is how it performs against itself. Intuitively, it could be reasoned that agreement would occur automatically by round five; both would be conceding two attributes, and both would accept offers different by two attributes. However, when negotiating against itself, AgreeProgressive requires that negotiation reach the final 15th round for agreement to occur. The reason for this is symmetry of behaviour. Since both Buyer and Seller follow the same strict pattern of attribute agreement, when the Buyer makes his second offer with one attribute in agreement with the Seller's initial offer, the Seller will agree on the same attribute. This effectively ensures that the Seller's next offer is the same as the Seller's first offer. This 'reflective' behaviour continues until round 15, where the Buyer will accept the Seller's offer regardless. Therefore, if negotiation does not reach round 15, both the Buyer and Seller representatives of AgreeProgressive are penalised, which is reflected in the rankings.

The relative success of JustAccept is a consequence of the nature of the other algo-rithms and the structure of the tournament. JustAccept does well because it always reaches a deal and because the opponents are behaving reasonably in that their offers are realistic. When JustAccept is acting as a Buyer, this is a close approximation to real life, where goods on sale are offered at a fair price and Buyers just accept that. It doesn't do quite as well as a Seller, but even there its behaviour is reasonable because Buyers open with reasonable bids. This algorithm is obviously open to exploitation, but the tournament has been structured to prevent this. Nonetheless, JustAccept has performed its role as a benchmark for calibrating the performance of others. The only algorithm to perform consistently better than JustAccept was AgreeProgressive, as-suming the number of negotiation rounds was at least 15.

The results presented here are typical. In other experiments, with different example sets and identical environment configurations, the ranking of the algorithms remains similar to the results displayed here. AgreeProgressive consistently does exceptionally well against the others, as a Buyer or Seller, as long as the round cut-off is at least 15 rounds. In its worst test it came third as a Buyer, but the tournament leader, Retreat in this case, was only ahead by a score of 0.006. Conversely, AgreeRandomOne per-forms very badly, always last in the rankings. The order of the middle rankings as shown in this paper is also representative. Of course, with some example sets, other algorithms do better than others, but in general big differences in the ranking are un-common.

5   Conclusions

We have described a series of experiments that have allowed us to compare various negotiating algorithms. Following Axelrod we have taken the view that an algorithm is best if it does well against a range of opponents. Although negotiation is a more complex behaviour to describe (and hence to measure) than simple collaboration, we have arrived at a similar result to Axelrod. One algorithm has performed better than expected, consistently doing well against a range of opponents. The algorithm is not the simplest in our set, nor is it the one we expected to be best. These are observations we have explained, to some extent. Further experiments which we plan, with these algorithms and with new algorithms, will lead, we hope to a greater understanding of negotiated agreement in an e-commerce context.

References

1. Axelrod, R.: The Evolution of Co-operation. Basic Books Inc., New York (1984)
2. Axelrod, R.: The Complexity of Cooperation. Basic Books Inc., New York (1997)
3. Burg, B.: Agents in the World of Active Web Services. To be published in Springer LNCS, see http://www.hpl.hp.com/org/stl/maas/pubs.html
4. Bichler, M., Segev, A., Zhao, J.L.: Component-Based E-Commerce: Assessment of Cur-rent Practices and Future Directions. ACM Sigmod Record: Special Section on Electron-ics Commerce, Vol. 27, No. 4 (1998) 7–14

5.   Binmore, K., Vulkan, N.: Applying Game Theory to Automated Negotiation. Netonomics, Jan. 99, see `http://www.worcester.ox.ac.uk/fellows/vulkan` (1999)

6.   Cranor, L.F., Resnick, P.: Protocols for Automated Negotiations with Buyer Anonymity and Seller Reputations. Telecommunications Policy Research Conference (TPRC 97), see `http://www.si.umich.edu/~presnick` (1997)

7.   Farhoodi, F., Fingar, P.: Developing Enterprise Systems with Intelligent Agent Technology. Distributed Object Computing, Object Management Group (1997)

8.   Fingar, P., Kumar, H., Sharma, T.: Enterprise E-Commerce. 1st edn. Meghan-Kiffer Press, Tampa FL (2000)

9.   Fogel, D.B.: Applying Fogel and Burgin's Competitive Goal-Seeking through Evolutionary Programming to Coordination. Trust and Bargaining Games. Proceedings of the 2000 Congress on Evolutionary Computation (CEC 2000), IEEE Press Piscataway NJ (2000) 1210–1216

10. Henderson, P.: Laws for Dynamic Systems. Proceedings of the Fifth International Conference on Software Reuse (ICSR 98), IEEE Computer Society Press, (1998) 330–336

11. Henderson, P., Walters, R.J.: Behavioural Analysis of Component-Based Systems. Information and Software Technology, Vol. 43, No. 3 (2001) 161–169

12. Henderson, P.: Asset Mapping - Developing Inter-enterprise Solutions from Legacy Components. In: Systems Engineering for Business Process Change - New Directions, Springer-Verlag UK, (2002) 1–12 see `http://www.ecs.soton.ac.uk/~ph/papers`

13. Sesseler, R.: Building Agents for Service Provisioning out of Components. Proceedings of the Fifth International Conference on Autonomous Agents (2001)

State-Based Modeling Method for Multiagent Conversation Protocols and Decision Activities

Ralf König

Department of Process Automation
Technical University of Hamburg-Harburg
21073 Hamburg
(+49) 40 42878 – 3547 (– 2493 for fax)
r.koenig@tu-harburg.de

Abstract. In this paper an approach to model conversations and agents decision activities by using state-based models is presented. These models are parts of a layered architecture which separates the aspects regarding the communication from the decision activities of the agents. By this, application independent conversation protocols are defined. The application dependent parts of the model are formed by a sequence of activities an agent has to perform to solve a problem. The different layers of the architecture are separated from each other by well-defined interfaces.

Keywords: multiagent communication, conversation, decision activity, conversation specification

1   Introduction

Research in multiagent communication is still an ongoing process. Topics of this research are communication languages, either for semantics of messages as well as for representation of contents, representation of communication patterns or coordination descriptions, to name only a few. In the area of agent communication languages (ACL), speech act based languages (like KQML) have shown, that the basic idea to adopt the speech act theory [20] to model multiagent communication is seminal. Using these speech act typed messages as "building blocks" for multiagent communication, the structure of message exchange between two or more agents (termed as *conversation*) can be specified. A good definition of *conversation policy* resp. *conversation* can be found in [9, p. 123]

> Any public declaratively-specified principle that constraints the nature and exchange of semantically coherent ACL messages between agents can be considered a conversation policy.

For these specifications of conversations different approaches have been made. Depending on the focus of the approaches, different representations are used, including formal grammars, petri nets, temporal logic, state machines, state transition diagrams, to name only a few. Up to now there is no agreement among researchers which is the best way to define specifications for conversations. In this paper an approach is presented to model conversations as well as the agent internal decision activities by using

R. Kowalczyk et al. (Eds.): Agent Technology Workshops 2002, LNAI 2592, pp. 151–166, 2003.

state-based models. The approach bases on a layered architecture build upon a speech act based ACL for defining conversation specifications between agents (on the lower level), for defining communication services using these conversations (at the medium level) and for defining the sequence of the activities of the agent solving a problem (on the upper level). Doing this, the agent uses the communication services provided by the medium level. Thus communication is a means which gives the agent the opportunity to reach a goal by making use of help of other agents.

This approach aims to form the base for a framework, which helps the designer of agents to develop cooperating multiagent systems by using the communication services provided by the framework. As a means for modeling the well known notation of state charts defined in the UML is used.

2   Fundamentals

2.1   Speech Acts and Conversations

The speech act theory forms one of the fundaments of research in multiagent communication today. A speech act is an utterance which is pronounced by a (human) speaker to cause a reaction at the (human) hearer. The utterances are classified by their types, e. g. question, assertion, request [20].

Concerning computer sciences the speech act theory gave the motivation to define multiagent languages like the KQML (currently the most used ACL [7, p. 3]) and FIPA-ACL. To be not determined to one ACL, only the type of the speech act will be declared in the following.

This paper follows the definition made in [25, p. 80] saying that a conversation is a structured exchange of messages. For the following explanation it is assumed that only two agents are involved in the conversation.

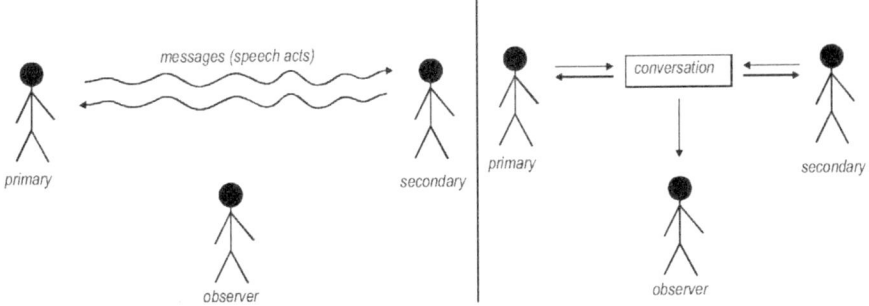

Fig. 1. Perspectives for modeling a conversation

The conversation can be modeled from three different viewpoints. The first one is the viewpoint of the agent who starts the conversation (it will be called the "primary"). An other viewpoint is the one of the agent who is addressed by the primary (the "secondary"). As a third party the observer is introduced (Fig. 1). It is not included in the conversation.

Thinking of a spoken conversation between humans, the primary and the secondary play the roles of speaker and hearer whereas the observer is still a hearer. This means that primary and secondary are making inputs and outputs to the conversation whereas the observer only gets the outputs.

2.2 Related Work

In the last years a lot of research in modeling conversation policies has been done. These works can be separated into two main directions [26, p. 384]. One direction understands conversations between agents as a process emergent from within the society, which can not be defined by hard-wired social laws (like fixed protocols). The other direction (termed as the *off-line approach*) understands a conversation as a predefined process, which can be ruled by a fixed protocol used by the participating agents. This paper follows the off-line approach in which three different representations have been used mainly ([21, p. 155]). The most significant are state transition diagrams (finite state machines) and petri nets. Beside this the Dooley Graph (as well as the Enhanced Dooley Graph) can be found as a state-based representation for conversations [21, p. 155].

In a petri net, the places can be marked by a number of identical tokens. Colored petri nets (CPN) and predicate transition nets (Pr/T-Nets) are extended petri nets using individualized tokens. Approaches using petri nets are made from von Martial (Pr/T-Nets) [15], Fallah-Seghrouchni et al. (CPN) [8], Cost et al. (CPN) [5] and Lin et al. (CPN) [14].

In this paper, state transition diagrams (STD) will be separated from finite state machines (FSM) in the manner that a STD models the states of a system and the transitions between them. These transitions can occur of a multitude of reasons which can have their causes inside or outside the modeled system. Following the idea of Winograd and Flores to model a conversation as a state-based process, the following approaches base on state transition diagrams (STD): Winograd and Flores [27], Belakhdar and Ayel [3], Bradshaw et al. [4], Greaves et al. [9] and Pitt and Mandami [19]. An extension of STD, the state charts [10], were used by Moore [17].

In contrast, for a FSM it is claimed that transitions only occur in case of an input (represented by an input symbol) from outside the modeled system. Models using FSMs or extensions of them where presented by Barbuceanu and Fox (FSM) [1], Barbuceanu and Lo (FSM) [2], von Martial (FSM) [15], Martin et al. (PDT) [16], Nodine and Unruh (FSM) [18] and Wagner et al. (extended FSM) [24].

Comparing these approaches, it can be seen that all approaches modeling conversations from the viewpoint of an observer are using either STD or petri nets, whereas the approaches using FSM (or state charts) are representing the conversation from the viewpoint of a participating agent. For modeling a conversation from the point of view of a participating agent who receives and sends messages, a model supporting input and output operations is more suitable, because the interaction between the agents can be expressed by these inputs and outputs. When a conversation should be modeled from an overall view (the observer's view), it is sufficient to use a model which is able to express, that a message has been transmitted from one agent to another, like a transition in a STD or in a petri net.

In contrast to STD, petri nets are able to express concurrency, which is important if the overall system behavior (including the agents internal decisions) is modeled. Models of this kind can be found in [14] for instance.

Most of the approaches focus on modeling the communication between two agents, without providing the opportunity to model communication among a greater number of participants. Furthermore, only a few approaches include the agents internal activities (the decision activities) in the model.

The approach presented in this paper models the decision activities of the agents as well as the communication between one agent and a group of agents. It bases on state-based modeled components which interact by inputting / outputting symbols (see Sects. 3.1 and 3.5 for details). The components are modeled by STD. The use of interacting, state-based modeled components ensures that these model components can easily be implemented as objects, which is helpful for the development of a software framework using this model.

2.3 Protocol Design

The presented approach uses the conversation as one layer of an architecture for agent communication. As agents are (at least from the authors point of view) software programs running on computers, it is helpful to take a look at the protocol design of communication protocols for computer networks.

Communication between computers is regulated by protocols which are organized in different layers (protocol stack). A protocol is a collection of rules and conventions defining how a dialogue has to take place [23, p. 33]. The ISO-Model defines a protocol layer as two connected protocol machines.

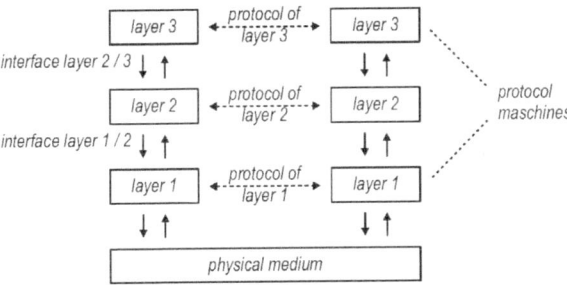

Fig. 2. Protocol stack

Each protocol machine consists of two interfaces, one to provide services to the user of the protocol (*user interface*), e. g. the next higher architecture layer, the other for sending and receiving data to / from the lower architecture layer. To simplify the understanding it is often assumed, that the protocol machine communicates directly with its peer (using its *peer interface*) by omitting the protocols on the lower levels (indicated by the dashed lines in Fig. 2).

To avoid confusion, this paper distinguishes between the *protocol* and the *protocol machine*. Whereas the protocol defines the way <u>how</u> the communication takes place, the protocol machine is the <u>realization</u> of the protocol which can be used to transfer data between two computers or agents.

3   State-Based Modeling Method

This section presents a state-based model for multiagent communication. The first section introduces the underlying architecture and it illustrates the purpose of its layers in short. The following sections describe the layers in depth. Section. 3.6 describes roughly the more complex contract net protocol and Sect. 3.7 dwells on the framework implemented by using this model.

3.1   Structure of the Layered Architecture

In the layered architecture, communication services (which are defined in the lower part of the hierarchy) are used by the decision activities of the agent, which are forming the upper part of the hierarchy.

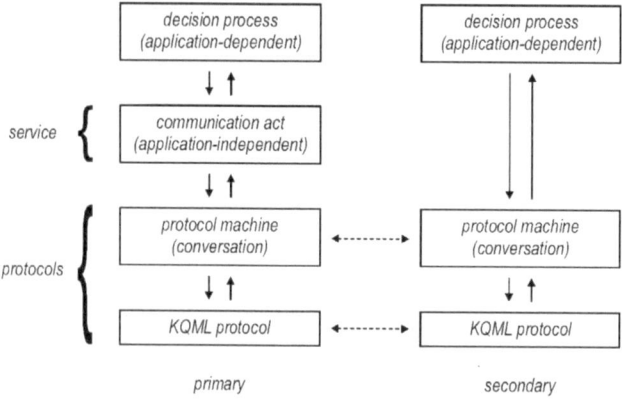

Fig. 3. Architecture

These decision activities will be termed as decision processes. A *decision process* consists of a number of activities (time consuming acts) performed by the agent to solve a problem. Hence the state of a decision process is defined by the execution of an activity. The communication service is defined in a *communication act*. Such an act is a means to serve a purpose, making use of the help of other agents. The message exchange among the agents is organized by making use of conversation policies, defined in a conversation protocol.

Considering the protocol stack the decision process uses the communication act which makes use of the protocol machine (which implements the conversation protocol). Underneath the conversation, the KQML protocol and transport protocols of the network are placed. Whereas the conversation policy and KQML are protocols (because either on the primary as well as on the secondary side a layer exists) the communication act is no protocol, because there exists no corresponding layer on the secondary's side. The separation between the decision process and the communication act is made, to have the opportunity to define communication acts independently from any application.

To summarize, every protocol stack layer is seen as a service provider to the upper layer. Because the services on an upper layer are build upon the services of the lower layer, these lower layers form "building blocks" for the upper ones. As an example, a conversation consists of a set of KQML-performatives. A communication act consists of conversations and so on. On the upper level of this hierarchy, the decision process makes use of communication acts to fulfill its task.

This approach separates the aspects concerning the communication (which are independent from a specific application) from the aspects regarding the decision activities of the agent (which are dependent from the application). By this, application-independent conversation protocols and communication acts can be defined and modeled in a framework. Using this framework, the agent developer only has to model the (application-dependent) decision activities of the agent as decision processes (which extends predefined patterns (skeletons)) and can make use of the framework to give the agents the ability to communicate.

The remainder of this section will show how to define conversation protocols, protocol machines, communication acts and decision processes.

3.2   Conversation Protocols and Protocol Machines

In the following, the conversation will be modeled as a FSM from the point of view of the observer. This FSM will serve as a base to represent the viewpoint of two participating agents. The resulting two FSM define the protocol machines (primary / secondary). A conversation is defined as a structured exchange of speech act typed messages between two agents. Furthermore the conversation is restricted to be synchronous, that means that the roles of speaker and hearer are defined for every message. It is not allowed, that in one state both agents are speakers, for instance.

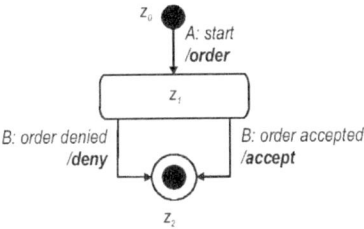

Fig. 4. Order Protocol

Looking from the viewpoint of the observer, the heard speech acts are outputs of the conversation. Thus they form the output alphabet of the FSM. As an example, a simple kind of conversation is used. In this conversation the primary gives the order to perform a task to the secondary which answers with a denial or an acceptance (Fig. 4). Depending on the result of an agents decision, the conversation chooses one of the transitions. As a result, the agents decisions serve as inputs to the conversation. In state z_1 the secondary (agent B) decides to fulfill the order (indicated by the event *order accepted*) or to deny it (indicated by *order denied*). Depending on this decision, the appropriate speech act type is selected.

After defining the conversation policy from the viewpoint of the observer, the protocol machines (modeled by FSMs) for the primary and the secondary are created. The conversation protocol is a means to transform the agents decisions into speech act typed messages. When an agent inputs its decisions into the conversation, the protocol machines convert them into speech act typed messages (on the senders side) and re-convert them from speech act typed message into agents decisions (on the receivers side). This is shown in Fig. 5.

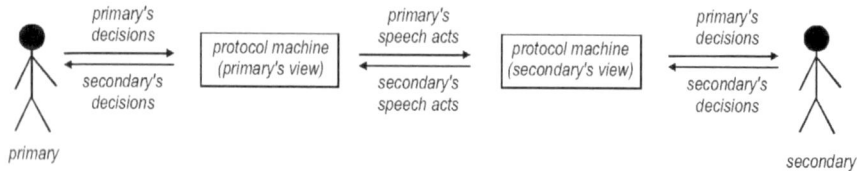

Fig. 5. Interaction of protocol machines and agents decisions

Both protocol machines consist of all states defined in the conversation protocol from the observers perspective. Every time the agent got control over the conversation, its decision result forms the input for its protocol machine. If the peer agent got the control, the speech act type of its message is used as input. In this conversation, the control over the conversation changes with every utterance from one agent to the other.

Fig. 6. Primary's protocol machine   **Fig. 7.** Secondary's protocol machine

To clarify this, the order-protocol (defined in Fig. 4) is regarded. The protocol machines for the participating agents are created by changing transition labels. Figure 6 shows the protocol machine for the primary, Fig. 7 the one for the secondary.

The primary starts the conversation by the decision result *start*. Its protocol machine changes to z_1 by outputting an *order*-typed message. This message type serves as the input of the secondary protocol machine which changes from z_0 to z_1 by outputting the decision result of the primary agent (*start*) to the secondary agent. Depending on its decision regarding the order (remember that the message got a content, in which the secondary agent was told to do something) either an *order accepted* or an *order denied* is input into the secondary protocol machine which changes to z_2 (final state) by outputting either an *accept*-typed or a *deny*-typed message. The primary protocol machine receives the message via its peer interface and moves to z_2 either by outputting *order accepted* or *order denied* depending on the speech act type of the received message.

3.3   Communication Acts

As the conversation describes the message exchange between two agents, an agent needs to be able to communicate with more than one agent. For that a "component" is needed which manages several conversations.

This component is termed *communication act*. In general a communication act fulfills a special purpose by making use of a conversation policy. Examples for a communication act are the answer to a question by a number of agents, the assignment of a task to a number of agents or to exactly one agent out of a group of agents. Please note that a communication act is not a protocol because there exists no peer layer at the peer agent (see Fig. 3).

This part of this section defines the communication act for the order-protocol defined in Fig. 4. Such a communication act can be seen as a general schema of using a protocol when communicating with several partners. Considering the order protocol, the following situation can be assumed.

Agent A needs to assign a task to other agents by using the order-protocol. As the agent A can not be sure that the first agent it gives the order to execute the task is able to fulfill it, it has to continue giving orders until one agent states that it is able to fulfill the task. On the other hand, if agent A would give the order to all agents at the same time, possibly more than one agent would fulfill it and execute the task. But only one agent should fulfill it. Hence agent A needs to give orders to the agents one by one, until an agent reports that the order can be fulfilled by executing the task in question.

The communication act modeled as a STD is shown in Fig. 8. It interacts with more than one object (a decision process and a set of conversations). Therefore the events and actions are extended by a prefix defining the affected object. The prefix *dec* stands for the decision process, *conv* for a conversation.

After the occurrence of the event *start* (*dec.start* means that the event was caused by the action *start* from the decision process *dec*) the act changes to z_1, creates the order for the first agent and moves to z_2 by performing the action *start* which affects the conversation (*conv.start*). The act waits in z_2 until an external event, caused by the conversation, occurs. In case of the event *order accepted* the communication act moves to z_3 (final state) by performing the action (outputting) *task assigned* to the decision process. In the other case (event *order denied*) the act moves to z_1 or to z_3.

When all available agents have been addressed (ordered), the act moves to z_3 by outputting *task not assigned* to the decision process, otherwise to z_1 without producing an output.

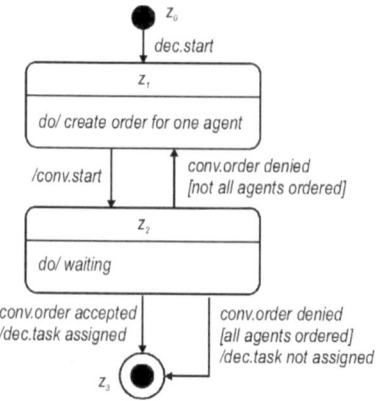

Fig. 8. Communication act (order-protocol)

The communication act doesn't matter what kind of task is assigned, it is independent from the message content and the application. It describes only how conversations all using the order-protocol have to be managed, to make sure that <u>one</u> task is assigned to <u>one</u> agent out of a group of several agents.

3.4   Decision Processes

The highest layer of the architecture is formed by the application dependent decision process. The decision process is modeled by a STD. In this section only the events, actions and essential states of the process are mentioned. These "skeletons" will be expanded in Sect. 4.2. to application specific decision processes for planning agents.

Like shown in Fig. 3 different decision processes for primary and secondary agent are needed. The decision process on the primary's side interacts with a communication act, the one on the secondary's side with a protocol machine (conversation).

Starting with the decision process on the primary's side, Fig. 9 is shown. This process interacts with the communication act, indicated by the prefix *comm* at the events and actions.

After creating the task to assign, the process moves to z_2 (*waiting*) by starting the communication act with the action *start*. Depending on the behavior of the addressed agents (whether they fulfill the order or not) the communication act causes either the event *task assigned* or *task not assigned*, which lets the decision process continue to z_3 (in case of a successful assignment) or z_4 (in case of failure) to execute activities mentioning the success of the assignment. The process finishes in z_5.

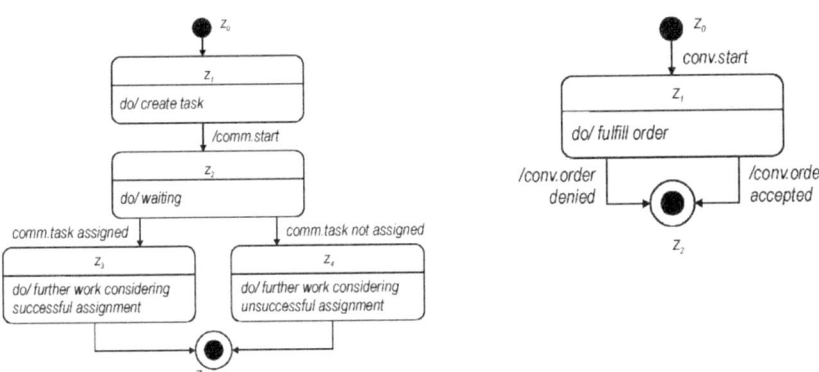

Fig. 9. Decision process on the primary side

Fig. 10. Decision process on the secondary side

On the secondary side, the decision process interacts with the protocol machine of the conversation (indicated by the prefix *conv*). This process is shown in Fig. 10. The process is started by the event *start* caused by the conversation. In z_1 the process tries to fulfill the order and continues to z_2 by causing the events *order accepted* or *order denied*, depending whether the order could be fulfilled or not.

3.5   Interaction between Layers

To explain the interaction between the layers of this architecture, an example is assumed. Considering the case that a secondary agent receives a message, a decision process is started to work on this message. Assuming that the agent needs the help of others, the decision process selects one or more appropriate communication acts which starts one or more conversations. Figure 11 shows the resulting structure.

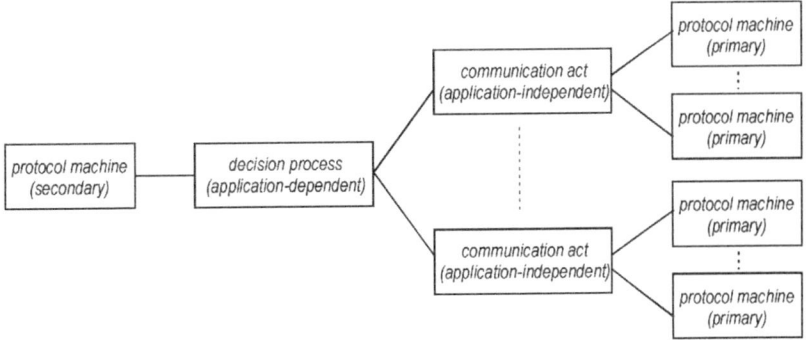

Fig. 11. Structure of protocol machines, decision processes and communication acts

3.6   Contract Net Protocol

The intention of Sects. 3.1–3.5 was to explain the structure of the presented model for multiagent communication as well as the function of the different layers of the architecture. Therefore, a simpler protocol was used to be able to present the approach in detail. By using this model, several protocols were modeled, including protocols for task assignment using the contract net approach [6,22]. Figure 12 shows the conversation protocol from the perspective of an observer.

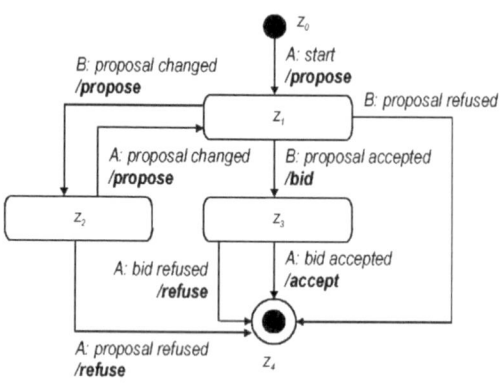

Fig. 12. Contract Net Protocol

Here the agent B is told to react on a proposal send by agent A. Receiving the proposal, agent B has in z_1 the opportunity to create a bid (*proposal accepted*), to vary the proposal by sending a counterproposal (*proposal changed*) or simply to leave the conversation by doing nothing (*proposal refused*). In case of a bid send by agent B (conversation is in z_3), agent A can accept or refuse the bid. In case of a proposal send by agent B, the conversation moves to z_2, where agent A can refuse the proposal or vary it by sending a counterproposal.

3.7   About the Framework

The model presented in this paper has been implemented in a framework using the object oriented language Smalltalk (VisualWorks 2.5.2). The framework provides communication services (communication act, protocol machines for primary and secondary) for task assignment and for information seeking. Due to its structure, the framework can easily be extended by further communication services. By this a tool for designing application independent communication services is provided.

4   Application Example

The developed framework was used to build up an agent-based distributed planning system for production planning. The structure of the planning system as well as the types of agents used in it are explained in Sect. 4.1. Section 4.2 clarifies by an example how to extract the skeletons of the primary and secondary decision process (defined in Figs. 9 and 10) to application dependent decision processes.

4.1   Short Description of the Planning System

The planning system consists of different types of agents, two of them will be described in detail. The first one is responsible for planning the manufacturing process of products (*product planner*), the second one for creating schedules of workstations (*workstation planner*). Figure 13 shows the structure of an example planning system containing two product planners and three workstation planners. The arcs connecting the agents show the communication structure. Two connected agents communicate with each other while solving a problem.

The planners can be seen as service providers able to perform services by executing processes on workstations or by producing products. In its planning process a planner solves its planning task by working on it up to the point where it needs the help (information or resources) of other planners. By using communication, the planner requests for service, which is given by the fulfillment of tasks. This help enables the planner to fulfill its own task successfully.

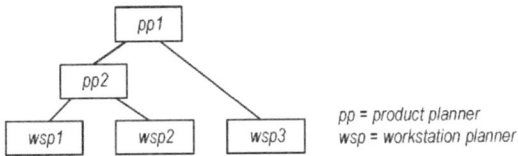

Fig. 13. Structure of an example multiagent system

The product planner conducts conversations with other product planners or with workstation planners (like *pp1* in Fig. 13). Every product planner needs the help of workstation planners to fulfill its task. In contrast to, workstation planners are able to fulfill their tasks without the help of other agents. Therefore, a workstation planner still acts as a secondary in a conversation because every time it is involved in a communication, a product planner asks it for its help. A product planner can be either a primary as well as a secondary in a conversation, because it can be asked by an other product planner for its help (secondary) or it can ask other planners for their help (primary).

In the following example in Sect. 4.2, the skeletons of the decision processes for the order-protocol (defined in Figs. 9 and 10) will be expanded to an application specific decision process.

4.2   Decision Process for Planning Agent

The decision process belongs to the product planner. This agent receives messages telling it to manufacture products in a specified quantity and time window. Its decision process is shown in Fig. 14.

The agent acts as a secondary. Solving its problem, it has to conduct conversations, which makes it also to a primary. The decision process expands the skeleton of the secondary decision process (Fig. 10). In z_1 the agent acts as a primary, so within z_1 the skeleton of a primary decision process (Fig. 9) is expanded.

In case of an event *start* from the conversation, the decision process moves to z_1 and onwards to $z_{1,1,1}$ to define the subtasks which are needed to manufacture the product in question. A subtask is either a product order for a subproduct, needed to manufacture the product or a process order for a process, needed to be processed on a workstation. For these subtasks planners have to be found, which are able to fulfill these tasks. If the product planner wasn't able to specify all needed subtasks or if it couldn't find at least one agent for each subtask, the decision process leaves to $z_{1,1,3}$. Otherwise the decision process moves to $z_{1,1,2}$ and creates one communication act (described in Fig. 8) for each subtask. After leaving $z_{1,1}$ the process starts the created communication acts by the action *start* and enters $z_{1,2}$ to wait in $z_{1,2,1}$ until all communication acts have been finished.

Depending whether all subtasks were assigned, the process moves to $z_{1,3}$ or $z_{1,4}$. In $z_{1,3}$ the process stores the assigned subtasks and the agents responsible for them. If not all subtasks could be assigned, the process withdraws made assignments in $z_{1,4}$. By leaving z_1, the process moves to z_2 by outputting *order accepted*, if the product could be manufactured indicated by assigning all needed subtasks to other agents. Otherwise *order denied* is output.

5   Conclusion

This paper presented an approach making use of a layered architecture, which ensures the reusability of the defined protocols and the communication acts by defining a clear separation between the application-independent and the application-dependent parts. The approach models conversation policies by making use of finite state machines. These policies were transformed into protocol machines, which provides communication services to agents. The modeled conversation protocols are application-independent as well as the communication acts, which are a mean for agents to fulfill their goals by making use of communication. The sequence of activities an agent has to perform to reach its goal (which is a state-based process), was called a decision process. It was modeled by a state transition diagram. The conversation, the communication act and the decision process are parts of a layered architecture. The layers are separated by well-defined interfaces. The interaction between the layers is modeled by the exchange of input/output symbols between the participating layers.

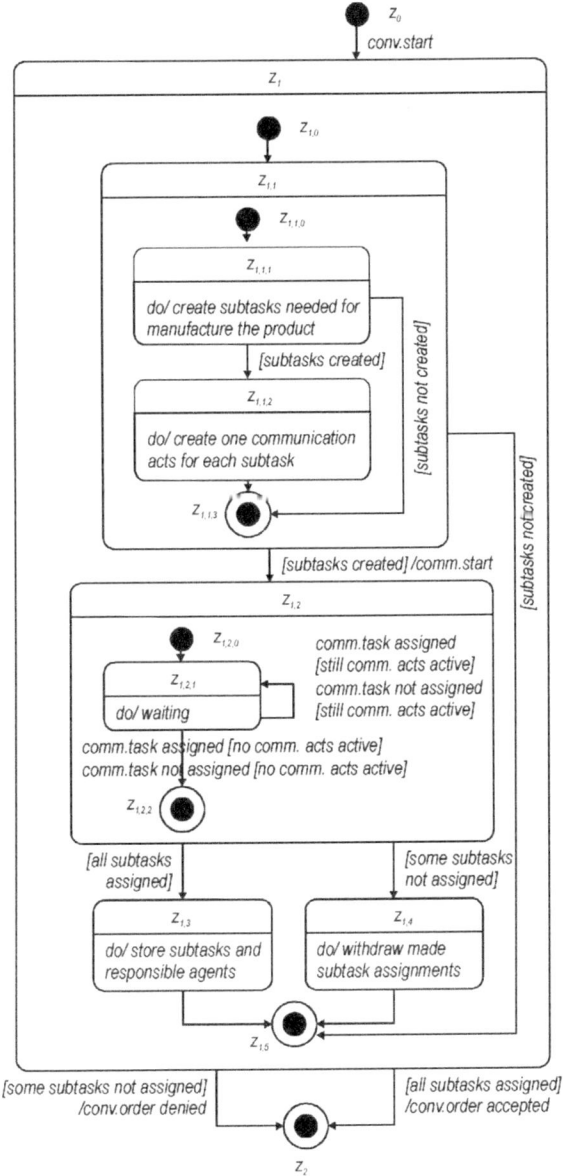

Fig. 14. Decision process of product planner

Acknowledgements. The author thanks the anonymous reviewers for their thoughtful comments and suggestions to improve this paper.

References

1. Barbuceanu, M.; Fox, M. S.; COOL: A Language for Describing Coordination in Multi Agent Systems; in: First International Conference on Multi-Agent Systems – ICMAS '95; pp. 17–24; The MIT Press, California, 1995
2. Barbuceanu, M.; Lo, W.-K.; Conversation Oriented Programming for Agent Interaction; in: [7], pp. 220–234
3. Belakhdar, O.; Ayel, J.; Modelling Approach and Tool for Designing Protocols for Automated Cooperation in Multi-agent Systems; in: Agents Breaking Away (MAAMAW '96), LNAI 1038; Van de Velde, W.; Perram, J. W.; pp. 100–115; Springer, Berlin, 1999
4. Bradshaw, J. M.; Dutfield, S.; Benoit, P.; Woolley, J. D.; KAoS: Toward An Industrial-Strength Open Agent Architecture; in: Software Agents; pp. 375 Bradshaw, J. M.; The MIT Press, California, 1997
5. Cost, R. S.; Chen, Y.; Finin, T.; Labrou, Y.; Peng, Y.; Using Colored Petri Nets for Conversation Modeling; in: [7], pp. 178–192
6. Davis, R.; Smith, R. G.; Negotiation as a Metaphor for Distributed Problem Solving; in: Artificial Intelligence, Vol. 20, pp. 63–109; Elsevier Science Publishers, North, 1983
7. Dignum, F.; Greaves, M.; Issues in Agent Communication, LNAI 1916; Springer, Berlin, 2000
8. Fallah-Seghrouchni, A. E.; Haddad, S.; Mazouzi, H.; Protocol Engineering for Multi-agent Interaction; in: Multi-Agent System Engineering (MAAMAW '99), LNAI 1647; Garijo, F. J.; Boman, M.; pp. 89–101; Springer, Berlin, 1999
9. Greaves, M.; Holmback, H.; Bradshaw, J.; What is a Conversation Policy?; in: [7], pp. 118–131
10. Harel, D.; Statecharts: A visual formalism for complex systems; in: Science of Computer Programming, Vol. 8, pp. 231–274; Elsevier Science Publishers, North, 1987
11. Holzmann, G. J.; Design and validation of computer protocols; Prentice Hall, New Jersey, 1991
12. Jennings, R. J.; Woolridge, M. J.; Agent Technology, Foundations, Applications and Markets; Springer, Berlin, 1998
13. König, R.; Modeling Conversations and Decision Processes by using state-based models; Technical Report ET8-2002-01; Department of Process Automation Techniques; Technical University of Hamburg-Harburg, Hamburg, 2002
14. Lin, F.; Norrie, D. H.; Shen, W.; Kremer, R.; A Schema-Based Approach to Specifying Conversation Policies; in: [7], pp. 193–204
15. von Martial, F.; Coordinating Plans of Autonomous Agents, LNAI 610; Springer, Berlin, 1992
16. Martin, F. J.; Plaza, E.; Rodríguez-Aguilar, J. A.; Conversation Protocols: Modeling and Implementing Conversations in Agent-Based Systems; in: [7], pp. 249–263
17. Moore, S. A.; On Conversation Policies and the Need for Exceptions; in: [7], pp. 114–159
18. Nodine, M. H.; Unruh, A.; Constructing Robust Conversation Policies in Dynamic Agent Communities; in: [7], pp. 205–219
19. Pitt, J.; Mamdani, A.; Communication Protocols in Multi-agent Systems: A Development Method and Reference Architecture; in: [7], pp. 160–177
20. Searle, J. R.; Speech Acts; Cambridge University Press; Cambridge, 1969
21. Shen, W.; Norrie, D. H.; Barthès, J.-P., A.; Multi-agent systems for concurrent intelligent design and manufacturing; Taylor & Francis, London, 2001

22. Smith, R. G.; The Contract Net Protocol: High-Level Communication and Control in a Distributed Problem Solver; in: IEEE Transactions on computers, Vol. C-29, No. 12, pp. 1104–1113, 1980
23. Tanenbaum, A. S.; Computernetzwerke; Pearson Studium, München, 2000
24. Wagner, Th.; Benyo, B.; Lesser, V.; Xuan, P.; Investigating Interactions between Agent Conversations and Agent Control Components; in: [7], pp. 314–330
25. Weiss, G.; Multiagent Systems, A Modern Approach to Distributed Artificial Intelligence; The MIT Press, Cambridge, 1999
26. Walker, A.; Wooldridge, M.; Understanding the Emergence of Conventions in Multi-Agent System; in: First International Conference on Multi-Agent Systems – ICMAS '95; pp. 384–389; The MIT Press, California, 1995
27. Winograd, T.; Flores, F.; Understanding Computers and Cognition: A New Foundation for Design; Ablex Publishing Corp., New Jersey, 1986

A Framework for Inter-society Communication in Agents

R. Abbasi, F. Mitchell, and S. Greenwood

School of Computing and Mathematical Sciences
Oxford Brookes University
Oxford, OX3 0BP
UK
{rabbasi,frmitchell,sgreenwood}@brookes.ac.uk

Abstract. Normally a society of agents is developed to cope with and solve a particular set of problems or to achieve some particular goals and objectives. Different people, organizations and groups have developed and deployed agent societies for their own particular needs. These societies have to communicate, coordinate and compete with each other. In this paper an explanation is provided as to what is meant by inter society communication in agents, why it is needed and some basic problems in the area of inter-society communication are identified. A framework for inter-society communication in agent systems is also proposed.

1 Introduction

A typical agent society is composed of more than one agent, who work together to achieve a shared purpose or goal, through a division of work, and are integrated by decision processes. In order to achieve its goal, a society has to provide some mechanism for its component agents to communicate with each other.

Communication between agents can be direct or indirect. Direct communication is based on theory of communication, which emerged from the telecommunications research of Shannon and Weaver in the 1940's. In this model [20,27] an act of communication consists of the sending of some information from one party (sender) to another party (receiver), while the information being passed is normally encoded in some form, sent on a medium and decoded on reception. This is the most commonly used method of communication in agent societies on which many models are based. [10,12,16,17,21]. A Model of communication in agent societies is normally known as Agent Communication Language (ACL)[23]. Most of these ACL's are semantically based on speech act theory [18,24]. KQML [12] and FIPA ACL [10] are the most commonly used ACL.

R. Kowalczyk et al. (Eds.): Agent Technology Workshops 2002, LNAI 2592, pp. 167-178, 2003.

2   Inter-society Communication

It is possible to have different agent societies, each using different ACL's for their internal communication purposes. Normally agents communicate only or mainly with other agents within their society. However, there is gradual recognition of the need for these societies to interact with each other. Agent societies can learn from each other, can use services provided by other societies and can offer services to other agent societies. As human societies collaborate, coordinate and also compete with each other, similarly agent societies will also need to collaborate, coordinate and compete, either for their own purpose or for achieving the goal set for them by users, in order to achieve this, agent societies have to communicate with each other.

From software engineering viewpoint just in the same way that there is a need for different distributed object systems, written in different programming languages to communicate, it is becoming increasingly apparent that there is a need for different agent societies, with different communication methods, to be able to communicate. CORBA [5] and SOAP [22] are possible solutions for distributed objects, but due to the differences between objects and agents, they are currently unsuited for the challenges posed by Inter Society Communication (ISC). What is needed is a CORBA like solution for agent societies.

3   The Framework

There are many issues and problem associated with providing a mechanism for Inter-Society Communication in Agents (ISCA). In [1] five major issues and problems have been identified and a framework has been proposed and partially implemented and tested. This framework consists of five layers as described in Fig. 1. Each layer addresses one issue or problem. A detailed description of each layer follows.

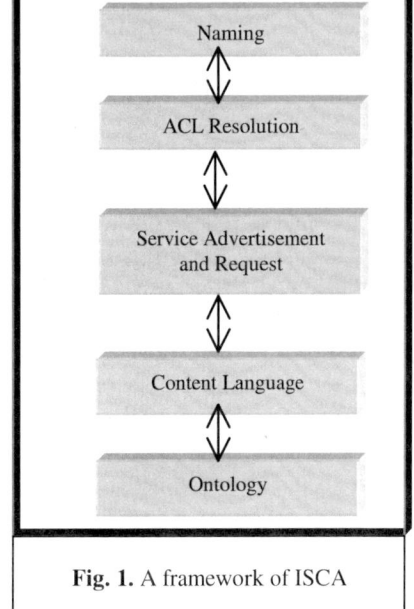

Fig. 1. A framework of ISCA

3.1   Naming

Every agent society can be considered as an organizational unit. These societies cannot communicate with each other on a society level unless and until they recognize each other's existence. Another, related problem is discovering the existence of other societies which is not possible without a unique identification for each society.

It is worth mentioning that at an intra-society level there are few communication methods which do not require unique naming scheme for agents within society. These methods include communication via environment, blackboard or broadcasting. We do not consider these as suitable for inter-society communication for the following reasons. Communication through the environment is infeasible simply because agent societies do not have to share a common environment. A blackboard is potentially more useful, however blackboards can become a bottleneck and a central point of failure. One solution to this is to provide a mechanism for providing replication of the blackboard, however, reliable replication is a known hard problem in distributed systems (and most solutions require a naming scheme for the replicants, !) therefore we believe blackboards are not suitable as a mechanism for inter-society communication. Similarly broadcasting may serve at an intra-society level where agents are on same network or subnet and also the number of agents within a society is normally not in thousands or millions, but this is not suitable for inter-society communication[1]. Finally although we have not considered system security in this framework, it is an important concept, and security is severely weakened (if not removed), if there is no form of identification available. Therefore at an inter-society level we have to rely on a direct communication method, which will require the unique identification of each individual society.

There are different schemes and standards available for naming of agents within a society. The research in this area does not highlight any suitable naming standard for agent societies, which can guarantee a unique name for each society. The required characteristics for a naming scheme for agent societies are listed below.

- It should generate a unique name for each society.
- It should be to able handle an arbitrary number of names.
- It should be able to deal with an arbitrary administrative organization for names.
- It should be dynamic i.e. able to handle changes made to names, attributes associated with the names and the distribution of the naming database.
- It should be efficient, providing speedy retrieval of names and their attributes.
- It should provide transparent distribution.
- It should be reliable i.e. providing fault-tolerance.
- It should be secure.
- It should be distributed
- It should be hierarchal

Using the above requirements a naming system for agent societies has been designed and implemented, which forms the first layer of the framework. In this system any society, which wants to be a part of the global environment, has to acquire a unique name. There are three main components of this naming system.
1. **Name Resolution Language**: A language used for name resolution
2. **Societies Naming Agent (SNA)**: An agent responsible for the generating a unique name for the societies

---

[1] What would had happened if broadcasting was being used for email? It might work in the beginning when emails addresses are few but later with billions of email addresses and millions of email servers it would not be workable.

3. **Local Naming Agent (LNA):** an agent responsible for acquiring a name for
    the society

The process of acquiring a unique name for the society requires a dialogue or communication between the SNA and the LNA. Different LNA's belonging to different societies will be speaking different ACL's. It is currently infeasible to design an SNA that could understand all the ACL's. However the dialogue between the SNA, and the LNA will always be very specific hence there are only a few different ACL messages used during this dialogue. Keeping these two things in mind a Name Resolution Language (NRL) has been proposed. This NRL is designed as a native language for the SNA. There is no set of performatives in the NRL. A performative is a speech act, and if it was used we would have to use some content language as well to code the message. As the case under consideration is quite simple we can omit both the performative and the content language. In NRL a field called 'status' is used, that is not a speech act. Status just depicts what is the status of the dialogue between the SNL and the LNA. Other fields of the message are 'sender', 'receiver' and 'name'. The name, which is being negotiated, is coded in the field 'name'.

The main function of the SNA is to keep the information about the names of all the existing societies in its domain[2] and make sure that all the new societies in that do main are allocated a unique name. In order to avoid a bottleneck and also for the purposes of reliability, efficiency and robustness it is proposed to have more than one SNA for a specific domain. These SNA's work together as a team to ensure that all the societies in one domain have a unique name. Whenever a SNA receives a 'check' message for a name, it checks in its local knowledgebase to see whether the name has already been allocated or not. If the name has been already allocated, it proposes another name to the requesting party, which might be of interest to the requesting party, but is not being used at that time. Currently a simple algorithm for proposing a new name is being used, but this could be modified at a later stage if required. If the name is not being used according to local information of the SNA then it tells the requesting party that the name is available by sending them a message with status 'available'. If the SNA receives a request 'request-allocate', it means that the requesting party wants a name to be allocated to them. Now the SNA has to make sure that this name is not being proposed to some other party by another SNA. In order to do that the SNA talks with it's fellow SNA's and checks the availability of name by sending a 'can-i-offer-it' message. Then it waits for the responses from all other SNA's. If the responses from all the other SNA are positive (can-offer) then the name is allocated to requesting party and all other SNA's are informed that the name has been allocated. If the response from even one SNA is negative (do-not-offer) then a new name is proposed to the requesting party. If, after a certain time period, no message is received from one or more fellow SNAs, a time out will occur and the message will be sent out again.

The Role of the LNA in this system is to act on the behalf of the agent society, and to obtain a unique name for the society. This is a simple process and may only need to be carried out once. As work progresses it might be desirable to allocate other func-

---

[2] We assume that the societies of the agent are well organized into domains. Each society belongs to at least one domain and there could be many domains of the societies. One domain can have more than one sub-domain.

tionality to the LNA. In the test implementation this job is assigned to the Name Server/Facilitator agent.

The Naming system described above has been implemented for a single domain of societies and tested successfully. For the purpose of inter-society communication all the societies at the global level have to be uniquely named so that communication between agent societies at the global level is possible. A hierarchical naming system is proposed which can be based on geography and functionality.

3.2   ACL Resolution

Within a society of agents the same ACL is normally used for communication. Most ACL's are based on speech act theory. In a typical ACL, agents communicate with each other using messages. Each message starts with a speech act (sometimes called a performative in some ACL's) followed by some other predefined fields. These messages normally travel from one agent to another in the format of a string. When an agent receives a message from an agent belonging to another society without prior knowledge of the ACL used by the other society, the outcome is unpredictable and can cause the following results:

1. Agent is not able to parse the message because it does not contain some mandatory fields.
2. Agent is able to parse the message but the speech act or performative is not understood.
3. Agent is able to parse the message and is even able to identify the speech act but there is a chance that the meaning of speech act is different to both ACL's so in this case the agent would misunderstand the message.
4. Agent does not understand the ontology used in the message
5. Agent does not understand the language used for the content field of the message.

In all these cases the agent is not able to understand the message properly resulting in failure of possible communication between the agents. Also the agent cannot tell other agents that it is not able to understand the message, as any message send may cause similar kinds of problems for the agent belonging to the other society.

It is known that translation between different ACL's is a non-trivial problem. But without tackling this problem, the dream of inter-society communication cannot be achieved. Agent societies normally do (or will need to) communicate with each other mainly for information sharing, service provision and service request

This means that all the communication between agent societies is generally limited to the types of communication outlined above, hence it is possible to propose some basic performative set which can be used to design messages. This limited scope allows the proposal of a reduced set of fields in message. This reduction in fields in a message and the limited number of primitive performatives will simplify the problem and enable the design of a mechanism for translation between different ACL's. Three performatives have been proposed as primitive performatives, which are Inform, Query and Request.

The next step is to propose a generic set of fields. In any ACL a message is sent from one agent to another so sender and receiver fields are needed. The content field is needed for the content of the message, language is needed to tell the recipients what language is used for the content and ontology is needed to tell what ontology the sender is using. These fields have to be supported by every ACL to be able to use the framework. It is possible to envisage a situation where fields like reply-with and in-reply-to may be advantageous, especially when multiple conversations are going on at the same time between two societies. Some people prefer to use conversation-id instead of reply-with and in-reply-to therefore this field is also included in the list.

Here generic names for fields have been used and it is possible that different ACL's will use other names for those fields. As long as any message is being coded using only the above-mentioned fields or their equivalent fields the translation mechanism will be able to translate the message in other ACL's.

All the messages for inter society communication will use any of the above listed performatives and should be coded using the above-mentioned fields. In different ACL's there could be different equivalents for these performatives and fields. It is not possible, or even desirable to force every one to use exactly the same performatives and fields because this would be the equivalent to introducing a new ACL. The idea is to let every society use its own ACL, with its own fields and performatives for communication within that society, and also let it communicate with external societies using the same language. It is the responsibility of a translator to ensure that each society receives a message in a format, which it understands. When a society wants to become part of a global environment in order to provide services to other societies or wants to make use of services provided by other societies, it will have to inform the translator what ACL it is using and what are the equivalent of the primitive performatives and generic fields. For the purpose of encoding this an ACL Description Format (ADF) is proposed. The fields of the ADF and their description are given below.

- **SocietyName:** States the name of the society using the ACL being described.
- **BaseLanguage:** States the base language for the ACL being described.
- **NoOfFields:** States how many fields are used in the message of the ACL.
- **NoOfPerformatives:** States how many performatives are supported by the ACL.
- **FieldsList:** Includes all the fields being used in ACL with the equivalence information. If the ACL is using any field for which there is no equivalent generic field name then slot for generic name is filled with the reserved word "NULL". The syntax is (field name1: primitive field name1, field name2: primitive field name2,....)
- **PerformativeList:** Includes all the perfomatives supported by the ACL being described, along with the equivalence information. The reserved word "NULL" is used instead of the primitive performative where there is no primitive performative available. The syntax of this field is similar to the one of FieldList.

The job of the translator is to translate a message from an ACL to a generic format and vice-versa. Each society can have its own translator. The translator always receives a message in the generic format from the outside world and then translates it into its own native language. Whenever anyone from within the society has to communicate with an agent in another society, it has to do it via the translator. The translator will translate the message into the generic format before sending it out.

In this scenario the translator is not only responsible for translating the messages but also acting as a Foreign Minster[3].

Another approach can be to divide the above-mentioned role into two parts, where one agent is the translator while another is responsible for communicating with the external world, the Foreign Minster (FM).

Yet another option is to move the translator outside the society and making them available globally. A FM is still needed who will be responsible for the communication with the outside world. All the communication with the outside world is achieved by this FM.

In all cases the address of the FM has to be made available to other societies, so that any agent from another society can discover how that society can be accessed. When an agent from within a society wants to communicate with the outside world it sends its message to the FM. The FM finds some suitable translator who understands its ACL. Then the FM sends this message to that the translator. If the translator understands the ACL spoken by the receiver society it translates the message for the agent and sends it to the respective FM of the receiver society, which then receives the message and passes it to the appropriate agent. If the translator does not understand the ACL used by the receiver's society it can make use of the services of another translator, who does understand the ACL used by the receiver's society before forwarding it to respective FM.

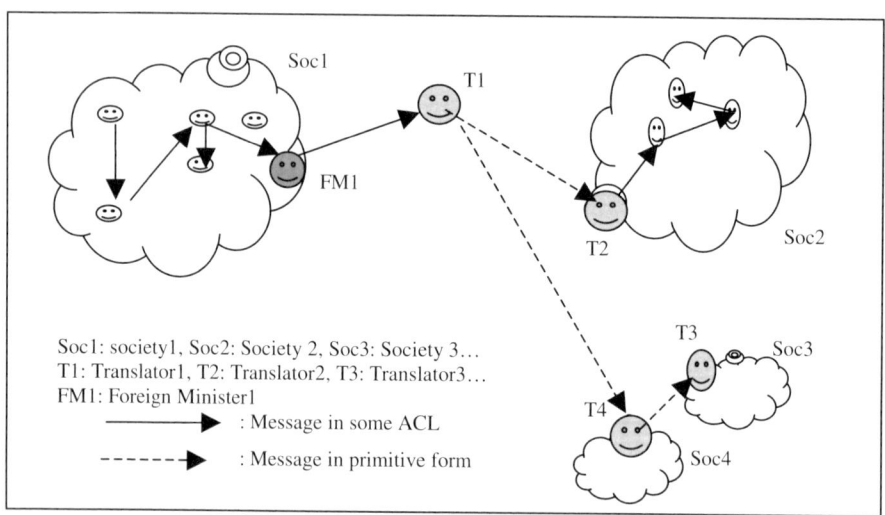

Soc1: society1, Soc2: Society 2, Soc3: Society 3...
T1: Translator1, T2: Translator2, T3: Translator3...
FM1: Foreign Minister1
⟶ : Message in some ACL
----⟶ : Message in primitive form

Fig. 2. Translators and foreign ministers

The above mentioned translation mechanism has been implemented and tested successfully for translation between FIPA-ACL and KQML and the societies using these languages were able to communicate with each other with the facilitation of the trans-

---

[3] In our Framework a Foreign Minister is an agent who is a link with other societies. It is able to discover information about other societies such as their location, services provided by them and the ACL, content language and the Ontology being used.

lator. In future this mechanism will be tested on other ACL's, which are substantially different from KQML and FIPA ACL in order to demonstrate its generality for ACL's based on speech act theory.

3.3   Service Advertisement and Request Mechanism

When a society needs a service, the first hurdle is to find a suitable society that can provide the required service. There is a requirement that there should be some available mechanism, which can be used by the societies to advertise their services. The societies should also be able to find suitable services for them using the same mechanism.

The main goal is to enable requester societies to obtain a response according to their needs and request. For this purpose we have proposed a service broker mechanism based on classification of problems and information.

Classification System

A classification system, which will enable societies to search for required services in some respective class is needed. With the provision of such a classification system, it will become easier for societies to advertise and request the services. Also the process of request resolution will be able to use class information and will become simpler. For the purpose of ISCA a classification system is needed that:

- can deal with classification of both information and services.
- is dynamic i.e. services could be added, deleted and modified dynamically.
- can deal with the fact that one particular service can belong to more than one class at the same time.
- can give a unique identification to each service with respect to classes it belongs to.
- can deal with the fact that one service can be provided by more than one society.
- provides some provision to relate services with the provider societies.
- can be used by agents without involvement of human users.

The design of a complete classification system is not the subject of this work but in order to demonstrate the principles, a prototype for classification system is proposed and used in the ISCA framework for a limited set of societies. However any classification system which had the desired properties listed above could be used.

The proposed classification system classifies the problems and information into a variable-level tree structure and each level in this tree is identified by a series of numerical values. Each main class represent a specific domain for example one for science, the second for industry, the third for business and so on. Each main class is divided into more than one sub-classes and each sub-class is divided into further sub-sub-classes. The depth of the classification tree can be extended to any reasonable level needed.

Every service or type of information provided by a society can be associated with a description based on its position in the tree. Because services can have multiple descriptions, the system allows a service to be listed in more than one part of the tree.

This information has to be supplied by the service provider society to the broker. For the purpose of service advertisement and request a language has been proposed called "Services Advertisement and Query Language for Societies" (SAQLS).

Services Advertisement and Query Language for Societies (SAQLS)
The provider of a service not only has to provide class information to the broker but also has to specify the input and output signatures of its service. Other information that may be provided includes the ACL, content language and ontology used. Based on this information the broker generates a token for each service. Some work, like LARKS [25], has been carried out in this regard but it is not primarily designed for inter-society communication, and also it does not address the concept of classification. Therefore a language has been proposed called Services Advertisement and Query Language for Societies (SAQLS). Societies can also use this language to find a service and can specify the required upper limit on the expected number of results. The brief description of fields in a SAQLS message is provided in Table 1.

Table 1. Description of different fields used in a typical SAQLS message

Name of Field	Descriptions	Message Directions
Performative	It is used to describe type of message it can have one of these value:-Advertise, Query, Advertised-ok, Query-response, Unadvertise, Unadvertised-ok	Society→Broker Broker → Society
Sender	Tells who is the sender of message. It will have either name of broker or name of FM/Translator of society	Society→Broker Broker → Society
Receiver	Tells who is the receiver of message. It will have either name of broker or name of FM/Translator of society	Society→Broker Broker → Society
Classlist	Tell to which classes a service belongs to	Society→Broker Broker → Society
Input	Used for input specification of services	Society→Broker Broker → Society
Output	Used for output specification of services	Society→Broker Broker → Society
ACL	Tells ACL being used by the society	Society→Broker Broker → Society
CL	Tells CL used being by the society	Society→Broker Broker → Society
Ontology	Tells ontology being used by the society	Society→Broker Broker → Society
Descriptor	Is used for encoding descriptor of services	Society→Broker Broker → Society
Provider	Is used to encode name of provider society	Broker → Society
Threshold	Tells what threshold is required	Society→Broker
Message-No	Tells the serial number of message if multiple messages are being send and is of the form 2/5.	Broker → Society
Weight	Describes the weight of proposed service as measured by the broker	Broker → Society

Service Broker

The service broker is an agent, which stores the information provided by the service provider, and also helps the requester societies to find a service suitable for their requirements. The service broker maintains a knowledgebase about the services provided by the agent societies and it stores all the information provided about the services by the societies. In order to accomplish this it has to transform the information provided by the societies via a SAQLS message into some internal form. In the test implementation tree and hash-table data structures have been used to maintain the information so that searching through the knowledgebase is made easier.

Matchmaking

The resolution of requests is an important problem. In the proposed solution the class information is the main parameter. A simple algorithm has been designed for distance calculation between classes. The information about input and output specification, ACL, content language and ontology is also used in the request resolution mechanism. The main goal is to generate a minimum number of results, which are as near to the request as possible. The broker agent also returns sufficient information to the requester along with the results so that the requester can utilizes its own matchmaking criteria if desired.

The above service advertisement and request mechanism has been implemented for two simple agent societies, each performing simple mathematical functions. The next step is to test it for a larger number of societies offering different types of services.

3.4   Content Language Resolution

One of the most important fields of any ACL message is the content field. This field, in combination with the speech act, determines the meaning of the message. Within a society agents normally use the same language for the content field and even if, as in some cases, multiple languages are used, they are understood by the agents in the society. However, in the area of inter-society communication there is a very strong possibility that content language used by one society is not understood by the other society. Another service, which can resolve this problem associated with use of different CL is required.

In the framework being proposed, the fourth layer will consist of some solutions for the problems associated with the use of multiple content languages. For this layer, a mechanism will be designed to facilitate the use of the formal description of content languages for translation of the messages from one content language into anther without any loss or alteration in the semantics of the messages.

3.5   Ontology Resolution

Agents use different symbols and words, which are understood by all the agents within a society as all the agents within a society share the same (implicit or explicit) ontology. However, when an agent tries to send a message to another agent belonging to another society that is using a totally different ontology then even use of the same ACL and content language does not guarantee that message will be properly understood by the receiving agent. There is a very good chance that the receiving agent will not be able to understand the message at all. In order to address this problem we need another service, which allows the different societies to share the ontologies being used.

When requesting a service it needs to be guaranteed that the ontology used by both the service provider and consumer is the same. We do not want to end up in a situation where a society looking for type of computer mouse ends up talking with a society that is dealing with different classes of species of mouse.

One idea is to classify ontologies on the basis of functional domain. A good starting point is to look at the work carried out by the semantic web initiative (W3C) [28,30] in this field. If every ontology is described formally, it is often theoretically possible to transform a message between different ontologies. The fifth layer of the framework will consist of a solution to the problems associated with the use of multiple onotologies. The aim for this layer is to use available formal descriptions of ontologies in order to resolve differences between them.

4   Discussions and Further Work

A framework for inter-society communication in agents has been proposed. This Framework is still under development, however the design and implementation of the first three layers has been completed. The initial testing of these layers has been carried out and the usability of the framework has been demonstrated for some simple domains. The next step will be the design and implementation of the remaining two layers. Once all five layers are completed the whole system will be tested together as an entity. The Framework is being designed and implemented in such a way that each of its layers will form a complete system and will work independent of the other layers. This will provide freedom for agent societies to make use of the layers as and when they wish.

References

1. Abbasi, R.; Mitchell, F.; Greenwood, S.; Interoperability in societies of agent, Proceedings of the International Workshop on Multiagent Interoperability, Aachen Germany, September (2002).
2. Agentcities: http://www.agentcities.org/ (May 2002)
3. Bertolini, P.; Busetta, A.; Molani, M. Nori,; A. Perini.: Designing peer-to-peer applications: an Agent-Oriented Approach, in this volume.
4. Breemen, A.: Integrating Agents in Software Applications, in this volume.
5. Common Object Request Broker Architecture (CORBA), Homepage: http://www.corba.org/ (2002)
6. Communication Working Group of Agentcities: http://www.agentcities.org/Activities/WG/Comms/ (May 2002)

7. Dale, J and Mamdani, E.: Open Standards for Interoperating Agent-Based Systems. In: Software Focus, 1(2), Wiley (2001)

8. DARPA Agent Markup Language (DAML) homepage: http://www.daml.org/ (2002)

9. FIPA Messaging Interoperability Service Specification: http://www.fipa.org/specs/fipa00093/ (May 2002)

10. FIPA, Agent Communication Language: http://www.fipa.org/repository/aclspecs.html (2001)

11. Jennings, N.; Wooldridge, M. (eds.): Agent Technology, Springer (1997)

12. Labrou, Y.; Finin, T.: Agent Communication Languages: The Current Landscape, Intelligent Systems, Vol. 14, No. 2, IEEE Computer Society (1999)

13. Labrou, Y.; Finin, T.: Semantics for an agent communication language. The fourth International Workshop on Agent Theories, Architectures, and Languages. Rhode Island, USA (1997)

14. Muller, P.: The Design of Intelligent Agents - A Layered Approach, volume 1177 of LNAI (1996)

15. Ontology Working Group of Agentcities http://www.agentcities.org/Activities/WG/Ontology/ (May 2002)

16. Open Agent Architecture: http://www.ai.sri.com/~oaa/whitepaper.html (2002)

17. Sadek, D.; Bretier, P.; Panaget, F.: ARTIMIS technology, http://www.cselt.it/fipa/torino/cfp1/propos97_015.htm (1997)

18. Searle, J.: Speech Acts. An Essay in the Philosophy of Language. Cambridge (1969)

19. Service Description and Compositions Working Group of Agentcities: http://www.agentcities.org/Activities/WG/SDC/ (May 2002)

20. Shannon, C.: A mathematical theory of communication, Bell System Technical Journal, vol. 27, (1948) 379-423 and 623-656.

21. Shoham, Y.: Agent Oriented Programming: an overview of the framework and a summary of recent research, Knowledge Representation and Reasoning Under Uncertainty (1994) 123-129.

22. Simple Object Access Protocol (SOAP) homepage: http://www.w3.org/TR/SOAP/ (2002)

23. Singh, M.: Agent Communication Languages: Rethinking the Principles. IEEE Computer, volume 31, number 12, (1998) 40- 47.

24. Singh, M.: Multiagent systems, A Theoretical Framework for Intentions, Know-How, and Communications, Lecture Notes in Computer Science, Volume 799, Springer-Verlag (1994)

25. Sycara, K. Widoff, S. Klusch, M and Lu, J.: LARKS: Dynamic Matchmaking Among Heterogeneous Software Agents in Cyberspace. Autonomous Agents and Multi-Agent Systems, 5, (2002) 173–203.

26. Vizine-Goetz, D.: Using Library Classification Schemes for Internet Resources. Proceedings of the OCLC Internet Cataloging Colloquium, San Antonio, Texas (1996).

27. Weaver, W.: Shannon, C: The Mathematical Theory of Communication, Urbana, Illinois: University of Illinois Press (1963).

28. Web Ontology Working Group of W3C: http://www.w3.org/2001/sw/WebOnt/ (May 2002)

29. Wooldridge M.: Verifiable Semantics for Agent Communication Languages, ICMAS'1998 Paris (1998).

30. World Wide Web Consortium (W3C): http://www.w3c.org/ (May 2002)

Action Recognition and Prediction for Driver Assistance Systems Using Dynamic Belief Networks

Ismail Dagli[1], Michael Brost[2], and Gabi Breuel[2]

[1] STZ-Softwaretechnik, Im Gaugenmaier 20,
73730 Esslingen, Germany
ismail.dagli@stz-softwaretechnik.de
[2] DaimlerChrysler, Research and Technology, Postfach HPC T728,
70546 Stuttgart, Germany
{michael.m.brost,gabi.breuel}@daimlerchrysler.com

Abstract. The design of advanced driver assistance systems always aims at enabling the driver to master today's traffic in a more safe and comfortable way. In order to judge the risks in a situation and initiate precautionary actions, future systems have to possess the capability to predict the behavior of surrounding traffic participants. This paper outlines an approach to predictive situation analysis for driver assistance systems and discusses one key issue in more detail - namely the predictive action recognition. In this context, a situation representation formalism will be introduced that exploits time as a compact physical measure. Furthermore, it will be shown how probabilistic networks can be used for reasoning about driver (action) intentions and how such networks can help to cope with uncertainty resulting from inaccuracy in models and sensor data. First results are shown in simulation for highway overtake scenarios. In the situations presented the prediction for an upcoming lane change can be made by the assessment of the time gaps to the nearest neighbors of that specific vehicle.

1 Introduction

The challenging vision of research in the field of driver assistance is to develop a scene understanding methodology for advanced assistance systems that can cope with complex traffic scenarios. The aim is that the next generation of driver assistance systems is able to optimize their information, warning and control strategy by considering driver preferences, driver intentions and the overall traffic situation. As an example, early recognition of an intervening vehicle would help to adapt the control strategy of an ACC (Adaptive Cruise Control) system early enough to avoid that the safety distance is violated by a lane changing vehicle. Obviously, a strict control strategy that only considers the distance to the front vehicle is not capable to handle such a situation appropriately. Therefore, it becomes important that assistant systems are provided with concepts and techniques that enable prediction of future situations.

R. Kowalczyk et al. (Eds.): Agent Technology Workshops 2002, LNAI 2592, pp. 179-194, 2003.
© Springer-Verlag Berlin Heidelberg 2003

This article presents a general scheme for reasoning about driver motivations, goals, plans and actions that facilitates prediction at the planning and action level. According to this scheme, one aspect - the action (and action intention) at action level is discussed in more detail. The approach taken at this level is to take typical driver behavior from driver models and field experiments and model this knowledge within a probabilistic network, that can deal with uncertainty resulting from inaccuracy in models and sensor information.

1.1 Context

In a complex traffic situation with many vehicles, taking all possible actions into account would lead to combinatory explosion and is intractable. Therefore, prediction must be concentrated on relevant and probable driving behavior. Plan recognition is a reasonable concept to define plausible behavior that can help to focus the prediction of future situations. Looking at the variety of actions that a vehicle can perform in a dynamically evolving traffic situation, it is very difficult to predefine sequences or patterns of actions which match the behavior of drivers. Therefore, a plan recognition approach with pre-defined behavior patterns is less effective for driver assistance systems. Instead of modeling fixed plans for plan recognition, it proves better to predefine abstract motivations and to deduce goals and plans dynamically for all relevant vehicles in the scene.

Therefore, the general scheme introduced in the following suggests that plan recognition is based on dynamic, search-based planning, requiring only a set of abstract motivations which are defined beforehand by the designer.

Motivation-based behavior recognition and prediction, as in [3], is based on the idea that driving behavior is strongly affected by the motivations of the driver and that these motivations can be - at least for highway traffic - formulated with a few sentences. Therefore, the concept assumes that the drivers of all surrounding vehicles have a set of motivations. Furthermore, it assumes that the driver has situation dependent goals according to the given motivations. After setting the goals, the driver develops plans to achieve those goals and chooses the one with the highest probability of success, the highest convenience, or the minimum risk. The driver executes the plan, monitors the progress of the situation continuously and re-plans if the outcome becomes less attractive than in the initial situation.

The concept introduced here assumes a predefined set of possible and relevant motivations (relevant for the assistance system) for each driver in every situation. The system assigns possible goals to each of the vehicles in the scene and creates possible plans according to the goals and motivations. The plan recognition module takes these plans and assigns probability hypotheses according to the current observations. The early discovery of conflicts in plans of two or more vehicles require reliable prediction of future situations. Figure 1 gives an overview of this approach.

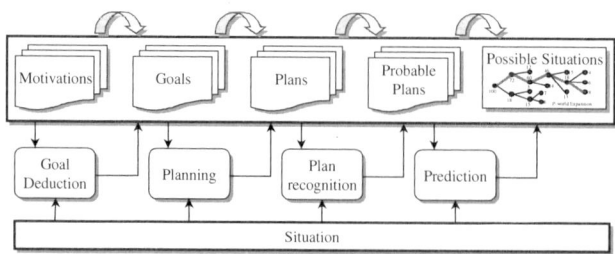

Fig. 1. Approach to behavior recognition and prediction

Plan recognition is one important aspect but early discovery of conflicts in plans of two or more vehicles that can lead to critical situations require also reliable prediction of situations in the future. Prediction as it is suggested here is performed in a possible situation/scenario structure, where possible futures are expanded within a tree and each node that represents a possible future is associated with a probability that is gained by the combination of the plan probabilities given by the plan recognition modules. This process is repeated for every vehicle in the scene until a probability has been assigned to each plan. The fusion of the most probable plans of all cars is then used to generate the possible situations - each of those situations is connected to a specific probability value.

The expanded possible situations are the basis of the situation analysis. Basically, situation analysis looks for relevant situations in the future that may affect the overall information, warning and control strategy of the assistance function. If the probability of a critical situation observed in the possible situations structure exceeds a certain threshold, the assistance system responsible for handling this situation becomes active and initiates a precautionary action in form of a warning or vehicle control.

1.2 System Architecture

Figure 2 shows the resulting software architecture: Situations and Motivations are stored in a global database. The whole world representation consists of a history of situations, motivations, goals, plans and possible worlds. The main component of the system is a heuristic planner that can cope with the variety of situations by exploiting heuristic search methods. Plans are created online according to pre-defined heuristics derived from motivations, the driver model and domain constraints. The plan recognition module takes the current set of possible plans and assigns probability hypotheses based on the observation of physical actions of vehicles. In order to make the prediction more efficient, prediction is focused on future situations resulting from combination of plans rather than considering each possible action. Furthermore, paths in the possible situations structure that are assigned a higher probability are expanded further into the future than less probable paths. The results (possible future situations) serve as basis for situation analysis. Please refer to [1] for a more detailed illustration.

The action recognition module that is discussed in the following is a self-contained unit that is responsible for the recognition of a single action within a plan description. This is done by observation of the physical parameters of the vehicle. The result of the action recognition module is a probability assignment for each possible action which is fed into the plan recognition module.

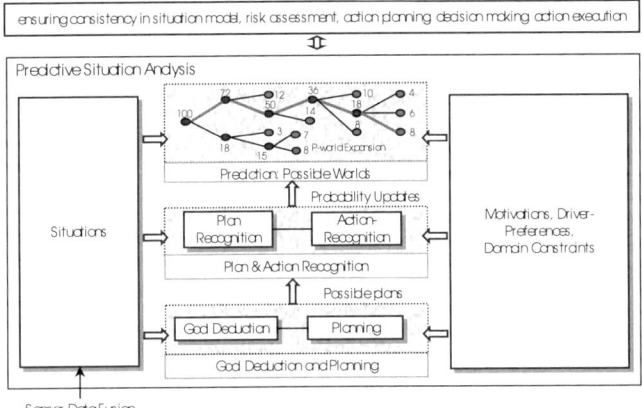

Fig. 2. Architecture overview

2 Modeling Principles

The following sections provide an introduction to the concepts and techniques which are the foundation for the realization of the action recognition model in Sect. 3.2. Firstly, an overview of the used domain knowledge is given. Secondly, the used representation formalism – namely dynamic belief networks (DBN) – is introduced, to illustrate the methodologies behind the model.

2.1 Behavior Models

The notion of predictive[1] action recognition in traffic is based on the idea that actions in traffic are initiated by typical driving behaviors, and hence, even though afflicted with uncertainty, can be predicted.

The following sections introduce vehicle following and lane change models supported by experimental investigations of actions in traffic that form the basis for modeling action recognition with DBNs. The key concept in the probabilistic network is the representation of the (local) situation in terms of time measures. This form of representation is chosen mainly because of the lack of scalability of DBNs. The representation of physical parameters like distance, (relative) velocity and acceleration in time measures saves random variables in the belief network and causal links which reduces the state space of probabilistic inference. The following measures are used in

---

[1] The term predictive is related to the fact, that the aim is to recognize the intention to an action before it actually occurs.

the dynamic belief network in Sect. 3.2: net-time-gap (T_N) and time-to-collision (T_{TC}) concept which are defined as follows:

Definition 1.
The net-time-gap (T_N) is defined as the time interval between two vehicles regarding an absolute position x_0. After the front vehicle (i-1) crosses the point x_0 it takes T_N for the following (ego i) vehicle to cross x_0, assuming that the physical parameters of i do not change during this interval:

$$T_{N,i} = \frac{-v_i \pm \sqrt{v_i^2 + 2\Delta x_i a_i}}{a_i} \tag{1}$$

where Δx_i is the free space between both vehicles, v_i and a_i are the velocity and acceleration of the ego vehicle, respectively.

Definition 2.
The time-to-collision (T_{TC}) is the time that is needed by vehicle (ego i) to reach (or collide with) the front vehicle (i-1), assuming that physical parameters of both vehicles do not change during this interval:

$$T_{TC,i} = \frac{-\Delta v_i \pm \sqrt{\Delta v_i^2 + 2\Delta x_i \Delta a_i}}{\Delta a_i} \tag{2}$$

where Δx_i is the free space between both vehicles, Δv_i and Δa_i are the relative velocity and relative acceleration, respectively

2.1.1 Vehicle Following

The Optimum Velocity Model (OVM)
In [2] a microscopic vehicle following model, called Optimum Velocity Model (OVM[2]), is presented where the following behavior is stated with a simple differential equation:

$$a_i = k\left(V_{i,optimal}\left(\Delta x_i\right) - v_i\right) \tag{3}$$

The parameter k reflects the sensibility and must be adjusted to the specific driver (see [2] or [10] for further detail). The optimum velocity $V_{optimal}$ is dependent on the distance and is given by a hyperbolic function:

$$V_{i,optimal}\left(\Delta x_i\right) = c_1\left(\tanh\left(c_2\left(\Delta x_i - c_3\right)\right) + c_4\right) \tag{4}$$

---

[2] The model assumes an optimal velocity for vehicle following. The introduced control equation seems to be dependent only on this *"optimal velocity"* (the names origin). Since this optimal velocity is also dependent on the distance, over another equation, the control strategy considers both: the distance and the relative velocity. In which sense the optimal velocity is optimal is not stated.

In [10] the initial model for Japanese highway traffic is adapted to city traffic in Germany, by the adjustment of the parameters c_1, c_2, c_3 and $c_4{}^3$. Several tests with this adjusted model that where performed to tune the parameters of the DBN have shown that typical values of T_N are between 1.3 and 1.4 seconds where the time-to-collision T_{TC} oscillates at high levels (>50sec).

The Vehicle Following Model According to Wiedemann
A frequently cited vehicle following model is given in [14] which considers thresholds in human perception and reaction. The model explicitly considers inaccuracy and resolution of human perception for distances and relative velocity.
The author defines four states for vehicle following scenarios:
1. *Unaffected:* Front vehicle is too far and there is no rapprochement observable. The driver drives at desired speed.
2. *Consciously affected:* The driver recognizes the rapprochement and starts braking.
3. *Unconsciously affected:* For $\Delta v_i \approx 0$ the driver can not recognize whether the distance increases or decreases. The driver keeps the last chosen velocity, which leads to a hysterlsls effect.
4. *Aversion of danger:* If the distance falls below the minimum following distance, the driver brakes harder than in case 2.

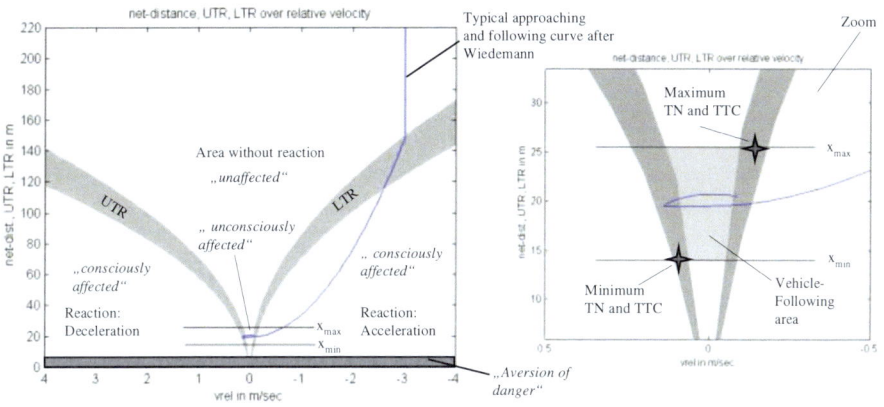

Fig. 3. Vehicle following model after Wiedemann, [14]

The interesting states are those which arise in the stationary vehicle following state – consciously and unconsciously affected. Here, the relative velocity and distance oscillate within certain bounds that are defined as follows:

UTR is the threshold where the following vehicle decelerates:

---

^{3} Manstetten et. al. define $c1 = 7.84$; $c2 = 0.184$; $c3 = 10.4$ and $c4 = 0.913$. The desired velocity for driving without front vehicle is the desired velocity which is $V (\Delta x \rightarrow \infty) = 15m/s$ for the defined parameters.

$$UTR : \Delta x = \sqrt{k_2 \Delta \dot{x}} \text{ for } \{\Delta \dot{x} \leq 0\} \tag{5}$$

LTR is the threshold where the following vehicle accelerates:

$$LTR : \Delta x = \sqrt{k_3 \Delta \dot{x}} \text{ for } \{\Delta \dot{x} > 0\} \tag{6}$$

Δx_{min} is the minimum following distance and is given by where Δx_s is the distance to stop.

$$\Delta x_{min} = \Delta x_s + \sqrt{k_1 \Delta \dot{x}} \tag{7}$$

The maximum following distance Δx_{max} is roughly twice the minimum distance.

$$\Delta x_{max} \approx 2\Delta x_{min} \tag{8}$$

The properties that have to be satisfied in the stationary vehicle following (VF) state can be derived from the above statements, which leads to:

$$\exists i : \Delta x_i > \Delta x_{i,min} \wedge \Delta x_i < \Delta x_{i,max} \wedge \left(\Delta x_i < \sqrt{k_2 \Delta \dot{x}_i} \wedge \Delta \dot{x}_i \leq 0 \vee \Delta x_i < \sqrt{k_3 \Delta \dot{x}_i} \wedge \Delta \dot{x}_i > 0 \right) \rightarrow VF(i) \tag{9}$$

As shown in Fig. 3, typical values for maximum and minimum T_N and T_{TC} can be concluded from this equation. Simulations with adapted parameters show T_N in the range of 1.45-1.55 seconds for the stationary case and T_{TC} oscillates at high levels like in the OVM model. Figure 3 depicts the results for adapted parameters[4] gained by simulation.

Experimental Results
Experimental results given in [7] and [8] show that T_Ns and T_{TC}s tend to be smaller as those gained by simulation with the models above. The authors identify a maximum of roughly 1 second for T_N. The authors further claim that drivers accept foreseeable risk, e.g. very small T_Ns are given in scenarios where the front vehicle changes its lane or the following vehicle intends to overtake.

2.1.2 Lane Change Models
Most lane change models divide the lane change process in phases, [5], [4]:
- The decision process of a lane change, where this process can be further subdivided in phases (see below).
- The initiation and announcement of a lane change.
- The realization of a lane change.
- The end of a lane change.

The last two points are important for prediction as defined in 1.20 and play a minor role for action (intention) recognition and will not be discussed further. Interested readers can refer to [5].

Lane Change Desire and Decision

---

[4] k_1=10sm, x_s=2m, k_2= 0.75 10^4 and k_3=-0.5 10^4

The decision process for lane change can be sub-divided in further phases,[4]:

1. Perception of the situation
2. Evaluation of satisfaction on lane→lane change desire
3. Determine feasibility of lane change →lane change decision

Synthesis of lane change desire: There can be many motivations leading to a lane change intention, [3]. Following a certain route or overtaking slower vehicles are two named examples. Moreover, reactive behavior can also motivate lane changes, e.g. if a fast vehicle approaches from behind. [4] defines a measure that summarizes the lane change desire: contentment on lane. This approach is the basis for modeling action recognition with DBNs, where contentment is expressed in terms of T_Ns and T_{TC}s. Feasibility of lane change: The feasibility of a lane change can be expressed in terms of T_Ns and T_{TC}s to the nearest neighbors, as done in the experimental investigations of [4] and [5].

Table 1. Minimal and maximal time gaps for lane change, [4]

	Minimum Value	Maximum Value
T_n behind	0,3 sec	1,7 sec
T_n ahead	0,3 sec	3,0sec

[5] introduces subjective values T_N=0.6sec and T_{TC}=3sec, where most drivers categorize the lane change as dangerous. Those values show that the evaluation of the feasibility of a lane change and also the lane change decision are driver (and also traffic type) dependent, which is an issue for the DBN model.

Lane Change Initiation
Action intention recognition requires the evaluation of typical behavior that initiates the action. Hence, the following variants for the initiation of a lane change will be exploited in the DBN model in Sect. 3.2.

- In most European countries the initiation of a lane change using a turn indicator is prescribed by law. Unfortunately, this is not applicable for vehicles around because reliable detection of the vehicle turn signal is unsolved and a major issue in environment perception. Assuming inter-vehicle communication the turn signal can be transmitted to concerning vehicles, but can not be the single source of evidence.
- In many cases, the obligation to initiate a lane change by using a turn indicator signal is omitted. A second source of evidence, that can be used in the DBN, is the offset and also the relative velocity to the center of the lane.
- Further evidence can be collected by evaluating the acceleration and decelerations processes of the vehicle. As shown in the preceding section there is a relation between T_Ns and T_{TC}s to the target lane that influences the lane change decision. The thesis utilized here is that the probability for a lane change increases if the conditions for vehicle-following models are fulfilled for the neighboring lane.

2.1.3 Summary

Models and experimental results presented so far are the basis for designing an action recognition model for driver assistance systems. As a matter of fact human driving behavior is subject to variances, and hence predicting driver-vehicle interactions is afflicted with uncertainty. The approach presented here consists of applying concepts and techniques of probability theory to tackle the uncertainty in sensor data acquisition and behavior models.

2.2 Dynamic Belief Networks (DBN)

Belief networks are directed, acyclical graphs, which illustrate the causal dependencies between probabilistic variables. An associated inference engine provides the necessary adjustment of each variable's probability distribution including the dependencies and possible a priori knowledge, called evidence. Many time-series models, e.g. Hidden-Markov-models or Kalman-filters, are examples of dynamic belief networks. The precondition for such models is, that an event at *t-1* might result in an event at *t* but never vice versa. A simple one-state Markov-Model illustrates such a precondition in the following statement.

$$P(X_1, X_2, X_3, \cdots, X_t) = P(X_1)P(X_1 \mid X_2)P(X_2 \mid X_3) \cdots P(X_t \mid X_{t-1}) \qquad (10)$$

This precondition provides a great reduction of possible events X_{t+1} and makes the whole concept feasible. The order of the models can be adapted, i.e. more than two time-steps can be included in such a model.

A first order dynamic belief network is composed of two stationary models for each time step and a time transition model comprised of links describing the dependencies between two time steps. The resulting model can be seen as nothing more than a new stationary model.

Please refer to [12] and [13] for a more detailed discussion on belief networks. The inference engine used in this work is described in [9].

3 Action Model and Action Recognition

3.1 Action Model

Elementary (atomic) actions represent non-dividable operations at the planning level and also realistic activities that a driver can perform. Unlike discrete, fully observable and deterministic domains, as in user interfaces, an action in traffic at the physical level represents a process that requires time and can only be determined with uncertainty, because these physical processes are subject to variances. Hence, an action model for traffic must satisfy requirements raised by the planning level algorithms (see [3]) and also by the peculiarities of the traffic domain.

Several action model concepts have been proposed in the context of autonomous vehicles and driver assistance systems, see [11], [15] and [6]. The dynamic (online) planning approach for plan recognition has additional requirements, so that a slightly different model will be proposed here.

The model satisfies the following requirements:

- Actions represent physical processes that continuously change the vehicle (and situation) state. These processes, even though afflicted with uncertainty, must be observable through sensors.
- The completion of an action leads to a change in the discretely represented situation model at the planning level.
- Actions are univocally distinguishable in the formulation and the degree of abstraction from higher level concepts as motivation, goal and plans.

In [3] an action model is introduced in the context of search-based action planning for highway traffic. The model depicted in Fig. 4 excels by the chosen level of abstraction that facilitates efficient planning. The model is based on the considerations in Sect. 2.1. Hence, the two actions beginning with the term *Remain* represent the vehicle following intention (if a vehicle in front exists). Therefore, evidence for lateral or longitudinal action which is intended to stabilize the vehicle, either to adjust the offset to the center of the lane or adjust the speed and distance to a vehicle in front, are seen as evidence for the *Remain* actions. Moreover, there is always a plan generated for plan recognition and also a goal or motivation, which is expressed through the *Remain* actions. The following definitions illustrate the properties of the action model.

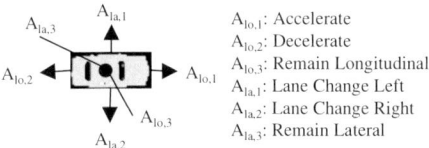

$A_{lo.1}$: Accelerate
$A_{lo.2}$: Decelerate
$A_{lo.3}$: Remain Longitudinal
$A_{la.1}$: Lane Change Left
$A_{la.2}$: Lane Change Right
$A_{la.3}$: Remain Lateral

Fig. 4. Action model for action recognition

The following definitions illustrate additional properties of the action model.

Definition 4.
The actions are grouped in longitudinal (lo) and lateral (la) actions. Actions within a group are exclusive, i.e. if an action in a group is recognized and belongs to the evidence set E, then all other actions in the same group are not a member of E.

$$\exists i : A_{g,i} \in E \rightarrow \bigwedge_{i \neq j, g \in \{lo, la\}} A_{g,j} \notin E \qquad (11)$$

Definition 4, means e.g., that an accelerating car cannot break simultaneously or a lane change to the left cannot take place while the car changes lane to the right. This definition also means that sum of probabilities of a group is always 1.

Definition 5.
Actions belonging to different groups may be executed at the same time or observe a certain time relation. These relations can be qualitatively formulated with 13 time relations, [1][5].

---

[5] [1] defines 7 time relations: $op = \{<, =, m, o, d, s, f\}$, corresponding to smaller, equal, meets, overlaps, during, starts, finishes. A total number of 13 arises by considering also the inverses $op = \{>, =, mi, oi, di, si, fi\}$.

$$A_{g,i} \ op \ A_{f,j} \ \text{for} \ \{g,f \in \{lo,la\} | \ g \neq f\}; \ i \in g ; \tag{12}$$
$$j \in f ; op = \{<,>,=,m,mi,o,oi,d,di,s,si,f,fi\}$$

Definition 5 explains time relations between specific actions; e.g. $A_{lo,1} \ m \ A_{la,2}$ defines a period of acceleration followed by a lane change to the right. Definition 4 is reflected in the structure[6] of the DBN. Definition 5 plays a role in the expansion of the possible future situations. Regardless of time dependencies, there are dependencies between actions from different groups, for e.g. a lane change to the left is mostly associated with an acceleration action. Dependencies between groups are modeled in the dependency graph (see Sect. 3.2).

3.2 Action Recognition

The emphasis of the DBN illustrated in the following is early recognition of driver actions (intended actions of the driver) which is possible by considering time gaps to the nearest neighbors of the vehicle and also by expanding this static relationship over a number of time steps (prediction[7]).

The model in Fig. 5 is not expanded over time for the sake of readability. The dashed links represent the time transition model that links the stationary models while the solid links represent the causal connections of the network.

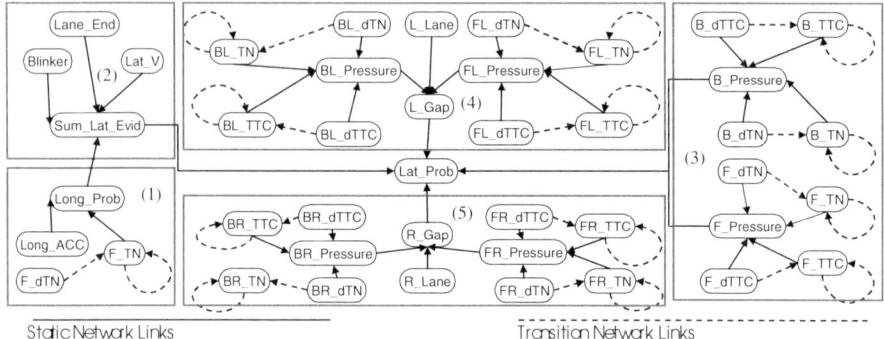

Fig. 5. Resulting DBN

This model can handle traffic on roads with more than two lanes. Information is only needed and gathered as evidence from the neighboring lanes to the left and to the right.

In Fig. 5 the nodes' names consist of certain abbreviations. The first letter(s) indicates what neighboring car is the source of possible evidence; e.g. the node FR_TN represents a probability distribution over the net-time-gap to the vehicle which is right

---

[6] Actions within a group are allocated to one node in the probabilistic network.
[7] Prediction here refers to the short term prediction of an action and must not be confused with prediction as defined in Sect. 1.2.

and ahead of the concerned vehicle as does B_TTC for the time-to-collision in respect to the vehicle in the back. The nodes X_dTTC represent the change of the time-to-collision in respect to the vehicle X. The X_PRESSURE nodes sums up the evidence concerning the neighboring vehicle X and represents the pressure that this car emits. The pressure a neighboring car emits increases the closer the car is to the current position of the ego car and the higher its relative velocity is.

The nodes Lat_Prob and Long_Prob represent query variables which indicate the action probabilities for the action model in Sect. 3.1. The following table explains the nodes of the network:

Table 2. Description of the nodes

Node	Description
X_TTC	Time-To-Collision values to the vehicles of the same and neighboring lanes. The abbreviations B, F stand for back, front and R, L for right, left, respectively.
X_TN	Net-Time-Gap values to the vehicles. Same as X_TTC.
X_Pressure	These nodes are introduced to reduced the amount conditional probabilities within the CPT (see equation 13 below). The abstract measure summarizes the T_n and T_{tc} and their derivations to a single boolean value.
L_Gap, R_Gap	The two pressure values regarding the neighboring lanes are again summarized to a L_Gap and R_Gap value
Blinker	Evidence node representing the turn signal indicator state. This signal is only available for the vehicle equipped with the system.
Lane_End	This node states whether the lane on which the concerning vehicle ends, and if it ends, in which direction the vehicle has to perform merging. This evidence is only available if appropriate navigation information is provided by (GPS) sensors.
Lat_Evid	The lateral offset of the vehicle together with the relative velocity relative to the center (time to lane crossing as a derived measure) of the lane are incorporated in this value.
Long_Evid	Similar as Lat_Evid, but for the longitudinal actions. Here, the acceleration or deceleration of the vehicle is taken as node value.
Sum_Lat_Evid	Summarizes the lateral evidences (except the derived pressure values) for a lateral action within one node. This reduces the amount of conditional probabilities that have to be specified.
Long_Prob	Query node that state whether the vehicle performs a motivated acceleration or deceleration process. Motivated in this context means that, e.g. an acceleration of a vehicle to adjust the gap to the front vehicle is taken as evidence for the Remain_Long action.
Lat_Prob	Query node that states the probability distribution over the lateral actions.

In the following the network is further divided in 5 subnets as depicted in Fig. 5. The subnets and their functions are explained in the following.

1. The Long_Prob node is causally dependent on the time gap (F_TN) and also on the evidence given by the sensed longitudinal acceleration (Long_ACC). The acceleration is evaluated with respect to the time gap which is an indication for the vehicle following intention. Acceleration or deceleration to adjust the gap to the front vehicle is taken as evidence for the vehicle following intention and hence for the Remain_Long action.

2. This subnet summarizes the evidence for a possible lane change: the turn indicator signal (Blinker), the lateral velocity (Lat_V), longitudinal action evidence (Long_Prob) and the lane information (Lane_End).

3. This subnet models the contentment on lane as introduced in Sect. 2.1.2: The little the time gaps on the same lane the less the contentment is and a lane change becomes more probable, where small T_Ns and T_{TC}s to the back increase the probability for a lane change to the right and small T_Ns and T_{TC}s to the front increase the probability for a lane change to the left. The pressure (F_PRESSURE) emitted from the car in front is determined by the net-time-gap (F_TN) and the time-to-collision (F_TTC) as well as their variation over the last time steps (F_dTN, F_dTTC). The same approach is taken for the car in the back.

4. The Lat_Prob is causally dependent on time gaps and also T_{TC}s on the neighboring lanes, where the models given in Sect. 2.1 are expressed in the conditional probability table of the nodes. The more the vehicle following model expressed in terms of T_Ns and T_{TC}s is satisfied for a particular lane, the more likely a lane change is in the corresponding direction. This works well for free traffic but in the case of crowded traffic these conditions are satisfied without being an indication for a lane change. Therefore, the T_Ns and T_{TC}s can not be the single source of evidence for a lane change, so that the evidence provided in subnets 2 and 3 is also needed. The subnet works just like subnet 3 but for the lane left of the occupied lane. The pressure each car emits is generated using knowledge about the relevant time gaps. Again the car in front and in the back on the specific lane are the ones monitored. The key values are the net-time-gaps (BL_TN, FL_TN) and time-to-collision (BL_TTC, FL_TTC) and their respective derivatives (BL_dTN, FL_dTN, BL_dTTC, FL_dTTC).

5. Same as subnet 4 for lane change to the right.

The time transition model applies only for continuos nodes representing T_Ns and T_{TC}s. The network is expanded over time whereby T_Ns and T_{TC}s are recalculated for each step with the aid of the derivations of the time gaps and TTC and fed into the network as new evidence for the next time step. With this, a small horizon prediction can already be performed within the model, based on the assumption that the physical parameters of the vehicle will not change significantly within a short period of time.

All other nodes (except evidence and query nodes) represent intermediate nodes that are included to reduce the amount of assignments in the conditional probability tables. This beneficial property of belief networks is due to the conditional independ-

ence assumption inherent to those models which states that any node is independent
of all ancestors given its parents, which results in following relationship:

$$P(X_i \mid X_{i+1}, \ldots, X_n) = P(X_i \mid Parents(X_i)) \tag{13}$$

One remaining issue is adaptability of the model to drivers and traffic flow types,
where an adaptability for the driver is only applicable for the driver of the own vehicle. In fact two approaches have been taken for adaptation of the network: Bayesian
learning and re-instantiation with adapted and predefined parameters.

The first approach exposed to be infeasible because of the required learning data
sample size and the inherent computational complexity in learning algorithms which
makes real-time online learning infeasible. Further investigations of bayesian learning
will be conducted as soon the required learning data is available. The second approach provides good results, when the parameterization of the network is tuned according to predefined driver and traffic types.

The following points are contemplating the properties of the model:

- DBNs are not scalable[8], and therefore reducing complexity in models is a major
  issue in designing such networks. Hence, the physical parameters distance, velocity, acceleration that are relevant are expressed in terms of "compressed" physical
  parameters of time.
- Representing the situation in terms of nearest neighbors reduces complexity and is
  sufficient for early action recognition based on physical parameters of vehicles and
  vehicle relations.
- DBNs provide concepts and techniques that can be used for short horizon prediction. The approach here proposes a continuation of physical parameters (T_Ns and
  T_{TTC}s) that facilitate reasoning about future beliefs.
- The adaptation of the network for different driver and traffic types is done by re-
  instantiation of the network with different parameters. Nevertheless, the major aim
  is to apply learning algorithms for the adaptation of the network shown here. Major
  benefits are expected regarding parameterization of the network when representative data for driving behavior is collected and used for learning.

4   Simulation Results

Simulations have been carried out for various highway scenarios and the applicability
to future assistance systems has been assessed. Figure 6 shows an excerpt of the results gained for overtaking scenarios. The prediction horizon in the DBN is set to 3
time steps where one time slice is 0.5 seconds. The figure shows that a probability of
0.8 is assigned to the change left action roughly 1.5 seconds before the actual lane
change is performed. This is because before a lane change (or a turn indicator) is
observed, the time gaps to the neighbors on the target lane must be adjusted, which
provides additional evidence for early action recognition.

More important is the fact that the T_Ns and T_{TC}s are predicted within the 1s prediction horizon. This means the probability distributions within the T_N and T_{TC} nodes are

---

[8] Inference in belief networks is NP-Hard

recalculated by probabilistic inference given the assumption that the derivations of T_Ns and T_{TC}s do not change within the short prediction horizon. With that, the probability distribution of the Lat_Prob node for the predicted T_Ns and T_{TC}s can be extracted from the network. Further investigations have shown that a prediction horizon larger than 2s is not feasible because of the inherent assumption that the change of the T_Ns and T_{TC}s over each time step will not change within the prediction interval (the hypotheses become to inaccurate, when the behavior if the driver changes).

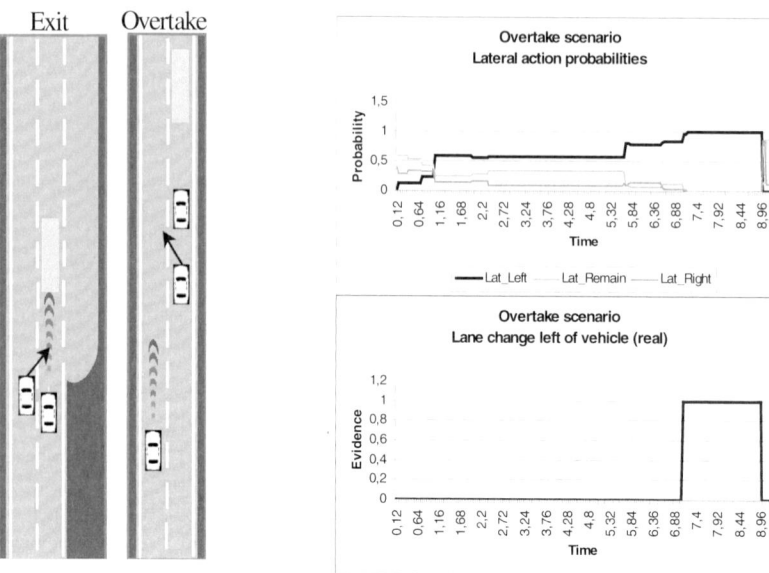

Fig. 6. Highway scenarios and results of lane change probabilities for overtake scenarios. The prediction horizon is 3 time steps.

The advantages of the early recognition of a single action are twofold: Firstly, employing this approach to the general scheme introduced in Sect. 1 enables early and reliable updates of the plan probabilities in the plan recognition module and enhances the prediction process. And secondly, the action (driver intention) recognition module can be used as self-contained unit and applied without realizing the whole framework, because the short horizon prediction achieved here can already bring enhancements for driver assistance systems. For instance, applying the approach to a lane change assistant would mean that necessary warnings can be identified earlier so that the reaction time of the driver can be bridged.

5 Conclusion and Outlook

Next generation assistance systems as adaptive cruise control (ACC) or lane change assistants should be able to perform predictive situation analysis to increase comfort an safety. The important aspect which makes up the quality of an assistance system is the interference strategy which determines when and how the assistance system informs, warns or even actively controls the vehicle. Without reasoning about driver intentions and predicting future situations this issue can hardly be solved.

The general scheme introduced at the beginning provides a framework for predictive situation analysis where action (intention) recognition is one major part. The approach for action recognition could give significant improvements to today's assistance systems (without applying the whole framework), especially to those coping with the driver in the loop. An appropriate information and warning strategy e.g. for a lane change assistant is conceivable, if the lane change intention of the driver can be determined with a certain probability. Hence, designing such a system requires looking beyond reactivity and control.

References

1. J.F. Allen, Towards a theory of action and time, Artificial Intelligence 23, 123-154, 1984
2. Bando, Masako, Hasebe, Nakanishi, Nakayama, Shibata, Sugiyama Phenomenolical Study of Dynamical Model of Traffic Flow Physical Review E, 51:1-18, 1995
3. I. Dagli, D. Reichardt, Motivation-based Approach to Behaviour Recognition, In Proceedings of the IEEE Intelligent Vehicle Symposium, Versailles, 2002
4. D. Ehmanns, Simulation Model of Human Lane Change Behaviour, Technischer Bericht, Institut für Kraftfahrwesen, Aachen, 2001
5. W. Fastenmeier, H. Gstalter Ulf Lehnig, Analyse von Spurwechselvorgängen im Verkehr, Zeitschrift für Arbeitswissenschaft, (1/2001), Darmstadt, 2001
6. J. Forbes, N. Oza, R. Parr, S. Russel, Feasibility study of fully autonomous vehicles using decision-theoretic control, Final Report, Computer Science Division, University of California, Berkley, 2000
7. A. Hiller, Mikroskopische Modellierung der Längsdynamik von Fahrer-Fahrzeug-Einheiten im Stadtverkehr Diplomarbeit, Institut für Systemdynamik und Regelungstechnik, Universität Stuttgart, 1999
8. J. Kienzle, Analyse von Einzelfahrzeugdaten, Diplomarbeit, Institut für Systemdynamik und Regelungstechnik, Universität Stuttgart, 2000
9. U. Kjaerulff, A computational scheme for reasoning in dynamic probabilitic networks, Proceedings of the Eights Conference on Uncertainty in Artificial Intelligence, p. 121-129, 1992
10. D. Manstetten, W. Krautter, T. Schwab, Traffic Simualtion Supporting Urban Control System Development, 4th World Congress on Intelligent Transportation Systems, 1997
11. R. Moeck-Hecker, Wissensbasierte Erkennung kritischer Verkehrsituationen, Dissertation, Düsseldorf, VDI Fortschritt-Berichte, Reihe 12, Nr. 209, VDI-Verlag, 1994
12. J. Pearl, Probabilistic Reasoning in Intelligent Systems, Systems Networks of Plausible Interference, ed. Morgan Kaufmann, 1988
13. D. Spiegelhalter, P. Dawid, S. Lauritzen, R. Cowell, Bayesian Analysis in Expert Systems Statistical Science, Vol. 8, No. 3, 1993
14. R. Wiedemann, Simulation des Straßenverkehrsflusses, Band 8 der Reihe Schriftenreihe des Instituts für Verkehrswesen an der Universität Karlsruhe, Universität Karlsruhe, 1974
15. M. Zeller, Planerkennung im Straßenverkehr, Dissertation, VDI Fortschritt-Berichte, Reihe 12, Nr. 286, VDI-Verlag, 1996

Collaborative Agent System Using Fuzzy Logic for Optimisation

Dayou Li and Jie Zhang

Computing and Information Systems Department, Faculty of Creative Arts and Technologies
University of Luton, Park Square, Luton, LU1 3JU, United Kingdom
{dayou.li,jie.zhang}@luton.ac.uk

Abstract. Collaborative agents are able to find global solutions to complex systems using local knowledge rather than a complete global knowledge set. The complexity, on the other hand, brings uncertainty. Global optimisation in a uncertain environment can be undertaken using fuzzy collaborative agents. Negotiation, involving generating, evaluating and modifying action plans, under fuzzy circumstance is a key to construct fuzzy collaborative agents and fuzzy multi-agent systems. This paper presents an approach of negotiation using fuzzy reasoning, in particular *fuzzy backward reasoning*. A fuzzy collaborative agent has local knowledge in the form of a fuzzy model. Fuzzy backward reasoning is performed based on the fuzzy model to generate and modify an action plan in each agent.

1 Introduction

Global knowledge representing the behaviours of a complex system is often hard to acquire and therefore incomplete in practice. This makes it difficult to perform global optimisation for the system. Collaborative agents, each of which has sufficient local knowledge, have been developed for the global optimisation without using a global knowledge set. The collaboration is often achieved through negotiation which involves generating, evaluating and revising action plans (Wangermann and Stengel, 1998; Moree and et al, 2000; Dospisil, 2001). The complexity of such a system can also mean uncertainty. Knowledge about the system contains ambiguity and vagueness. It is convenient to represent the knowledge using fuzzy sets and fuzzy implications. The collaborative agents in this situation, then, must be able to apply the fuzzy knowledge to the production, evaluation and modification of action plans.

Negotiation between agents requires two categories of fuzzy implications. Those used to produce a plan reflect knowledge about decision-making, as the plan is a sequence of actions an agent may take. These implications are mappings from the state and auxiliary variables to the actions. Whilst, because an agent is committed to a task which associates to a part of a system, called a sub-system, fuzzy implications used to evaluate an action plan should describe the behaviours of the sub-system. They are mappings from the actions and the auxiliary variables to the state variables of the sub-system. The fuzzy implications regarding decision-making are difficult to acquire. As a fuzzy system in general is the mimic of an existing system, the fuzzy implications regarding decision-making have to be developed based upon a successful

R. Kowalczyk et al. (Eds.): Agent Technology Workshops 2002, LNAI 2592, pp. 195-210, 2003.

decision-making mechanism. This requires the existence of such a successful mechanism in the first place. The fuzzy mappings will not be able to construct when the successful mechanism is not available.

To revise a fuzzy action plan, an agent must possess the ability to perform constrained optimisation in a fuzzy domain where both the cost function and constraints are fuzzy. The fuzzy constrained optimisation aims to find an intersection between the cost function and the constraints (Bellman and Zadeh, 1970; Wang, 1983; Zimmermann, 1990). The intersection is then used to modify the action plan. In the case where the intersection cannot be found, a new method is required for the purpose of reversing.

In this paper, an approach of negotiation using fuzzy knowledge was presented. This includes the production, evaluation and modification an action plan. When generating and revising an action plan, *fuzzy backward reasoning* is applied to the fuzzy implications that describe the behaviours of a sub-system. Hence, it avoids the problem of the lack of decision-making knowledge. Fuzzy forward reasoning is used to evaluate the action plan.

This paper is organised in the following way. Section 2 defines the collaborative agents, including variables reflecting its state, functions being able to provide and knowledge which sustains the functions. Section 3 introduces the concept of fuzzy backward reasoning and its usage to the negotiation between the collaborative agents. Section 4 discusses the negotiation process. Section 5 shows the construction of a collaborative agent system for controlling liquid levels in a three-tank system and discusses experimental results.

2   Model of Fuzzy Collaborative Agents

A collaborative agent is self-interest, has its own commitment and is able to negotiate with other agents. The negotiation procedure contains generating action plans, evaluating and revising these plans and finally forming a plan that is acceptable by all involved agents.

Definition 1:
An action plan $\mathbf{u}_i = [(u_{i1}, u_{i2}, \cdots, u_{in}), (\mathbf{x}_{i1}, \mathbf{x}_{i2}, \cdots \mathbf{x}_{in})]$ consists of a sequence of actions an agent A_i may take as well as its effects on auxiliary variables, where u_{ij} stands for the jth action in the sequence of actions and $\mathbf{x}_{ij}$ is a set of auxiliary variables influenced by u_{ij}.

There is normally a time constraint τ associated with a task, in which an agent must complete the task. That is the corresponding sub-system should reach its steady-state after τ. This time constraint applies a limitation onto the length of an action plan. Let T be the sample time when an action is applied to the sub-system and feedback is observed. The length of the action plan n should satisfy $n \leq \tau / T$.

Definition 2:
A plan $\mathbf{U}$ is the disjunction of revised action plans.

Actions and their effects on auxiliary variables form two subsets of an action plan, whilst $\mathbf{u}_i$ ($i = 1, 2, \cdots$) are themselves the subsets of the plan $\mathbf{U}$. The subsets of actions of different action plans may have intersection if two or more agents share the same facility to fulfil their commitments. The subsets of effects may also have intersections.

Definition 3:
A fuzzy collaborative agent A is a triple A = $\{\mathbf{V}_A, \mathbf{F}_A, K_A\}$, where $\mathbf{V}_A$ is a set of variables, $\mathbf{F}_A$ is a set of functions and K_A stands for local knowledge.

The model of a collaborative agent is shown in Fig. 1. In $\mathbf{V}_A$, there are two types of variables, namely *state variables* and *auxiliary variables*. The former is the state of the task an agent is committed to. These variables reflect the extent to which the agent pursues its commitment. The agent will generate an action plan according to the values of these variables. The latter represents environment that surrounds the agent. Changes in the environment may be caused by disturbances and by other agents when they take actions for their own commitments. The environment changes may have influences on the state of the task the agent committed to. As the collaborative agent is self-interest, it must be able to perceive the environment changes and make decision on whether or not to response to the environment changes. The auxiliary variables enable an agent to capture these changes.

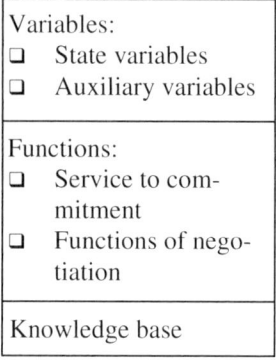

Fig. 1. Model of a collaborative agent

$\mathbf{F}_A$ contains functions of creating an action plan $\mathbf{u}_i$, evaluating action plans received from other agents and revising its own action plan. These functions are supported by the local knowledge K_A stored in the agent. As K_A is formed with a set of fuzzy implications, various fuzzy reasoning methods are used to implement these functions.

Local knowledge K_A within an agent is actually a fuzzy local model that describes the dynamics of a sub-system of a complex system. A system usually contains measured variables, manipulated variables and external disturbances. The measured variables reflect the state and the output of the system. The manipulated variables are the inputs of the system. They both are stimuli of the system. Giving values to the inputs, the system responses to the stimuli according to its dynamics, which can be represented as input-output relations. A local model is the mathematical description of these relations of a sub-system. A local fuzzy model is employed when uncertainty exists in the dynamics of a sub-system. In a fuzzy model, the values of input and out variables are fuzzy sets and the relations are represented using a collection of fuzzy implications. The local fuzzy model also contains fuzzy implications representing changes on the auxiliary variables because of the stimuli.

3   Action Plan Generation and Evaluation Using Fuzzy Reasoning

3.1   Fuzzy Reasoning

Fuzzy reasoning is performed based on fuzzy implications either to find a logical consequence for a given condition or to deduce a sufficient condition for a specified conclusion, called a given goal.

Definition 4:
Giving a fuzzy implication "if x is A then y is B" and a condition "x is A'", the procedure of searching for a logical consequence of the given condition is called fuzzy forward reasoning. That is

$$
\begin{array}{ll}
premise\,1: & x \text{ is } A' \\
premise\,2: & \text{if } x \text{ is } A \text{ then } y \text{ is } B \\
\hline
consequence: & y \text{ is } B'
\end{array}
\tag{1}
$$

The performance of fuzzy forward reasoning is straightforward because the fuzzy implication implies that "y is B" is a necessary condition of "x is A". By applying the composition rule of inference to premise 1 and premise 2, consequence can be deduced.

Definition 5:
Giving a fuzzy implication "if x is A then y is B" and a goal "y is B'", the procedure of searching for a sufficient condition that will logically lead to the given goal is called fuzzy backward reasoning. That is

$$
\begin{array}{ll}
premise\,1: & y \text{ is } B' \\
premise\,2: & \text{if } x \text{ is } A \text{ then } y \text{ is } B \\
\hline
consequence: & x \text{ is } A'
\end{array}
\tag{2}
$$

Performing fuzzy backward reasoning involves solving a fuzzy relational equation. The solutions are the sufficient conditions for the given goal (Wang, 1983; Pham and Li, 2001). The fuzzy relational equation is formed with the given goal, unknown variables and a fuzzy relation from the unknown variables to the given goal. The fuzzy relation consists of a collection of fuzzy implications. A fuzzy relational equation has the form:

$$X \circ R = Y_d \tag{3}$$

where X stands for a set of unknown variables that must be sufficient to the given goal, R is a fuzzy relation and Y_d is the given goal.

The algorithm used to solve a fuzzy relational equation in this work is described as follows (Wang, 1983):

STEP 1. Construct matrix **U**, in which

$$u_{ij} = \begin{cases} y_j & \text{if } r_{ij} > y_j \\ \phi & \text{otherwise} \end{cases} \tag{4}$$

where u_{ij} can either be a set having only one element y_j, or an empty set.

STEP 2. Calculate the *infimum* of each row of **U**. If all the entries in one row are ϕ, the *infimum* is set to be 1. The *infimum* of every row forms a vector, denoted $S = (s_1 \quad s_2 \quad \cdots \quad s_n)$.

STEP 3. Form the matrix **U'** by adding more y_js of Y to **U**.

$$u'_{ij} = \begin{cases} y_j & \text{if } r_{ij} > y_j \text{ and } y_j < s_i \\ \phi & \text{otherwise} \end{cases} \tag{5}$$

where u'_{ij} can either be a set having only one element, y_j, or an empty set.

STEP 4. Check whether every column of **U'** resulting from the last step contains at least one non-empty entry. If this is the case, the fuzzy relational equation is solvable and the vector S is accepted as the maximal solution. Otherwise, terminate.

STEP 5. Form matrix **W**. Select one non-empty entry from each column of matrix **U'** and put it into the corresponding position in **W**. Fill in other positions in the obtained **W** matrix with ϕ.

STEP 6. Calculate one minimal solution, $S' = (s'_1 \quad s'_2 \quad \cdots \quad s'_n)$ from matrix **W** by applying the *supremum* operation to each row of **W**. The *supremum* will be 0 if all the entries in a row are empty entries.

STEP 7. Return to *STEP 5* if further selections are possible, otherwise, terminate.

3.2   Generating an Action Plan through Fuzzy Backward Reasoning

As being formed with a set of fuzzy implications, a local fuzzy model can be represented as the following:

$$\begin{cases} R_1 : \{U, X\} \rightarrow Y \\ R_2 : U \rightarrow X \end{cases} \tag{6}$$

where U stands for manipulated variables, X is a group of auxiliary variables and Y is a group of state variables. R_1 describes the dynamics of the sub-system and R_2 represents the effects that U applies on X.

Formula (6) is equivalent to the following fuzzy relational equations

$$\{U, X\} \circ R_1 = Y \tag{7.1}$$

$$U \circ R_2 = X \tag{7.2}$$

The commitment of an agent to a sub-system is that the agent must drive the sub-system to achieve its desired output. When this value, denoted y_d is given and is represented as fuzzy sets Y_d, Eq. (7.1) becomes $\{U, X\} \circ R_1 = Y_d$. Fuzzy backward reasoning can then be applied to the equation and action U can be deduced. By substituting U into Eq. (7.2), the effects of this action on X will then be obtained.

Note that U satisfies Y_d in fuzzy domain does not necessarily mean that u, U's counterpart in a non-fuzzy domain, can satisfy y_d. The reason of this is the physical constraints on the action in the non-fuzzy domain. In fuzzy domain, an extreme fuzzy set has an extreme linguistic meaning, e.g. "very big". A non-zero value of this extreme fuzzy set logically means an extreme action is required. However, the non-fuzzy value of this logical extreme action is limited by what can be achieved physically when this action is converted to the non-fuzzy domain. Thus, although an extreme non-fuzzy value is obtained through the conversion, it might be not as extreme as specified.

This leads to the fact that y_d cannot be achieved in one go. The above procedure of producing an action and its effects on auxiliary variables X needs to be repeated several times until y_d is achieved. The iteration will produce a sequence of actions and their effects on X. An action plan is therefore formed.

3.3   Action Plan Evaluating Using Fuzzy Forward Reasoning

When an agent A_j perceives the change in environment caused by the actions of agent A_i $(i \neq j)$, it will ask for the action plan of A_i and then examines the influences of these actions and of the environmental changes on its own commitment. If the influence is negative, it will make its own action plan $\mathbf{u}_j$ aiming to reduce the influences. This action plan will then be sent to the agent A_i for its evaluation.

In both agent A_i and agent A_j, the evaluation is carried out using fuzzy forward reasoning. In the case where these two agents share the same facility to fulfil their commitments, both the subset of actions and that of effects in an action plan will substitute into Eq. (7). In the case where they do not share the same facility, only the subset of effects will be used in Eq. (7). When fuzzy forward reasoning is performed, Eq. (7) yields a sequence of Y', the sub-system's output. If this sequence of output is far from the one generated by its own action plan, the action plan received from the other agent is unacceptable.

4 Negotiation

4.1   Profit Function

In the case where the intersection of the actions, proposed by both A_i and A_j $(j \neq i)$, can be found, the intersections can be taken as a plan and global optimisation will be achieved. In the case where the intersections do not exist, a compromise is required between A_i and A_j $(j \neq i)$ for the purpose of global optimisation. A *profit function* is defined to justify the compromise.

Definition 6:
A profit function of an agent is the inverse proportional to the summation of 1 and the normalised abstract error between the desired output and actual output of a sub-system to which the agent is committed in a certain period of time. That is

$$p = \frac{1}{1 + |\dfrac{y_d - y}{y_d}|} \tag{8}$$

where y_d represents the desired output, y is the actual output and p stands for profit. When $y = y_d$ within the certain period of time, $p = 1$, otherwise p takes a value between 0 and 1.

As mentioned in Sect. 3, there is a time constraint τ associated to a task and an agent should be able to complete the task within τ. This is the commitment of this agent. The more the agent fulfils its commitment, the much profit it gains. When the agent makes the task complete within τ, it gains its maximum profit. The error $|y_d - y|$ at the end of period of τ quantitatively indicates the degree to which the task is completed, or equivalently the extent to which the agent fulfils its commitment. If the system is still far away from its desired state at the end of τ, the error is large, which means the completeness of the commitment is poor. One the other hand, if the system is very close to its desired state, the error will be very small, which shows that the agent fulfils its commitment very well.

4.2   Negotiation Process

The compromise is achieved through negotiation. A negotiation process involves generating, evaluating and modifying action plans. An agent will generate its action plan according to the task to which it is committed. It will ask for and evaluate action plans from other agents if it perceives environmental changes. The agent will modify its action plan for the compromise. The negotiation process between the collaborative agents is illustrated in Fig. 2.

Agent A_i and agent A_j commit to Task I and Task J, respectively. A_i perceives the states of the sub-system Task I associates and generates an action plan, $\mathbf{u}_i$, for the task. Then, A_i calculates p_i that is the profit according to $\mathbf{u}_i$. At the same time, it applies the first action of the action plan to the sub-system, which causes the change in auxiliary variables. When A_j perceives the change, it will ask for the action plan from A_i and calculate profit p_j. If p_j is not satisfactory, A_j generates its own action plan $\mathbf{u}_j$, calculates p_j and then sends it to A_i where the evaluation of $\mathbf{u}_j$ will take place. A_i calculates profit p_i'. If the condition

$$p_i' + p_j' > p_i + p_j \tag{9}$$

holds true, $\mathbf{u}_j$ is accepted, otherwise, $\mathbf{u}_i$ will be revised and sent to A_j for evaluation. This process will be carried out iteratively until the revised $\mathbf{u}_i'$ and $\mathbf{u}_j'$ that hold Eq. (9) are found.

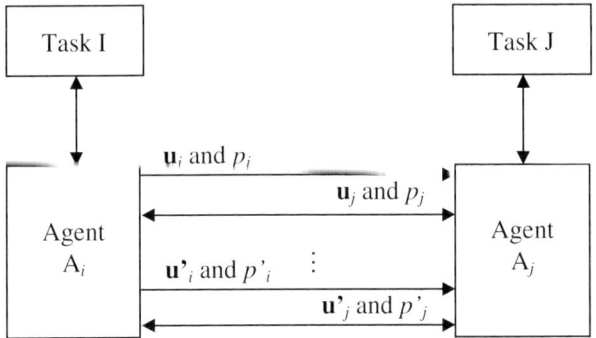

Fig. 2. Negotiation between collaborative agents

4.3   Calculation of Profit

An agent calculates its profit according to the profit function (Eq. (8)) using its own action plan $\mathbf{u}_i$. In this situation, all u_is and x_is in the action plan are substituted into Eq. (7.1) to deduce y at the end of τ for computing profit p. When evaluating an action plan of the other agent, it calculates the profit based on the action plan $\mathbf{u}_j$ received from another agent when evaluating the plan. If these two agents share the same facility for their commitments, both u_is and x_is are used, otherwise only x_is are taken into account.

4.4   Modification of Action Plan

The action plan that one agent, A_i, received from another agent, A_j, contains a sequence of values of auxiliary variables. These values should be taken into account when A_i modifies the action plan proposed by itself. Local knowledge and fuzzy backward reasoning are once again used for this purpose. The values of the auxiliary variables will be substituted into Eq. (7a) such as

$$\{U', X'\} \circ R_1 = Y_{dj} \tag{10}$$

where, X' are determined by considering the values x_j from the received action plan and U' stands for the new action in fuzzy domain. When fuzzy backward reasoning is performed, U' can be deduced. The non-fuzzy counterpart of U' is a sequence of new actions $(u'_{i1}, u'_{i2}, \cdots, u'_{in})$. If the new actions exist, the sequence of actions will be replaced by the sequence of new actions and a revised plan forms as the following:

$$\mathbf{u}'_i = \{(u'_{i1}, \cdots, u'_{in}), (x_{j1}, \cdots, x_{jn})\} \tag{11}$$

If, for some reasons, new actions for some of x_j cannot be found, the original actions will be retained in the action plan. Suppose u'_{im} is not exist, u_{im} will be retained and the revised plan will be in the form shown below:

$$\mathbf{u}'_i = \{(u'_{i1}, \cdots, u'_{im}, u'_{i(m+1)}, \cdots, u'_{in}), (x_{j1}, \cdots, x_{im}, x_{j(m+1)}, \cdots, x_{jm})\} \tag{12}$$

5   Control of a Three-Tank System Using Collaborative Agents

5.1   Three-Tank System

A three-tank plant is illustrated in Fig. 3. Three tanks, namely Tank-I, Tank-III and Tank-II from left to right, are installed on the top of a reservoir that is used to store liquid. The three tanks have the same cross section area and are coupled with connection pipes. Liquid levels in these tanks are regulated through manipulating the inlet flowrates. There are two inlets in this plant. One goes to Tank-I and the other to Tank-II. Two pumps, P_1 and P_2, pump liquid into Tank-I and Tank-II with flowrates q_1 and q_2, respectively. Liquid in these two tanks can enter into Tank-III through connection pipes. The flowrates of the connection pipes are manipulated using ball valves V_1 and V_3. The plant has one outlet pipe located at the bottom of Tank-II. Ball valve V_2 is installed on the outlet pipe to control the outlet flowrate. It can only be manipulated manually. There are three other pipes installed at the bottom of each tank, with ball valves V_4, V_5 and V_6, respectively, to simulate disturbances. Three pressure sensors are used to measure liquid levels in the three tanks, namely, h_1, h_2 and h_3.

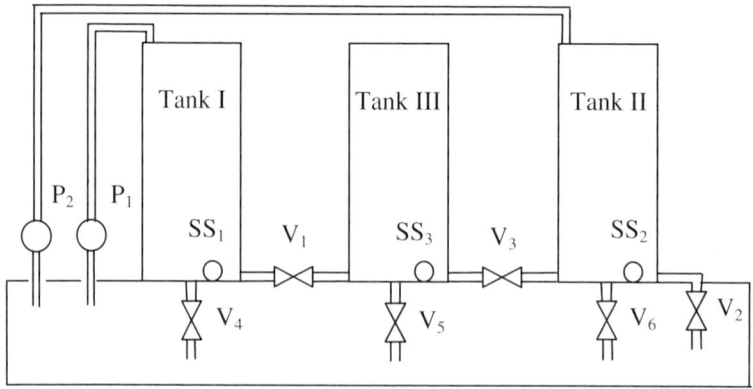

Fig. 3. A three-tank system

Global optimisation in this three-tank system is to find a sequence of optimal values of q_1 and q_2 that will drive h_1 and h_2 to their desired values h_{1d} and h_{2d}, respectively. As the tanks are coupled together, q_1 changes not only h_1 but also h_3, which influences h_2. q_2 has the similar influence. The determination of q_1 and that of q_2, therefore, tangle together. The dynamics of the three-tank plant can be described as follows:

$$(\dot{h}_1, \ \dot{h}_2) = f(h_1, \ h_2, \ h_3, \ q_1, \ q_2) \tag{13}$$

where $\dot{h}_1$ and $\dot{h}_2$ are the increments of h_1 and h_2, respectively, in the sample time T. q_1 and q_2 can be calculated for the given $\dot{h}_1$ and $\dot{h}_2$ when h_1, h_2 and h_3 are known. However, when the global knowledge is not available, the optimisation has to perform based on local knowledge.

5.2   Collaborative Agents

The global optimisation consists of two tasks. One is to decide a sequence of q_1 and the other to decide a sequence of q_2. Two agents, A_1 and A_2, are developed, each of which is committed to one of these tasks. The two agents have an identical structure as the two tasks are similar. Agent A_i ($i = 1, 2$) has four variables, namely q_i, h_i, $\dot{h}_i$, and h_3. q_i represents the action the agent will take, h_i and $\dot{h}_i$ are the state variables reflecting a degree to which a task is completed and h_3 is the auxiliary variable from which an agent can perceive the action taken by the other.

Agent A_i possesses local knowledge about the influences of an action q_i on $\dot{h}_i$ when h_i and h_3 are in certain levels. The local knowledge is also about the influence of q_i on h_3. This knowledge is imprecise and therefore is represented in the form of fuzzy relations, as shown in the following:

$$\begin{cases} R_{1i} : (Q_i, H_i, H_3) \rightarrow \dot{H}_i \\ R_{2i} : H_i \rightarrow H_3 \end{cases} \tag{14}$$

where Q_i, H_i, $\dot{H}_i$ and H_3 are the counterparts of $q_i, h_i, \dot{h}_i, h_3$ in fuzzy domain, respectively. The universes of discourse of Q_i, H_i, $\dot{H}_i$ and H_3 are partitioned into five fuzzy regions and five fuzzy sets, namely "n_big", "n_sml", "zero", "p_smal" and "p_big", are defined accordingly for each of them.

The fuzzy relation R_{11} is the relation from Q_1 to $\dot{H}_1$ for given H_1 and H_3. R_{11} contains a number of layers, each of which is corresponding to a condition where H_1 and H_3 have certain values. Four layers for the conditions of

H_1 and H_3 are "n_sml",
H_1 is "zero" and H_3 is "n_sml",
H_1 is "n_sml" and H_3 is "zero" and
H_1 and H_3 are "zero"
were used in this work and given in Eqs. (15.1) to (15.4).

$$R_{11}^{(1)} = \begin{pmatrix} 0 & 0.67 & 0.5 & 0 & 0 \\ 0 & 0.1 & 0.69 & 0 & 0 \\ 0 & 0.1 & 0.63 & 0.1 & 0 \\ 0 & 0 & 0.1 & 0.71 & 0.1 \\ 0 & 0 & 0 & 0.6 & 0.1 \end{pmatrix}_{H_1 = H_3 = "n\_sml"} \tag{15.1}$$

$$R_{11}^{(2)} = \begin{pmatrix} 0 & 0.5 & 0.69 & 0 & 0 \\ 0 & 0.1 & 0.58 & 0 & 0 \\ 0 & 0.45 & 0.56 & 0.1 & 0 \\ 0 & 0 & 0.1 & 0.5 & 0.1 \\ 0 & 0 & 0 & 0.4 & 0.4 \end{pmatrix}_{H_1 = "zero", H_3 = "n\_sml"} \tag{15.2}$$

$$R_{11}^{(3)} = \begin{pmatrix} 0 & 0.65 & 0 & 0 & 0 \\ 0 & 0.1 & 0.66 & 0 & 0 \\ 0 & 0 & 0.65 & 0.5 & 0 \\ 0 & 0 & 0.1 & 0.69 & 0 \\ 0 & 0 & 0.4 & 0.56 & 0.1 \end{pmatrix}_{H_1 = "n\_sml", H_3 = "zero"} \tag{15.3}$$

$$R_{11}^{(4)} = \begin{pmatrix} 0 & 0.61 & 0.5 & 0 & 0 \\ 0 & 0 & 0.52 & 0.41 & 0 \\ 0 & 0 & 0.62 & 0.1 & 0 \\ 0 & 0 & 0.56 & 0.67 & 0.1 \\ 0 & 0 & 0 & 0.63 & 0.57 \end{pmatrix}_{H_1 = H_3 = "zero"} \tag{15.4}$$

The fuzzy relation R_{21} is the relation from H_1 to H_3. R_{21} is shown in Eq. (16).

$$R_{21} = \begin{pmatrix} 0 & 0 & 0 & 0 & 0 \\ 0 & 0 & 0.3 & 0.8 & 0 \\ 0 & 0 & 0.7 & 0.2 & 0 \\ 0 & 0 & 0 & 0 & 0 \\ 0 & 0 & 0 & 0 & 0 \end{pmatrix} \tag{16}$$

R_{12} also contains a number of layers and the ones related to the work presented in this paper are show in Eqs. (17.1) to (17.4).

$$R_{12}^{(1)} = \begin{pmatrix} 0 & 0.69 & 0.5 & 0 & 0 \\ 0 & 0.55 & 0.68 & 0 & 0 \\ 0 & 0 & 0.66 & 0.44 & 0 \\ 0 & 0 & 0.5 & 0.78 & 0 \\ 0 & 0 & 0 & 0.71 & 0 \end{pmatrix}_{H_2 = H_3 = "n\_sml"} \tag{17.1}$$

$$R_{12}^{(2)} = \begin{pmatrix} 0 & 0.58 & 0.56 & 0 & 0 \\ 0 & 0.5 & 0.73 & 0 & 0 \\ 0 & 0 & 0.68 & 0.1 & 0 \\ 0 & 0 & 0.5 & 0.67 & 0.1 \\ 0 & 0 & 0 & 0.53 & 0.5 \end{pmatrix}_{H_2 = "zero", H_3 = "n\_sml"} \tag{17.2}$$

$$R_{12}^{(3)} = \begin{pmatrix} 0 & 0.64 & 0.5 & 0 & 0 \\ 0 & 0.55 & 0.57 & 0 & 0 \\ 0 & 0 & 0.43 & 0.44 & 0 \\ 0 & 0 & 0.5 & 0.63 & 0 \\ 0 & 0 & 0.37 & 0.1 & 0 \end{pmatrix}_{H_2 = "n\_sml", H_3 = "zero"} \tag{17.3}$$

$$R_{12}^{(4)} = \begin{pmatrix} 0 & 0.68 & 0.55 & 0 & 0 \\ 0 & 0 & 0.54 & 0.1 & 0 \\ 0 & 0 & 0.49 & 0.1 & 0 \\ 0 & 0 & 0.6 & 0.7 & 0 \\ 0 & 0 & 0.5 & 0.58 & 0.5 \end{pmatrix}_{H_2 = H_3 = "zero"} \tag{17.4}$$

The fuzzy relation R_{22} which is the relation from H_2 to H_3 is the same as R_{21}.

The agents have the capability of performing fuzzy forward reasoning and fuzzy backward reasoning and that of communicating to one another.

5.3   Negotiation

A typical situation where the two agents must collaborate is that A_1 lifts h_1 up to a new height h_{1d} ($h_{1d} > h_2$) (task 1) and meanwhile A_2 keeps h_2 at its original height (task 2). In this situation, when A_1 takes actions to fulfil task 1, h_3 increases as well. This will result the rise of h_2. To prevent h_2 from being affected, A_2 has to act. In turn, A_2's action may influence h_1 in the similar way. Therefore, the collaboration between A_1 and A_2 is required.

Simulation started from an initial condition where h_1 and h_2 are at their steady-states, i.e. $h_1 = h_2 = 200$ (mm) and a new height for h_1, $h_{1d} = 300$ (mm), was set.

A_1 generating action plan $\mathbf{u}_1$

A_1 created its action plan $\mathbf{u}_1$ based on fuzzy relations shown in Eqs. (14) and (15). To generate a control action q_1, the following fuzzy relational equation was adopted:

$$(Q_1, H_1, H_3) \circ R_{11} = \dot{H}_1 \tag{18}$$

For the first control action, $R_{11}^{(0)}$ was employed because both H_1 and H_3 are "n_sml" (as $h_1 = h_3 = 200$ (mm)). $\dot{H}_1 = (0\ \ 0\ \ 0\ \ 0\ \ 1)$ due to $h_{1d} = 100$ (mm). Equation (18) was then rewritten in the following:

$$Q_1 \circ \begin{pmatrix} 0 & 0.67 & 0.5 & 0 & 0 \\ 0 & 0.1 & 0.69 & 0 & 0 \\ 0 & 0.1 & 0.63 & 0.1 & 0 \\ 0 & 0 & 0.1 & 0.71 & 0.1 \\ 0 & 0 & 0 & 0.6 & 0.1 \end{pmatrix} = (0\ \ 0\ \ 0\ \ 0\ \ 1)$$

To make this fuzzy relational equation solvable, $\dot{H}_1$ has to be amended to $\dot{H}_1 = (0\ \ 0\ \ 0\ \ 0\ \ 0.1)$ (Pedrycz, 1990; Li, 1999). Fuzzy backward reasoning was performed by solving this equation with the method introduced in Section 3. This gave a solution, $Q_1 = (0\ \ 0\ \ 0\ \ 0\ \ 0.1)$. Defuzzifying this solution yielded $q_1 = 15$.

After having obtained the first control action, A_1 predicted the influence of this control action on h_3 which is the other part of an action plan. A_1 firstly calculated the actual value of $\dot{h}_1$ caused by q_1. This was done by performing fuzzy forward reasoning such that

$$(0\ \ 0\ \ 0\ \ 0\ \ 0.1) \circ \begin{pmatrix} 0 & 0.67 & 0.5 & 0 & 0 \\ 0 & 0.1 & 0.69 & 0 & 0 \\ 0 & 0.1 & 0.63 & 0.1 & 0 \\ 0 & 0 & 0.1 & 0.71 & 0.1 \\ 0 & 0 & 0 & 0.6 & 0.1 \end{pmatrix} = \dot{H}_1$$

It then computed h_3 by performing fuzzy forward reasoning based on R_{21}. That is

$$H_1 \circ \begin{pmatrix} 0 & 0 & 0 & 0 & 0 \\ 0 & 0 & 0.3 & 0.8 & 0 \\ 0 & 0 & 0.7 & 0.2 & 0 \\ 0 & 0 & 0 & 0 & 0 \\ 0 & 0 & 0 & 0 & 0 \end{pmatrix} = H_3$$

This yielded $h_3 = 210$.

The entire action plan $\mathbf{u}_1$ was worked out in this way and is shown below:

$$\mathbf{u}_1 = \{(15.0 \quad 15.0 \quad 15.0 \quad 5.4 \quad 4.5 \quad 3.7 \quad 2.9 \quad 2.1 \quad 1.3 \quad 0.3),$$
$$(210 \quad 227 \quad 245 \quad 260 \quad 274 \quad 284 \quad 292 \quad 297 \quad 299 \quad 299))\}$$

with a profit $p = 1.00$. A_1 took the first action immediately. As this affected h_2, A_2 joined the collaboration by starting negotiation with A_1. The negotiation process evolves.

A_2 evaluating action plan $\mathbf{u}_1$

A_2 evaluated $\mathbf{u}_1$ by performing fuzzy forward reasoning such that

$$(Q_2, H_2, H_3) \circ R_{12} = \dot{H}_2 \tag{19}$$

where Q_2 is "n_sml", H_2 is initially "n_sml" due to $h_2 = 200$ (mm) and H_3 comes from the second part of $\mathbf{u}_1$. H_2 was updated when new $\dot{H}_2$ was deduced through the fuzzy forward reasoning shown in Eq. (19). H_2 would eventually have reached 250 (mm) if $\mathbf{u}_1$ had been accepted and $p_2 = 0.8$.

A_2 generating action plan $\mathbf{u}_2$

A_2 generated its own action plan $\mathbf{u}_2$ to improve its profit. $R_{12}^{(1)}$ was employed to create such a control action q_2 that can keep $\dot{h}_2 = 0$ (the corresponding $\dot{H}_2$ is "zero"). This gave the following fuzzy relational equation:

$$Q_2 \circ \begin{pmatrix} 0 & 0.69 & 0.5 & 0 & 0 \\ 0 & 0.55 & 0.68 & 0 & 0 \\ 0 & 0 & 0.66 & 0.44 & 0 \\ 0 & 0 & 0.5 & 0.78 & 0 \\ 0 & 0 & 0 & 0.71 & 0 \end{pmatrix} = (0 \quad 0 \quad 1 \quad 0 \quad 0)$$

Performing fuzzy backward reasoning over this equation yielded $q_2 = 5.7$.

A_2 also predict the influence of the control actions on h_3 in the same way used by A_1. The action plan $\mathbf{u}_2$ is shown in below:

$$\mathbf{u}_2 = \{(5.7 \quad 5.7 \quad 5.7 \quad 5.7 \quad 5.7 \quad 5.7 \quad 5.7 \quad 5.7 \quad 5.7 \quad 5.7),$$
$$(200 \quad 200 \quad 200 \quad 200 \quad 200 \quad 200 \quad 200 \quad 200 \quad 200 \quad 200))\}$$

The profit increased to $p_2 = 1.00$.

A_1 modifying action plan $\mathbf{u}_1$

A_1 modified $\mathbf{u}_1$ through constrained optimisation in which the increase of h_3 was limited. 250 (mm) was chosen as the upper bound of h_3. The fuzzy counterpart, H_3, was accordingly somewhat between "n_sml" and "zero". All $R_{11}^{(i)}$ ($i = 1, \cdots, 4$) was adopted in fuzzy backward reasoning to deduce control actions q_1. R_{21} was used in fuzzy forward reasoning for predicting the influence in q_1. The revised action plan $\mathbf{u}_1'$ is shown in below:

$\mathbf{u}_1' = \{(15.0 \quad 9.0 \quad 5.1 \quad 4.0 \quad 2.9 \quad 2.5 \quad 2.1 \quad 1.3 \quad 0.3 \quad 0.3),$

$\qquad (210 \quad 227 \quad 245 \quad 250 \quad 250 \quad 250 \quad 250 \quad 250 \quad 250 \quad 250)\}$

The profit $p_1' = 1.00$.

Negotiation terminating

A2 evaluated $\mathbf{u}_1'$ and calculated the corresponding profit. For $\mathbf{u}_1'$, $p_2' = 0.93$. As $p_1' + p_2' > p_1 + p_2$, the negotiation terminated.

A$_2$ generated $\mathbf{u}_2'$ under the constraints of the second part of $\mathbf{u}_1'$ which A$_2$ accepted. The new action plan $\mathbf{u}_2'$ is

$\mathbf{u}_2' = \{(3.5 \quad 3.5 \quad 3 \quad 3 \quad 3 \quad 3 \quad 3 \quad 3 \quad 3 \quad 3),$

$\qquad (210 \quad 227 \quad 245 \quad 250 \quad 250 \quad 250 \quad 250 \quad 250 \quad 250 \quad 250)\}$

As the second part of $\mathbf{u}_1'$ is the same as that of $\mathbf{u}_2'$, a plan $\mathbf{U}$ was established for both A$_1$ and A$_2$ as shown in below:

$\mathbf{U} = \{(15.0 \quad 9.0 \quad 5.1 \quad 4.0 \quad 2.9 \quad 2.5 \quad 2.1 \quad 1.3 \quad 0.3 \quad 0.3),$

$\qquad (3.5 \quad 3.5 \quad 3 \quad 3 \quad 3 \quad 3 \quad 3 \quad 3 \quad 3 \quad 3),$

$\qquad (210 \quad 227 \quad 245 \quad 250 \quad 250 \quad 250 \quad 250 \quad 250 \quad 250 \quad 250)\}$

A$_1$ and A$_2$ performed task 1 and task 2 according to this plan.

5.4 Discussion

The action plan $\mathbf{u}_1$ is able to raise h_1 to its new desired height $h_{d1} = 300$ (mm). However, when it was created, its influence to h_2 was not taken into account. As the increase of h_1 can make h_2 deviate from its desired height, A$_2$ had to make its action plan $\mathbf{u}_2$. When A$_1$ modified $\mathbf{u}_1$, h_3 was assumed not exceed 250 (mm). Though h_1 would take longer time to reach h_{1d} according to the revised action plan, h_2 was less affected as the actual value of h_3 was below 250 (mm).

6 Conclusion

A collaborative fuzzy agent was proposed in this paper for the purpose of performing global optimisation with incomplete and ambiguous knowledge. The agent has local knowledge which is a fuzzy model describing the behaviour of a sub-system. It generates and modifies action plans through, in particular, fuzzy backward reasoning. A profit function was presented as a criterion for promoting the negotiation between two agents. A multi-agent system developed for a three-tank system illustrated the effectiveness of the proposed approach.

References

1. Bellman, R. E. and Zadeh, L. A.: Decision-making in a fuzzy environment, Management Science, Vol. 17, No. 4. (1970) 141-165
2. Dospisil, J.: Constraint agent can negotiate, Proceedings of International Conference on Trend in Computations, Euro Con's 2001, Vol. 1. (2001) 140-144
3. Li, D.: Fuzzy Control Using Backward Reasoning, PhD Thesis, University of Wales Cardiff, School of Engineering, Cardiff, UK, 1999.
4. Moree, B. J., Bos, A., Tonino, H. and Witteveen, C.: Cooperation by iterated plan revision, Proceedings on the 4th internation Conference on Multiagent Systems, (2000)
5. Pedrycz, W.: Inverse problem in fuzzy relational equations, Fuzzy Sets and Systems, North-Holland, Vol. 36. (1990) 277-291
6. Pham, D. T. and Li, D.: Fuzzy control of a three-tank system, Journal of Mechanical Engineering Science (Part C). Vol. 215, No C5. (2001)
7. Wangermann, J. P. and Stengel, R. F.: Principled negotiation between intelligent agents: a model for air traffic management, Artificial Intelligence in Engineering, Vol. 12. (1998) 177-187
8. Wang, P. Z.. Fuzzy Sets Theory and Its Applications, Shanghai Scientific and Technologic Publishers, Shanghai (1983)
9. Zimmermann, H. J.: Fuzzy Sets Theory and its Applications (1st Edition), Kluwer Academic Publishers (1985)

A Self-Organizational Management Network Based on Adaptive Resonance Theory[1]

Ping Jiang and Quentin Mair

Department of Computing, Glasgow Caledonian University
Glasgow G4 0BA, UK
{p.jiang, q.mair}@gcal.ac.uk

Abstract. This paper presents an organizational network for product configuration management within the context of Virtual Enterprise. Actors, from high level strategy making actors to low level physical devices, can advertise their own skill and knowledge and seek for partners to form dynamic alliances in a community. The network is organized based on Adaptive Resonance Theory(ART) which was originally used for unsupervised neural network learning and which allows the organization and cooperation of such product development alliances to be more flexible and adaptable. Some characteristics, which are inherent in real enterprises or society, such as self-organization, unsupervised learning, competition between actors are exhibited in the ART-based organization network and are the keys for evolution and development of enterprises.

1 Introduction

During long product life-cycles, there are huge numbers of actors involved in product design, development, and deployment. They can be distributed at different geographical sites and different information is shared between them. If such a system is designed from a global viewpoint it becomes inflexible, unchangeable and is too difficult to manage. However, take a look at each actor, such as a programmer, a manager, etc., who is involved in the product life cycle. There are not usually too many direct connections to control and usually a very explicit and clear task can be assigned to each actor. An actor-oriented distributed management structure may give a management system more flexibility, changeability, and interoperability, where the actor is any person or device involved in product management, manufacturing and customer service[1]. The UML analysis from an actor's perspective can directly guide people to realize their own cooperation agent that is capable of advertising their knowledge for both seeking for and cooperating with partners. A multi-agent system can then be constructed to realize collaborative work between actors during product life-cycles.

---

[1] This work is supported by DIECoM (Distributed Integrated Environment for Configuration Management), an IST project funded under the EC's Framework V programme (http://www.diecom.org)

R. Kowalczyk et al. (Eds.): Agent Technology Workshops 2002, LNAI 2592, pp. 211–225, 2003.
© Springer-Verlag Berlin Heidelberg 2003

A key technology of actor-oriented product management is service discovery from a group of actors, where the capability of each actor can be described, advertised, and discovered by other actors who are seeking for partners for a given task. It has a similarity to the service matchmaking among agents for information retrieval on the World Wide Web. However, for information retrieval, current research mainly focuses on passive matching of context and profile by the advertisement[2][3] with less focus on active learning and adaptive ability. Paper [4] attempts to endow WWW information retrieval with learning ability by using BP neural networks that can capture knowledge about users' interests and preferences; although the training of the networks might be quite slow and cumbersome. At the same time other drawbacks such as local optimum, missing semantic relation by hashing encode etc., exist. This kind of active learning and adaptive ability may be more important in product management in Virtual Enterprises. Within the context of a Virtual Enterprise, cross-organizational PCM (Product Configuration Management) faces dynamically changing environments and dynamically changing roles within organization. An actor with more adaptability and flexibility will be more powerful and useful in a competitive society. This requires that actors should have dynamic reorganization ability. The concept of dynamic reorganization allows agents to reconfigure and/or restructure their system in response to environmental or system changes.

This paper proposes an organizational network for dynamic partnership in a dynamic environment with adaptive ability, learning ability, competitive ability. It is based on Fuzzy Adaptive Resonance Theory (Fuzzy ART) that is a neural network model and is proposed by Carpenter, Grossberg, and Rosen[5] for clustering binary or analog data. The ART network is a self-organizational network and is based on a "winner-takes-all" competitive principle. It has unsupervised learning ability and adaptive ability for data clustering. In this paper, each actor involved in product configuration management is considered as a neuron in an ART network that is the proposed organizational network of product configuration management. On presentation of a task (input vector) from a contract provider, who is seeking for an appropriate partner for the task, the actor whose advertisement of its ability (connection weight) is closest to the input vector will become candidate of the contract, and will be allowed to learn the demands of the contract provider, i.e. modify its ability description (called connection weight in ART). After repeatedly advertising the task in the product management community, the network will adapt to the demand of the contract provider and store a prototypical element of each demand in the connection weights. Then, an actor with previous experience in a specific area will have more chance for a task in that area and will have more expertise to do it. In multi-agent system, agents should have information about their environment. Actors should talk each other based on an ontology of a community, where OIL(DAML-OIL)[6] can be used to represent the knowledge and information of each actor. In order to use numeric representation instead of symbolic representation of knowledge in ART networks, a feature vector of a local ontology is defined by using semantic distance of concepts and fuzzy inference is introduced for analysis of similarity of concepts. Simulations are carried out to verify features of the proposed management network.

2 Actor-Oriented Product Management Systems

In the life-cycle of product management, each actor goes through a sequence of part-nership creation, configuration, operation, dissolution, repeatedly. At the first stage, an actor may initiate a product development partnership and may advertise requirements of the task to a community. For instance, "I need a software engineer for a driver program coding" and "I need another hardware engineer for communication board design who has experience in IEEE 802.11b". The actor is called an "initiator" of the product development partnership. At the same time, actors as participators are adver-tising their ability in the community for getting the task assigned by the initiator. After finding a candidate for the desired task, the initiator will create a partnership with it. Both sides can modify the advertised ontology to adapt specific tasks through negotia-tion. Then the participator who gets the contract may decompose the task into subtasks and become a new initiator within its local community.

Therefore, each actor plays two roles in a system, on one side, it is an initiator of a partnership in a multi-agent community who seeks for qualified candidate to complete a given task, and on the other side, it is a participator in another multi-agent commu-nity who advertises its ability and seeks for tasks from other actors. Suppose an actor a_j belongs to two communities, as a task initiator in $C_I = \{a_{I1} \quad a_{I2} \quad ... \quad a_{IL}\}$ and as a task participator in $C_P = \{a_{P1} \quad a_{P2} \quad ... \quad a_{PM}\}$, respectively. An actor is capable of connecting two communities together and decomposes a given task $t_{input}(C_p)$ from C_p into a series of subtasks $t_{output}(C_I) = [t_{o1} \quad t_{o2} \quad \cdots]$ in C_I and local tasks t_{local} as

$$\{t_{output}, t_{local}\} = S_j(t_{input}) \tag{1}$$

where the task t_{input} is assigned by actors in C_p, i.e., $t_{input}=t_o(a_{pi})$, $i=1...M$, and t_{output} is the decomposed subtasks of the actors in C_I. The t_{local} is the tasks completed locally.

Within a community C, in order to facilitate cooperation and understanding amongst agents, there is an ontological definition Ω_C that defines the semantics of commonly used concepts and terminologies. The ontology can be defined using DAML-OIL(http://www.daml.org/), Suppose that the ontology in a community C can be defined as:

$$\Omega_C \equiv R(e_1, e_2, ..., e_N) \tag{2}$$

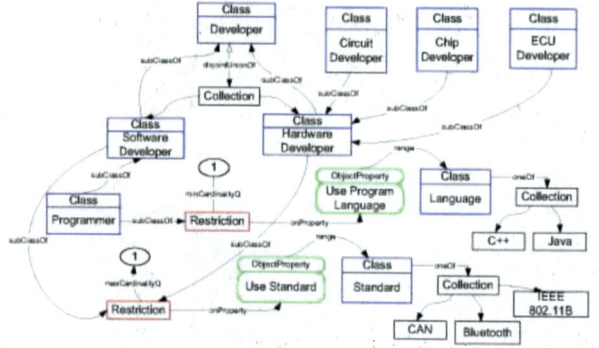

Fig. 1. Ontology of a community for electric device development

where $e_1...e_N$ are entities (concepts, terminologies, properties) related to the community and $R(\bullet)$ is the relationship between the entities.

An example of an ontology defined using DAML-OIL in the community of electric device development is shown in Fig. 1, where the entities of the society are {Developer, Software Developer, Hardware Developer, Programmer, Language, Standard, C++, Java, CAN, Bluetooth, IEEE 802.11b, ECU developer, Chip Developer, Circuit Developer} and reflect the content of the community.

As a contract provider, an initiator a_{Ii} can then advertise its demand for a subtask by using entities defined in the Ω_C. At the same time, the participators a_{Pj}, $j=1...M$, in the community expect a contract and advertise their capability for automatically creating alliances. The content of the advertisements of both initiator and participator is a semantic knowledge description or an instance of the local ontology based on the definition of Ω_C in a community C. It can also be written In OIL(DAML-OIL) and is called a local ontology.

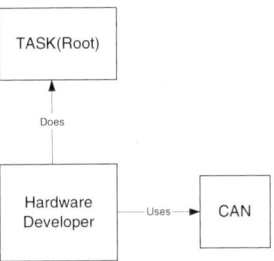

Fig. 2. Advertisement of a participator

An example of a hardware developer as a participator can announce himself as "An ECU developer using CAN for a given task" as shown in Fig.2, who is seeking for suitable task from the community of electric device development. In the same way, an initiator can announce a task to seek for a partner who can do it, for instance, who should be a programmer with knowledge and experience of Bluetooth and C++ as shown in Fig.3.

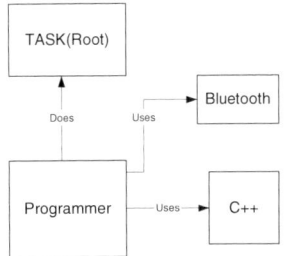

Fig. 3. Advertisement of an initiator

Note that, in the advertisements of both initiator and participator, there is a root object. For an initiator, the root corresponds to a subtask that an initiator can provide and its local ontology describes what kind of subtask it is. For a participator, the root corresponds to a task that an actor can do and the local ontology describes participator's ability to do the task. All the objects inside an actor and the relationship amongst objects should be defined around the root for its demand or capability advertisement.

For each actor, we define a feature vector for any task advertisement t as a numeric representation of the semantics of its local ontology:

$$V(t)=[s_1,s_2, ..., s_N]^T \qquad (3)$$

The component s_i of $V(t)$ has a one-to-one correspondence to the entity e_i of the ontology. The $s_i \in [0,1]$ is the semantic closeness between e_i and the root(task) in an OIL based advertisement. It is an inverse of semantic distance. One possible form of this inverse can be

$$s_i = \begin{cases} e^{-\alpha DIS(e_i,root)} & \text{if } e_i \text{ is appeared in the local advertisement} \\ 0 & \text{if } e_i \text{ is not appeared in the local advertisement} \end{cases} \qquad (4)$$

where the $\text{Dis}(e_i, root)$ is a semantic distance between entity e_i and the *root* calculated in the local ontology, e.g., in Fig.2 for a participator and in Fig.3 for an initiator. The α is a steepness measure[7], in fuzzy system, which is often selected to be $-7/\text{MAX(Dis)}$ because $e^{-7} \approx 0$ when $\text{Dis}(e_i, root)$ achieves its maximum.

In order to deal with automation of knowledge extraction, semantic distance or similarity between concepts has been researched in recent years, such as semantic web matchmaking[2][8] and conceptual clustering of database schema[9]. In [9], a general semantic distance is defined mathematically as an application of $E{\times}E$ into R^+, where E is a set of objects in a community ontology Ω_C, with the following properties:

i. $\forall x \in E, \forall y \in E,$ $\text{Dis}(x, y) = 0 \Leftrightarrow x = y$

ii. $\forall x \in E, \forall y \in E,$ $\text{Dis}(x, y) = \text{Dis}(y, x)$

iii. $\forall x \in E, \forall y \in E, \forall z \in E$ $\text{Dis}(x, y) \leq \text{Dis}(x, z) + \text{Dis}(z, y)$

The semantic distance can be use to characterize similarities or dissimilarities between two objects. Usually, a distance between two objects is the shortest path between them. The path description greatly depends on the view point of observation. Different types of semantic distances are proposed in [9]: visual distance, hierarchical distance and cohesive distance, etc.. For instance, a visual distance is very close to the graphical representation of a model. It is defined from a view-point that two objects semantically linked by a relationship (or a generation) are very often graphically close. In fact, anyone can define his semantic distance from a viewpoint which is of relevance to his problem. For instance, in a DAML-OIL ontology, two classes that are disjointWith should be given a large semantic distance but two that are sameClassAs should have a zero semantic distance.

Therefore, on the basis of definition of semantic distance, the feature vector can describe the requirements of an initiator and the ability of a participator from a designer's view. It gives the information both about how many entities are related to a task and about what kind of relationship is between the involved entities. Partner seeking becomes a matchmaking process between advertised feature vectors of initiators and participators in a community. An ART network can be constructed for automation of this matchmaking which considers adaptive capability, learning ability, and competitive properties.

3   Self-Organizational Management Networks

In the aforementioned actor-oriented product management system, how to organize the actors to form a partnership becomes a key for product management. In this section, a self-organizational ART network is proposed for adaptive and dynamic partnership seeking. Suppose that an actor a_i is a partnership initiator in a community C_I, who has decomposed a task t_{input} into a series of subtasks $t_{output}(C_I) = \begin{bmatrix} t_{o1} & t_{o2} & \cdots & t_{oK} \end{bmatrix}$ and is seeking for partnership in a community C_I. As a contract provider, it asks the community whether any actor has the desired capability and interest for the given tasks. The desired capability is described by the feature vectors $X_d(t_{oj})$, $j=1...K$, as defined in equation (3):

$$X_d(t_{oj})=[x_{d1}, x_{d2},...,x_{dN}]^T, \quad x_{di}\in[0,1], j=1...K, \quad i=1...N \qquad (5)$$

The vector $X_d(t_{oj})$ specifies what kind of abilities and contents are required for doing the subtask t_{oj}, where x_{di} is between 0 and 1; where a bigger x_{di} means the ability of e_i is more important for the subtask. Then, the series of feature vectors to describe the desired subtasks t_{oj}, $j=1...K$, are posted in the community to seek for qualified candidates.

At the same time, the participators a_j, $j=1...M$, in the community C_l are advertising their knowledge and capability to seek for a suitable task from initiators. The representation of the ability of each participator can also be written in OIL(DAML-OIL) and the corresponding feature vector can be calculated based on the semantic distance between entities and the root (the ability of a participator):

$$W(a_j)=[w_{j1}, w_{j2},...,w_{jN}]^T, \quad w_{ji}\in[0,1], j=1...M, \quad i=1...N \qquad (6)$$

This feature vector can be explained by the fuzzy cognitive map proposed in [10], which describes the relationship between concepts using connection weights. The w_{ji} in equation (6) describes a connection weight between the concept "ability of the participator a_j" and the concepts of e_i, $i=1...N$, defined in Ω_C. A higher weight indicates a stronger ability in the

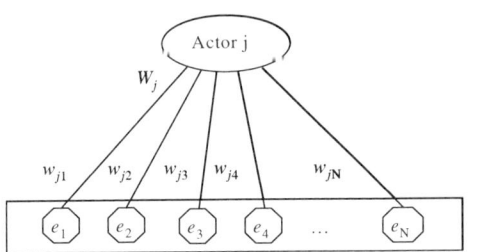

Fig. 4. Advertising ability of an actor

area related to the concept e_i. Then, the advertisement of each actor, equation (6), can be depicted as a fuzzy network shown in Fig.4. The weight or feature vector of an actor connects the "ability of Actor j" with the concepts of $e_1...e_N$ in the community. For instance, in the community of Fig. 1, a weight vector of {0.3, 0.5, 0.2, 0.6, 0.5, 0.2, 0.9, 0.8, 0.1, 0.4, 0.3, 0.1, 0.1, 0.1} implies that an actor is excellent at the coding using C++ or Java but relatively poor at hardware work.

Then, a 3-layers ART network can be constructed for adaptive and dynamic partnership seeking as shown in Fig.5. The layers F0, F1, and F2 are the input layer, comparison layer and recognition layer, respectively. The input layer F0 gets the demand $X_d(t_{oj})$ one by one from an initiator. Each participator advertises its capability in the comparison layer F1 for competence comparison. The nodes in F0 and F1 are composed of the entities of the ontology which is the same as the bottom layer of Fig.4. The corresponding nodes of layer F0 and F1 are connected together via one-to-one, non-modifiable links. The nodes in recognition layer F2 are the actors (participators) of the community C who are candidates for the given tasks. Altogether, there are M nodes corresponding to M participators in the community C, each one represents a characteristic competence of an individual actor and will take part in competition for the posted tasks from the initiator. There are top-down and bottom-up weight connections W_j between the nodes in the layer F1 and the nodes in the layer F2. Initially, the weight W_j is the advertised fuzzy weight vector of a participator a_j in Fig.4. During the

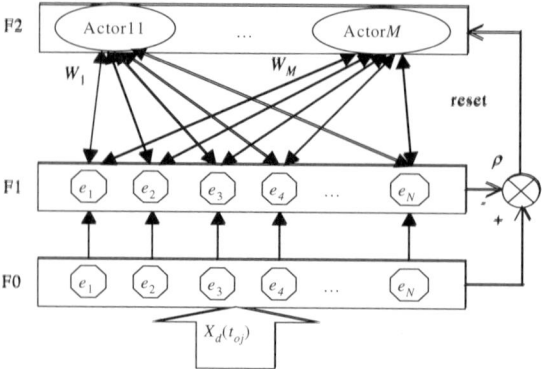

Fig. 5. ART network for self-organizational management

process of that the desired tasks $X_d(t_{oj})$, $j=1…K$, are posted by initiators and assigned to the participators iteratively, the weights will be adjusted based on a competitive principle such that the most competent candidate for a given task are selected as the winner. This competitive process is then conducted by examining the degree of match between the layer F0 and the Layer F1 of the winning candidate. If the degree of match between the demand of a task and the advertised capability is higher than a vigilance level ρ, a partner which fits is found and the task can be assigned to it for execution. Depending on performances of the execution, its weights can then be adjusted towards or backwards from the requirements for conducting the task. Therefore, it is a learning process based on the experiences of an actor. More experienced actors will be given more chance for similar jobs. After repetitive learning, the self-organizational ART network becomes a view of an initiator about the community i.e. which actor is competent for what kind of task.

The partner seeking process can be shown as below:

1) For each desired subtask t_{oj} from an initiator, the initiator posts the requirement vector $X_d(t_{oj})$ to the input layer of the ART network.

2) The comparison layer F1 attempts to classify it into one of participators based on its similarity to the advertised capability of each participator. A choice input of layer F2 is calculated for each participator a_j by bottom-up weight W_j:

$$T_j = \frac{|X_d \wedge W_j|}{\alpha + |W_j|}$$ (7)

where α is a positive real-valued number called the choice parameter, $X_d \wedge W_j$ is a vector that its ith component is equal to the minimum of X_{di} and W_{ji} and $|\bullet|$ is the norm of an vector, which is defined to be the sum of its components.

3) The actor with maximum choice input in the layer F2 will be selected as candidate for the task $X_d(t_{oj})$ (winner-takes-all competition):

$$T_J = \max\{T_i| i=1,…,M\}$$ (8)

4) The ability of the winner is sent back to the layer F1 by its up-down weights. The examination of similarity between the winner's ability and the required ability is conducted by a vigilance criterion of:

$$\frac{|X_d \wedge W_l|}{|X_d|} \geq \rho \qquad (9)$$

where $\rho \in [0,1]$ is the vigilance parameter given by the initiator.

There are the following possibilities after examination:

(1) If the criterion (9) is satisfied, the participator a_l may be competent enough for the task from the judgment of the initiator. Then go to step 5).

(2) Otherwise, the candidate a_l is reset and the next maximum participator is selected as a new candidate for the task. Then, repeat step 4 for vigilance examination.

(3) If no participators in the community C can pass the vigilance threshold, no competent candidate exists and a new actor should be recruited into the community, such as to recruit a new employee, to found a new department for the given task.

5) The initiator sends a contract to the winner a_l by the contract-net protocol[11], receives the bid and evaluates the bid from a_l. If the details can meet the requirement, awards the contract to a_l, otherwise resets a_l and selects the next maximum participator as candidate and go back to step 4).

6) Go back to step 1) for partner seeking of the next subtask $X_d(t_{oj})$, $j=1...M$, until all subtasks are assigned.

7) When any participator a_i completes its task and sends back results to the initiator, initiator evaluates the quality of the completed task in the interval, $\beta \in [-1,1]$, from "-1" very poor to "+1" perfect. Then, the advertised weight W_i will be modified by

$$W_i^{new} = (1-\eta)W_i^{old} + \eta\beta(X_d \wedge W_i^{old}) \qquad (10)$$

where $\eta \in [0,1]$ is the learning rate of fuzzy ART. A bigger η corresponds to a faster learning.

This is a learning process with self-organizational ability for multi-agent systems. It goes through a process from matchmaking to quality feedback. More qualified participators are given more chance to get a contract. Based on the quality of the execution, equation (10) updates the advertised ability of a participator. Better performance (bigger $\beta > 0$) will make the weight closer to what it has done and give this participator more chance to be a winner for the forthcoming similar jobs. On other hand, poorer performance (smaller $\beta < 0$) will result from the weight being far away from the given task described by X_d and may lose similar task assignment in the next round of competition. The learning rate is selected based on a tradeoff between new ability learning and previous experience forgetting. With a bigger η, the actor is more likely to do the job that it just did well. Usually, a small company or a person is likely in this case. With a smaller η, the change of the actor's expertise is slower, for instance, the capability centre of a bigger company is usually changed little but the company can be competent in a wider field around the centre, which can be implemented by setting a lower vigilance ρ. If no actor is competent at a task, a new actor called uncommitted node in ART network should be generated in step 4) and becomes the winner auto-matically. Therefore, the aforementioned organization of a community can begin from inception without a priori knowledge and grows up from its experience step-by-step. In addition, through cooperation between initiator and participator, their descriptions about the demand and actor ability, using the concepts defined in the community, will become closer although their understanding about the concepts of ontology might be different initially. Equation (10) always lets the advertisement of the participator adapt to the requirement of an initiator who can provide tasks to participators. This is a learning process such that the gap between participator and initiator becomes smaller and smaller and each participator expects to win the next round competition.

4 Semantic Relation of a Community

In the last section, a full space of ontology clustering for partnership seeking is pro-posed. The task assignment depends on the similarity between the demands of an initiator and the advertisement of a participator described in the space of ontology entities. Usually, this is a high-dimensional space. For high-dimensional clustering, most algorithms can not work efficiently because of the sparsity of data[12]. In a community, the subtasks decomposed by initiators have a degree of randomness and cannot be predicted exactly. Therefore, the participators can only advertise their abil-ity by a general means and from their own understanding about terminologies in the ontology. It is impossible to require both initiator and participator to use the same or similar description for their advertisements. Usually, for the advertised vectors, there might be a few dimensions on which the points are far from one another even though the essence is very close. This is because the components of the $(e_1, e_2, ..., e_N)$ space are highly correlated by the semantic relationship but the similarity comparison con-ducted in the last section thinks them irrelevant. Take an example, someone advertis-ing himself that he can do the work of "software development" but at the same time having not exactly said that he can program using "C++", might fail to get a work of "C++ programming" by the scheme of the last section. In fact, "software" and "C++" have a tight semantic relation. In this section, a fuzzy inference scheme is proposed considering semantic relationship inside the ontology. Then, the self-organizational network proposed in the last section can be used for partnership seeking in the sense of fuzzy matchmaking.

The concepts $(e_1, e_2, ..., e_N)$ used by an initiator to announce their demand is not nonfuzzy; any concept implies some aspects of other concepts, which can be defined by a grade of membership proposed by Zadeh in his fuzzy set theory[13]. The ontol-ogy Ω_C of a community C defines the semantic relation between concepts $(e_1, e_2, ..., e_N)$ and can be used to determine the grade of membership. Now, the prob-lem becomes a fuzzy inference from the initiator side to participator side such that a fuzzy matchmaking can be conducted on the layer F1 of Fig.5 and participators with relevant ability can be considered during competition. A fuzzy inference block can be added between layer F0 and layer F1 instead of direct connection in Fig.5. The conclusion of the fuzzy inference will replace X_d for similarity examination in equations (7) and (9).

Suppose the set of $\{e_1, e_2, ..., e_N\}$ forms a *universe of discourse* in the community. Any announcement t_{oj} of an initiator is a linguistic variable, such as "design *driver program* using *Java*". Then, the corresponding feature vector $X_d(t_{oj})=[x_{d1}, x_{d2},...,x_{dN}]^T$ in (5) is a fuzzy variable representing the subtask t_{oj} on the initiator side, where x_{di}, $i=1...N$, is a grade of membership corresponding to the ith variable in the universe. Then, fuzzy demand $X_d(t_{oj})$ on the initiator side should be transferred to a fuzzy variable $X(t_{oj})$ on the participator side, which considers fuzzy relationship between concepts$(e_1...e_N)$:

$$X_d(t_{oj}) \wedge (I \Rightarrow P) \Rightarrow X(t_{oj}) \tag{11}$$

where $(I \Rightarrow P) \in R^{N \times N}$ is a fuzzy relation between the initiator side and participator side. It reflects the relationship between entities such that similar concepts can be considered during matchmaking even though they are not explicitly declared in the demands of the initiator.

If the relation $R=(I \Rightarrow P)$ is known, for any given fuzzy variable $X_d(t_{oj})$, it is straightforward to get the fuzzy variable $X(t_{oj})$ based on fuzzy inference. The relation R that reflects correlation between concepts can be obtained based on an OIL description of the global ontology Ω_c, e.g., of Fig. 1, because both fuzzy variables of the initiator and participator are defined by the elements of the global ontology Ω_c or by the set of the universe $\{e_1, e_2, ..., e_N\}$.

In the ontology Ω_c, distances between each pair of concepts (e_i, e_j) can be calculated as stated in the section 2, such as by using *visual distance*[9].Thus, a concept distance matrix can be generated as

$$D = \begin{bmatrix} 0 & d(1,2) & \cdots & d(1,N) \\ d(2,1) & 0 & \cdots & d(2,N) \\ & & \cdots & \\ d(N,1) & d(N,2) & \cdots & 0 \end{bmatrix} \tag{12}$$

where each component of $d(i,j)$ is a semantic distance between concept e_i and concept e_j.

Then, the relation $(I \Rightarrow P)$ can be calculated accordingly based on the distance matrix:

$$(I \Rightarrow P) = \begin{bmatrix} 1 & r(1,2) & \cdots & r(1,N) \\ r(2,1) & 1 & \cdots & r(2,N) \\ & & \cdots & \\ r(N,1) & r(N,2) & \cdots & r(N,N) \end{bmatrix} \tag{13}$$

where $r(i,j)=e^{-\alpha d(i,j)}$ and α is a steepness measure.

Equation (13) reflects the relationship between concepts that are defined in the global ontology of Ω_c.

Then, for any linguistic demand t_{oj}, the fuzzy inference can be conducted based on equation (11):

$$X(t_{oj}) = X_d(t_{oj}) \vee . \wedge (I \Rightarrow P)$$

$$= \begin{bmatrix} x_{d1} & x_{d2} & \cdots & x_{dN} \end{bmatrix} \vee . \wedge \begin{bmatrix} 1 & r(1,2) & \cdots & r(1,N) \\ r(2,1) & 1 & \cdots & r(2,N) \\ & & \cdots & \\ r(N,1) & r(N,2) & \cdots & 1 \end{bmatrix} \quad (14)$$

$$= \begin{bmatrix} x_1 & x_2 & \cdots & x_N \end{bmatrix}$$

where $\vee . \wedge$ is an inner product of fuzzy relation, such as max-min composition in [13]:

$$x_i = Max(min(x_{d1}, r(1,i)), min(x_{d2}, r(2,i)), \cdots, min(x_{dN}, r(N,i))) \quad (15)$$

In the fuzzy inference (14), the input is a fuzzy set representing the demand of an initiator, the output X is also a fuzzy set reflecting which concepts might be required by the task where the semantic correlation between concepts has been considered.

Using the fuzzy set $X(t_{oj})$ instead of $X_d(t_{oj})$ in equation (7), (9), and (10), the ART network proposed in section 3 can realize fuzzy matchmaking during a process of self-organization.

5 Simulation Results

The proposed management network can give each contract initiator a view of actors participating in the contract competition and it can evolve gradually based on the experience of partnership execution. On the basis of this network, the community can be organized in a self-organizational and adaptive way. In order to verify the proposed scheme, the example shown in Fig. 1 is taken as background to the simulation, which is a community for electrical device development. Then, the concept distance matrix (12) of this community can be obtained by the definition of *visual distance*. The corresponding fuzzy relation between concepts can then be obtained based on (13) with $\alpha=1$, thereafter.

Initially, suppose there are two participators, Actor 1 and Actor 2, in the community, where Actor 1 is a work team for hardware development work and advertises itself by Fig.6, Actor 2 is a work team for software development work with its advertisement of Fig.7.

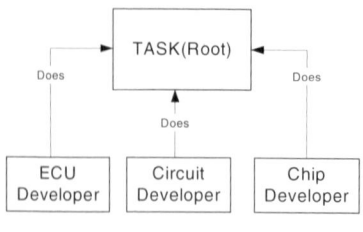

Fig. 6. Advertisement of Actor1

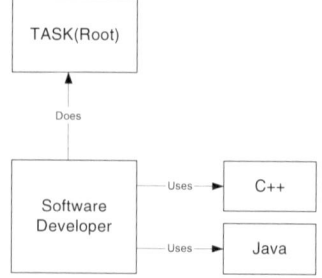

Fig. 7. Advertisement of Actor2

Accordingly, the feature vectors of both participators can be obtained by means of (3) and (4) with α=-4/MAX(Dis):

$$V_1 = \begin{bmatrix} 0 & 0 & 0 & 0 & 0 & 0 & 0 & 0 & 0 & 0 & 0 & e^{-\alpha*1} & e^{-\alpha*1} & e^{-\alpha*1} \end{bmatrix}^T$$

$$V_2 = \begin{bmatrix} 0 & e^{-\alpha*1} & 0 & 0 & 0 & 0 & e^{-\alpha*2} & e^{-\alpha*2} & 0 & 0 & 0 & 0 & 0 & 0 \end{bmatrix}^T$$

Suppose that a contract initiator from the automobile industry is seeking partners in the community for a telematic control unit (TCU) development. The TCU is a router which connects internal control area network (CAN) with external WLAN (Wireless Local Area Network, for example IEEE 802.11) and wireless WAN (Wide Area Network, such as the GPRS (General Packet Radio Service) or UMTS (Universal Mobile Telecommunication System), as well as with wireless radio networks such as bluetooth. The initiator has decomposed the development task into 6 subtasks with the feature vectors of:

Task 1: "C++ software work for Bluetooth application"

$$V(T_1) = \begin{bmatrix} 0 & 0 & 0 & e^{-\alpha*1} & 0 & 0 & e^{-\alpha*2} & 0 & 0 & e^{-\alpha*2} & 0 & 0 & 0 & 0 \end{bmatrix}^T$$

Task 2: "TCU programming for integration of Bluetooth, CAN, IEEE.802.11b"

$$V(T_2) = \begin{bmatrix} 0 & 0 & 0 & e^{-\alpha*1} & 0 & e^{-\alpha*2} & 0 & 0 & e^{-\alpha*3} & e^{-\alpha*3} & e^{-\alpha*3} & 0 & 0 & 0 \end{bmatrix}^T$$

Task 3: "Bluetooth communication chip design"

$$V(T_3) = \begin{bmatrix} 0 & 0 & 0 & 0 & 0 & 0 & 0 & 0 & 0 & e^{-\alpha*2} & 0 & 0 & e^{-\alpha*1} & 0 \end{bmatrix}^T$$

Task 4: "Bluetooth board design based on the designed chip"

$$V(T_4) = \begin{bmatrix} 0 & 0 & 0 & 0 & 0 & 0 & 0 & 0 & 0 & e^{-\alpha*2} & 0 & 0 & e^{-\alpha*1} & e^{-\alpha*1} \end{bmatrix}^T$$

Task 5: "TCU board design for connecting CAN, Bluetooth, and IEEE 802.11b"

$$V(T_5) = \begin{bmatrix} 0 & 0 & 0 & 0 & 0 & 0 & 0 & e^{-\alpha*2} & e^{-\alpha*2} & e^{-\alpha*2} & 0 & e^{-\alpha*1} & e^{-\alpha*1} \end{bmatrix}^T$$

Task 6: "Integration of hardware work with software work"

$$V(T_6) = \begin{bmatrix} e^{-\alpha*1} & e^{-\alpha*2} & e^{-\alpha*2} & 0 & e^{-\alpha*3} & e^{-\alpha*3} & 0 & 0 & e^{-\alpha*4} & e^{-\alpha*4} & e^{-\alpha*4} & 0 & 0 & 0 \end{bmatrix}^T$$

From the subtask announcements, we can find that T1 and T2 are software related programming work, T3, T4, and T5 are hardware related development work, T6 is integration work and requires both hardware and software knowledge.

Firstly, management networks are established for Actor1 and Actor 2 based on their initial announcement of V_1 and V_2. Both static and dynamic performances will be investigated by the following simulations:

1) Static competition

The subtasks from T_1 to T_6 are advertised in the community and are taken as inputs to the ART network. The subtasks can then be assigned to the best matching actors based on winner-takes-all competition principle. This is a static matchmaking process and the process is called unsupervised in ART. However, the initiator can control this matching process by adjusting vigilance. A higher vigilance means a stricter matching condition, a lower vigilance gives a looser matching condition. The following table shows the competition result for the given tasks with different vigilances:

Table 1. Competition results under different vigilance (where * means a new actor should be recruited into the community for the given task)

	T1	T2	T3	T4	T5	T6
$0 < \rho < 0.87$	Actor2	Actor2	Actor1	Actor1	Actor1	Actor2
$0.87 \leq \rho < 0.92$	Actor2	Actor2	Actor1	Actor1	Actor1	*
$0.92 \leq \rho < 0.95$	Actor2	*	Actor1	Actor1	Actor1	*
$0.95 \leq \rho < 0.96$	Actor2	*	*	*	*	*
$0.96 \leq \rho < 1$	*	*	*	*	*	*

It shows that, for a lower vigilance, $0 < \rho < 0.87$, the proposed network can map the hardware related tasks to the Actor1 and the software related tasks to the Actor2 automatically. Because of lower vigilance, Actor2 also gets the bid of T6 that is an integration work combining software and hardware development. On increasing the vigilance to 0.87, this integration task T6 cannot be assigned to any existed actors and a new actor should be recruited specifically for this kind of job. Further increasing vigilance necessitates that more specialized actors should be included into the community for the given tasks. Increasing vigilance means increasing precision but decreasing intelligence from the principle of IPDI [14]. In fact, each initiator can define a dynamic vigilance for partnership creation. It reflects the confidence of an initiator in the intelligence of specific participators. Participators with higher confidence can be given a lower vigilance or vice versa. This dynamic vigilance can also be adjusted through a learning process to form an adaptive vigilance based on the initiator's experience or other actor's evaluation.

2) Dynamic evolution:

This simulation aims to investigate the evolution ability of the ART network in a virtual enterprise. In a dynamic environment, participators with adaptability have to evolve and change themselves according to the demands from initiators for the purpose of winning the forthcoming contract competition. Therefore, competition can stimulate enterprise development. In the ART network, more competent participators are given more chance to win a contract and the winners are given a chance to learn the demands.

In order to examine the dynamic performance of the proposed network, we need to model the ability of each participator. Suppose the model of each participator can be expressed by a Gaussian function, which acts as a measurement of the actor's expertise and will feedback to the initiator as β in learning law (10) for weight updating:

$$\beta = 2e^{-\frac{\|X - W_i\|^2}{2\sigma_i^2}} - 1 \qquad (16)$$

where X is the desired task from an initiator, W_i is the advertised weight of the actor i, σ_i is an accepted field of the participator i, $\beta \in (-1,1]$ reflects the capability of the actor i, from "-1" very poor to "+1" very competent, for a given task X. Consequently, the actor i can perform a task X better when X is closer to W_i, the centre of its expertise. The accepted field σ_i reflects adaptability of an actor for a task with deviation from the actor's expertise centre. A bigger σ_i with wider range of acceptable tasks Now, suppose, in the community, there are intensive demands on actors with both hardware and software knowledge. The demands are expressed by the following feature vector series:

$$V(t_i) = V(T_6) + n(-0.1, 0.1) \quad , i=1...N, \tag{17}$$

where $n(-0.1, 0.1) \in R^{14 \times 1}$ is a vector with elements of uniform distributed noise between $[-0.1, 0.1]$.

Let the vigilance $\rho = 0.8$ and suppose the learning rate $\eta = 0.2$ and each actor has the same accepted field $\sigma_i = 1$, which implies a poor adaptability in fact. Advertise the new tasks (17) in the community, which require both hardware and software expertise.

Fig. 8 illustrates the evolution process of the ART network, where d-axis is the Euclidian distance between the feature vectors of initiators and the weights of participators, i-axis indicates subtasks sequence. As shown in Fig.8, due to the lower vigilance(0.8), Actor2 got the first 3 tasks. Modeled by (16) with $\sigma = 1$, each actor is able to complete the tasks satisfactorily only when they are close enough to its expertise centre. Due to the large difference, Actor2 lost the competition with Actor1 in the 4th round. However, the Actor1 has poor adaptability ($\sigma = 1$) too and it lost the contract in the 7th competition. A new actor, Actor3, had to be recruited into the community for the integration work of hardware and software development. This is an example of equal capability with $\sigma = 1$ for all actors.

Now, suppose Actor 1 is more competitive with $\sigma = 2$ but all other actors' accepted fields are kept to be $\sigma = 1$. As shown in Fig. 9, at the first 2 bids, Actor2 got the contract. Due to its poor performance, it lost the 3rd contract and cannot got back again because Actor1 with $\sigma = 2$ is more competitive than it. After Actor1 got the bid, the distance between its weights and the desired tasks is decreased when the tasks in (17) come to the community sequentially. It implies that the expertise of Actor 1 was changing from "hardware development" to "hardware and software integration" gradually because of continuous requirements from initiators.

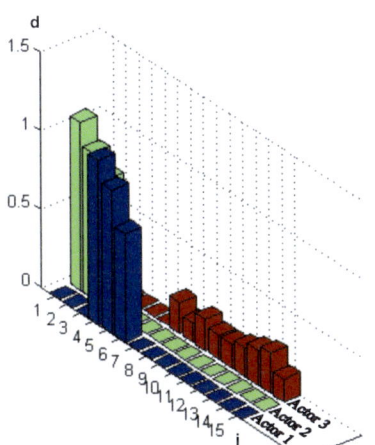

Fig. 8. Euclidian distances between demands and weights ($\sigma = 1$)

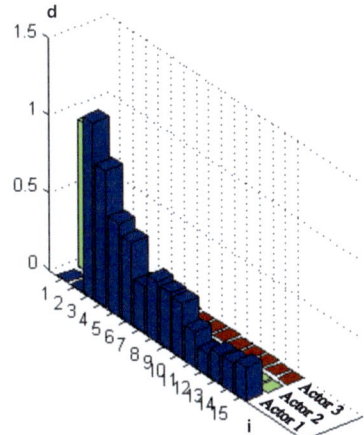

Fig. 9. Euclidian distances between demands and weights ($\sigma_i = 2$)

6 Conclusions

This paper presented a self-organizational management network for the organization of enterprise partnership. The connections between different actors who are seeking for partnership are adjusted based on fuzzy adaptive resonance theory so that the management network can exhibit unsupervised learning ability, adaptive ability, competitive ability, and self-organizational ability. These abilities usually exist in human society. Semantic difference is used to quantify distance between demand and provision; at the same time fuzzy inference is used to solve ambiguous expression from both sides. Working in this way Virtual Enterprises can evolve dynamically and force to improve product quality of each actor and organizational performance of partnerships.

References

1. Jiang P., Mair Q.: An Actor-oriented Approach to Distributed Product Management Systems. COMPSAC 2002, Oxford, August 2002
2. Sycara K., Klusch M., Widoff S., Lu J.: Dynamic Service Matchmaking among Agents in Open Information Environments. ACM SIGMOD Record 28(1) (1999) pp. 47–53
3. Jeng J.J., Cheng B.H.C.: Specification Matching for Software Reuse. ACM SIGSOFT Software Engineering Notes 20 (1995)
4. Choi Y.S., Yoo S.I.: Multi-Agent Learning Approach to WWW Information Retrieval Using Neural Network. Proceedings of the 4th international conference on Intelligent user interfaces, December, 1998
5. Carpenter G.A., Grossberg S., Rosen D.B.: Fuzzy ART: Fast Stable Learning and Categorization of Analog Pattern by an Adaptive Resonance System. Neural Networks 4(6) (1991) 759–771
6. Horrocks D. F., Broekstra J., Decker S., Erdmann M., Goble C., van Harmelen F., Klein M., Staab S., Studer R., Motta E.: The Ontology Inference Layer OIL. http://www.ontoknowledge.org/oil/TR/oil.long.html
7. Williams J., Steele N.: Difference, Distance and Similarity as a Basis for Fuzzy Decision Support Based on Prototypical Decision Classes. Fuzzy Sets and Systems, in Press.
8. Fankhauser P., Kracker M., Neuhold E.J.: Semantic vs. Structural Resemblance of Classes. Special SIGMOD RECORD Issue on Semantic Issues in Multidatabase Systems. 20(4) (1991) ACM Press
9. Akoka J., Wattiau I. C.: Entity-relationship and Object-oriented Model Automatic Clustering. Data & Knowledge Engineering. 20 (1996) 87–117
10. Miao C.Y., Goh A., Miao Y., Yang Z.H.: Agent That Models, Reasons and Makes Decisions. Knowledge-based Systems. 15 (2002) 302–211
11. Smith R. G.: The Contract-Net Protocol: High-Level Communication and Control in a Distributed Problem Solver. IEEE Transactions on Computers. 29(12) (1980) 1104–1113
12. Aggarwal C. C.: A Human-Computer Cooperative System for Effective High Dimensional Clustering. Proceedings of the seventh ACM SIGKDD international conference on Knowledge discovery and data mining, August 2001
13. Zadeh L.A.: Fuzzy sets. Information and Control. 8 (1965) 338–353
14. Saridis G.N.: Analytical Formulation of The Principle of Increasing Precision with Decreasing Intelligence for Intelligent Machines. Automatica, 25(3) (1989) 461–467

Mobile Software Agents for Location-Based Systems

Sebastian Fischmeister

Software Research Lab
Department of Computer Science
University of Salzburg
Fischmeister@SoftwareResearch.net
http://www.mobileshadow.net

Abstract. As mobile computing matures, location-awareness as part of context-awareness gains more attention; especially, location-aware services for mobile users. This paper concentrates on the software engineering issues of location-aware services and presents Mobile Shadow as an successful example of a design and an implementation of a scalable, fault tolerant, and component-based service infrastructure for location-aware services. The paper presents the basic concepts used in Mobile Shadow, the requirements, and the resulting component architecture.

Keywords: location-aware, context-aware, infrastructure, proactive services

1   Introduction

The continuous trend of miniaturization of hardware and the exponential growth of computation power has created a wide spectrum of small mobile computing gadgets. These gadgets enable the 'anytime and anywhere' communication paradigm, which has lead to the trend of wireless communication technology such as the global system for mobile communication (GSM) or the wireless local area networks (WLANs). These two trends form the basis of mobile computing.

Due to the high proliferation of mobile computing gadgets and technology, concepts such as context-awareness and especially location-awareness have regained interest. Only ten years ago, the location of a computer user did generally not change. Today, the user carries his notebook from one location to another, connects to the network, and continues using network-based services. A location-aware system would notice such a location change and would offer services specific to that new location.

An example of such a location-aware service is a reminder service. *John wants to carry books home with him for the weekend. Therefore, he wants to receive a reminder, when he is going home. Hence, he sticks a post-it on the exit door of the office building to remind him not to forget the books.* Instead of a post-it note, John could use an electronic location-aware service to remind him not to forget his books. John would enter the reminder text and most constraints (e.g.,

R. Kowalczyk et al. (Eds.): Agent Technology Workshops 2002, LNAI 2592, pp. 226–239, 2003.

building exit, after 5pm) and the service will send a reminder as soon as all constraints are fulfilled (i.e., John leaves the building after 5pm).

The Mobile Shadow project [10,12,9] at the University of Salzburg provides an infrastructure for such services. The system aims at three issues: proactive location-awareness, scalability/fault tolerance, and components & adaptivity. A location system basing on 802.11 WLAN technology enables proactive location-awareness. A decentralized infrastructure and the use of mobile code technology tackle the scalability and fault tolerance issues. And the use of agent technology and a component-based design support adaptivity by manipulation of the user.

Related research projects mainly concentrate on the locating issues and not on software engineering issues; Example projects are Cricket [20], Cyberguide [3], Active Bat [4], EventManager [18], or active badge [23]. Each system usually has one proof-of-concept service. The most closely related system is the stick-e document approach [5]. However, this system has never been implemented and only ran as a simulation at a workstation. Furthermore, the project did not concentrate on the software part of location-aware services. Another closely related project is the Lancaster Tour Guide [6]. The project provides a location-aware service for tourists in the Lancaster area. The project bases on an 802.11 WLAN and provides location-sensitive tourist information upon request. Thus, it is a reactive service only. The system bases on an extended version of HTML and uses thin client devices.

The remainder of the work is organized as follows. Section 2 introduces the concepts for proactive location-aware services and mobile code. Section 3 describes the Mobile Shadow system and presents a scenario and implementation details. Section 4 explains the different types of adaptivity available in the Mobile Shadow system. Finally, Section 5 concludes the paper.

2   Concepts

The following paragraphs provide an introduction to the core concepts of the Mobile Shadow architecture.

2.1   Locating versus Location

Systems combine the entities object and location in one of the two following relations:

1. Where is object o_1? – At location l_1.
2. Who is at location l_1 ? – The objects o_1 and o_2.

The first query refers to the mapping from objects to locations (see Fig. 1(a)) and the second from locations to objects (see Fig. 1(b)). We define systems using the first type of mapping (object $\rightarrow$ location) as locating systems and the second (location $\rightarrow$ objects) as location system. The key difference between locating and location systems is the fact that locating systems focus on the identification of

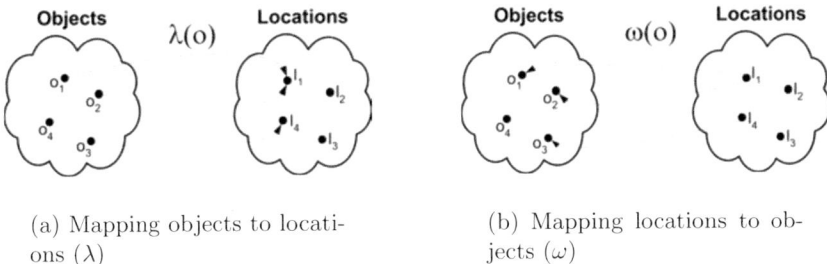

(a) Mapping objects to locations (λ)

(b) Mapping locations to objects (ω)

Fig. 1. Locating systems map objects to locations, whereas location systems map locations to objects

locations of objects, whereas location systems identify objects at locations. Each
system requires different methods and mechanisms to provide the result to the
query (for detailed discussion about the differences between these two mappings
see [11]).

2.2   Reactive versus Proactive Behaviour

Another important concept in the area of context-awareness is what we call pro-
activity versus reactivity. Currently, most location-aware systems offer reactive
services only. The technology to build such reactive systems is already available
[15] whereas current technology does not support proactive services out of the
box [8].

Reactive services rely on the prevailing request-response communication mechanism, where the service complements the client's incomplete knowledge. The
client actively "pulls" information from the service by issuing an explicit request. The most well known service, which uses this mechanism, is the World
Wide Web (WWW) and the Hypertext Transfer Protocol (HTTP). When a user
wants to visit a site, she enters the name, presses return (thereby issues the request), and will get the answer (the WWW server of that site will return the
requested document).

Proactive services deliver or "push" information to a client without explicit
request. Therefore, the user need not send an explicit request and will receive
information automatically. Such proactive services work autonomously as background processes and they prompt/inform the user as configured. To use such a
service, the client must subscribe to the service. During the subscription process,
the client creates a profile, which forms the basis for distinguishing wanted and
unwanted information. The reminder service mentioned in the introduction is a
typical proactive service. The service would miss its purpose, if the client always
has to query whether a reminder is set or not. Therefore, the reminder service
runs in the background and prompts the user, whenever a reminder is set.

2.3   Mobile Code

The advent of Java in 1994 and its built-in support for simple network programming revived mobile code paradigms. Although, prior projects existed, many researchers started researching mobile code at that time [14]. In general, mobile code is about moving data and/or code. There are four different mobile code paradigms: code-on-demand, remote evaluation, and mobile agents.

The *Code-on-demand* paradigm got widespread with the advent of Java applets. The client requests code from a server, the server returns the binary code, and the client runs this code. In this paradigm, only code is transferred via the network. The *remote evaluation* paradigm bases on transferring code. The client transfers code to the server, the server executes the code and returns the result to the client. Finally, the *mobile agent* paradigm is a mixture of the previously mentioned paradigms. A mobile agent consists of code, data, and a program state (i.e., it resumes operation at the remote host where it has left off before). The key property of it is "autonomous". The client sends a mobile agent to the server and the server executes it. After the execution, the mobile agent can autonomously decide to move to another server or to return to the client. In this paradigm, the program code, the data, and the program state are transferred through the network. In this work we use mobile agent and mobile code component synonymously.

3   Mobile Shadow

The University of Salzburg has built an architecture for proactive cell-based location-aware services. It is called *Mobile Shadow* [2]. In contrast to related work, Mobile Shadow bases on a locating system; it is optimized to answer queries such as who is nearby the building exit.

In this project concerning the software architecture, we aim at particular goals: modifiability/adaptivity, scalability, and fault tolerance. Modifiability is important, because we want to provide a platform for location-based services and thus want to add new services later on. To ease the user interaction, we also need to support adaptivity. By adaptivity we mean adaptation by manipulation by the user (see [16] for a discussion about the different concepts). The other goal is scalability; scalability is important, because most location-based services do make sense only when they cover a large area and to cover a large area and many users, the service must be scalable and the service platform must provide scalability features to services. Another aim is fault tolerance; if the system covers a large area, the system must not fail completely, if one service crashes or one cell crashes.

3.1   System Architecture

Several cycles of architecture design and evaluation lead us to a decentralized and localized architecture (see Fig. 2). Each cell has its own communication

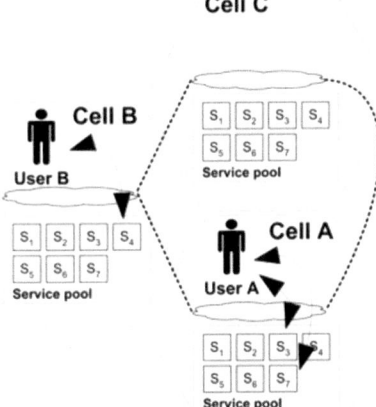

Fig, 2. The decentralized and localized architecture of Mobile Shadow

technology and a connection link to the adjacent cells. In the figure each cell includes only a service pool which runs a replica of a local service (i.e., S_1 to S_7). Each user accesses the services directly in his cell. For instance, user A stands in cell A and accesses the service S_5 and S_7 in Cell A through the cell's own communication system. User B stands in cell B and also accesses the services through the cell's own communication system.

This architecture satisfies our requirements. Concerning scalability, different users in different cells access different replica of the same service. Therefore the service access will not become a bottleneck. Also, we can multiply the number of service pools in one cell, so we also can perform dynamic load balancing. Concerning fault tolerance, if the communication technology in one cell fails, it does not affect the other cells. For example, if the communication technology in cell A fails, User B in cell B can still access the services.

3.2   Computation Model

Mobile Shadow meets the stated requirements and implements the suggested scalable, fault tolerant, and modifiable architecture through different models. What we call space model describes how we map the reality onto the system architecture (see below for a description of all models). The user model and the service model show how we represent users and services in this virtual reality. The user-space and the service-space model explain how we handle user and service movements. And finally the user-service model shows how we model user access to services.

Space Model. In the Mobile Shadow system, each physical place also has exactly one logical counterpart. The physical environment is split into several separate small cells and each of these cells has a virtual representation. For example, the location "building exit" exists as "shadow building exit" location, too.

Furthermore, the space model allows adding a hierarchical structure as a tree on top of it. Several cells can be virtually merged into one larger cell and several cells can be treated as one larger cell but still exist separately. For example, our computer science building consists of several cells, however, the whole building also represents a virtual location "computer science building".

User Model. Each user owns a virtual "alter ego" (a mobile code component). This component, called user agent, always resides at the logical counterpart of the current user's location. Therefore, if a user agent resides at the virtual "building exit" location, then the real user also resides at that location.

Additionally, each agent is associated with several roles. Each role defines the specific set of services available for the user. For example, staff members can access different services than students or visitors.

Each service consists of a trigger and a service implementation. The user agent runs the trigger and the local infrastructure runs the service implementation. A trigger consists of trigger constraints, a personal configuration, and a trigger action. The user can configure what triggers his agent has and the user can also set the personal configuration of each trigger via a small command line application or a web interface. The trigger constraints define conditions in order to call the trigger action such as time, location, or available services. Once all these constraints are satisfied, the trigger action activates the local service and transfers the personal configuration of the user for this service.

Figure 3 shows two user agents, their roles, and the included triggers. Although some roles may share the same triggers such as the local MessageBoard (see below), they also offer a different set of services (e.g., see Fig. 3(b) the developer role), or have different access permissions (e.g., the lecture reminder service is equivalent to the reminder service, however, the user has predefined reminders).

User-Space Model. To provide location-aware services, Mobile Shadow transforms user movements into virtual movements. Thus, if the user moves from location A to location B, then the user agent will move from the virtual location A to the virtual location B; Once the user agent arrived at the new place, it accesses the local resources such as a local database, other local user agents, and local available services.

Service Model. Services, similar to users, have a virtual representation. For example in a tour guide service, a museum painting offers a specific service; e.g., an insight into the art of the previous century. This painting owns a virtual service at its location. The service offers this information to user agents who visit this location. The service can even notify the user of its existence, depending on the trigger configuration of the user agent.

Real world services and services – i.e., services that are only available in the virtual user space – are available in Mobile Shadow. Such services are often

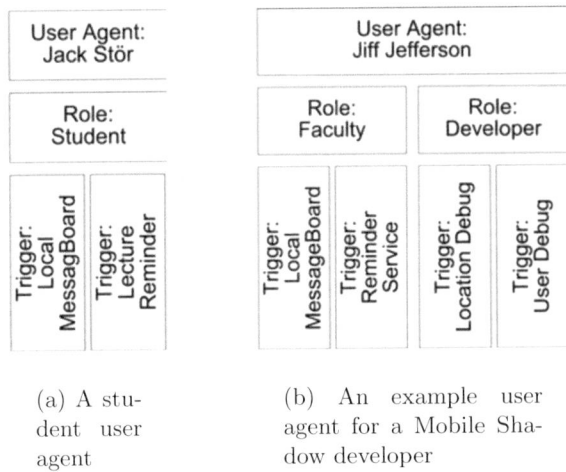

(a) A student user agent

(b) An example user agent for a Mobile Shadow developer

Fig. 3. Example user agents. Each user can have several roles and different triggers for each role

real world services transformed and extended into digital services. For example, a local message board service (see GeoNotes [7]); the service is similar to a blackboard, however, the notes are in an electronic form and the blackboard provides different notes at different locations. A user can read, add, or remove notes at any location she wants to.

Service-Space Model. The service-space model is static in contrast to the user-space model. So far, the Mobile Shadow project only consists of real and virtual services that do not move. However, one could think of moving services such as services that chase viruses. Service such as the painting service have a specific location and work with localized or personalized data.

User-Service Model. When a user agent moves to a new location, this location change activates the triggers of each service. The trigger tries to fulfill the trigger constraints. In case it succeeds, the trigger action activates this service and transfers the personal configuration parameters. Afterwards, the activated and personalized service processes the local data. If the service finds some information that may be interesting for the user, it will notify her according to the personal setup (e.g., via short message service or via display message or via email).

3.3   Mobile Shadow Component Architecture

Each location-aware service consists of two parts: the service and the trigger. The *trigger* contains the user specific configuration parameters, the trigger constraints, and the trigger action. In simple cases, the configuration cannot be

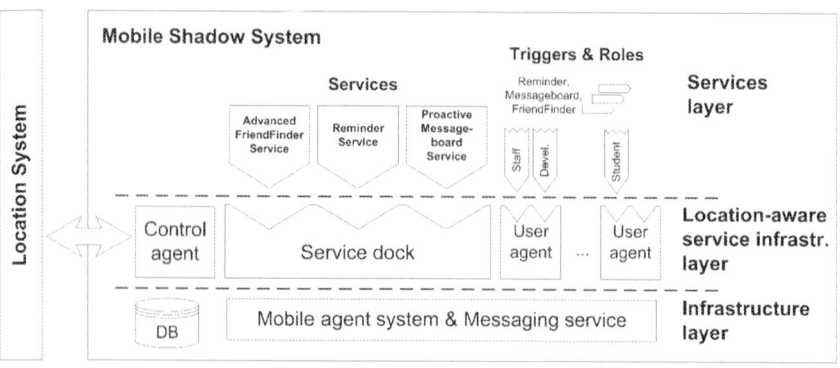

Fig. 4. Component concept of Mobile Shadow. The system has a modular design, which enhances support for adding and exchanging components

changed and the action is a simple method invocation. The *service* contains all server-side business logic such as database access or data processing. The *user agent* manages the roles and triggers (e.g., deactivating them before moving to a new location or activating them after moving) and provides the basic functionality for the triggers (e.g., finding the service dock). The *service dock* controls the services (e.g., registration of new services or starting and shutting down services) and provides some basic functionality (e.g., finding a service or gaining database access). The *control agent* manages the communication between Mobile Shadow and the location system and provides interfaces to external resources such as partner research projects. Furthermore, the control agent manages the user agents, e.g., it is responsible to move user agents to their new location. Finally, the *mobile agent system* provides the basic infrastructure for mobile code, which also includes an inter-agent messaging system. Finally, the *location system* tells the control agent, when an agent did move from one place to another.

3.4   Example Scenario

A location-aware service is the reminder service. The introduction already presented the objectives of the service. Among other services, Mobile Shadow provides this reminder service.

   User John wants to carry books home with him for the weekend. Therefore, he wants to receive a reminder, when he is going home. In the Mobile Shadow system, he opens the WWW interface to add location-based reminders. He enters/selects following data: cell id (pull down list), reminder text (maximum of 140 characters[1]), reminder start time and start date, reminder end time and end date, how often he wants to be reminded of this one event (e.g., John could wish to be reminded every day to lock his door, then the number of reminders is

---

[1] Messages may be delivered via the short message service to for cellular phones. Such messages have a limit of 140 characters.

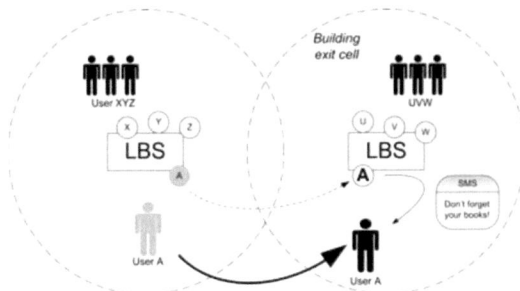

Fig. 5. Example scenario "John and the books.". John must not forget to carry books home for reading, thus he did set a location-based reminder in the exit cell of his building

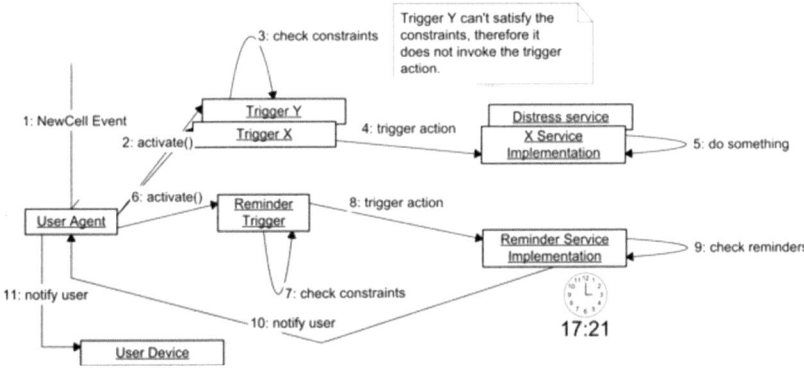

Fig. 6. Each time the agent arrives at a new place, it receives an event from the agent system and then it activates the triggers

infinite). The cell identification is equivalent to the location. The correct cell for this example is the building exit cell (i.e., *AP Hall*). The reminder text is "Hey! Don't forget the books for the weekend!". The reminder start time is "5pm" and the start date is "today". The reminder end time is "11pm" and the end date is "today". The reminder is only active between the start time and the end time. This prevents for instance, that John also receives the reminder, when he leaves for lunch. Finally, John can also define the number of reminders that he wants to receive. In this example, John sets the reminder count to "1". Therefore, John will receive only one reminder and then the system will deactivate the reminder. After the John filled out all the required fields, he submits the data and thus configures the trigger carried around by his user agent.

Figures 5 and 6 show what is going to happen, when John is leaving the building after 5pm. John moves through the building and thus enters and leaves several locations. At each location the user agent receives the *NewCell* event from the Mobile Shadow platform. Finally, John and his user agent enter the building exit cell. Arriving in this cell, the user agent again activates the triggers. The reminder trigger checks its constraints and executes the trigger action. Because it is past 5pm and before 11pm, the date is today, and the reminder is still active (the reminder count is greater than zero), the service will return a valid reminder. Now, the user agent uses the short message service (SMS) and sends the reminder to the John's cellular phone.

3.5   Implementation

The Mobile Shadow research project has consisted of two phases. Researchers at the University of Constance built the first prototype of the Mobile Shadow system. This first prototype implemented the whole functionality except the location system. Therefore, user agents had to be moved manually to demonstrate the system. The first prototype was implemented in Java, used the Aglets mobile agent system as basic infrastructure, and MySQL as database management system. Researchers at the University of Salzburg built the second prototype of the Mobile Shadow system. This prototype now includes the location system, the interface to Mobile Shadow, and three running services. The second prototype is implemented in Java but uses the Grasshopper Agent system [22,21,1,17] (to increase independence from the underlying system) as basic infrastructure and MySQL as database management system.

Based on an evaluation of wireless communication technologies [8], the second prototype was built on top of an 802.11b wireless local area network [19]. Figure 7 shows the ground floor of the computer science building. In addition the figure shows five access points of the WLAN and the integrated Mobile Shadow system. There are a total of 11 access points distributed on three levels of the building. However, the component architecture in Sect. 3.3 showed that the

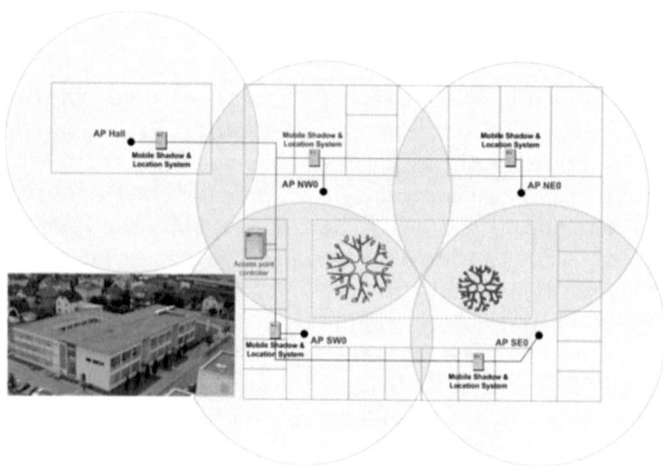

Fig. 7. Computer science building ground floor and integrated Mobile Shadow infrastructure The building plan shows five WLAN access point and how the Mobile Shadow system is integrated into it

location system is only loosely coupled, so we can easily exchange the WLAN location system by commercial ones or research projects such as Cricket [20], Cyberguide [3], or Active Bat [4], which would provide a more exact locating system than the WLAN system.

To realize the location system, the WLAN hardware vendor Proxim cooperated with the university and provided information about the implementation of the hand-off protocol (a client switches from one access point to another). In order to detect location changes, the locating system of Mobile Shadow analyses the network traffic between the access points and a central access point controller. When a client switches from one access point to another, then the new access point registers the client at the central controller. Therefore, the location system knows when to move the user agent. The system determines the destination of the user agent by querying adjacent access points. Once it has found the destination, it notifies the control agent of the location change.

In addition to a virtual place for each location, the Mobile Shadow system also requires a place to store 'homeless' user agents. This place is called *garage*. Once the user turns off the WLAN device, the locating system can no longer track the user. After a timeout, the location system issues a location change to the garage place. Then, the Mobile Shadow system moves the user agent to the garage. Besides easier maintenance, this simplifies the restore process once the user again turns off the device. In case of a restore (the user joins the WLAN again), the system knows where it can to find the agent[2].

4   Adaptivity

The computation model on which Mobile Shadow bases and its modular design allow adapting the system in three simple and efficient ways: (1) the user can define an specific configuration for each service (e.g., the user can define the list of friends for the FriendFinder service or the kind of notification of reminders), (2) the user can add triggers and thus subscribe to new services, and (3) the system designer can change user roles so they suit the new requirements.

Manipulation of the User Configuration. The easiest way of adapting the system to the personal needs of a user is to change the configuration of the trigger. Each service offers different parameters and each user can adjust this parameters to her personal needs via a web service. For example, the main intention of the FriendFinder service is to automatically notify the user whenever one of his friends enters the same location. The basic setup is to define this list of friends (i.e., enter the phone numbers of them). But there are also general settings, which are not specific for a certain trigger. For example, Mobile Shadow uses SMS as default notification service, but some users prefer other notification services. Therefore, users can also configure the notification mechanism (e.g., show a message on the device or send an email).

---

[2] To provide a scalable solution for the *garage*, we use a peer to peer model that is beyond the scope of this paper.

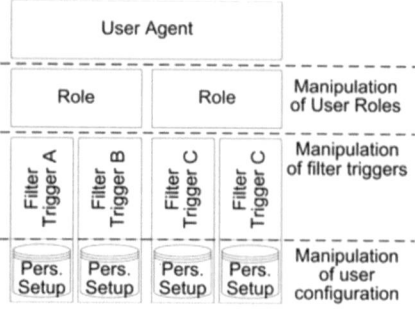

Fig. 8. Manipulation of the user agent setup. The user can adapt her user agent at three different layers

Manipulation of Triggers. The second possibility for a user to adapt the system is to add or remove triggers to and from his user agent. A separate WWW interface allows configuring these components and thus allows the user adapting her subscription setup. Due to use of component technology and Java, the user can execute the changes during runtime and no restart is necessary.

Manipulation of User Roles. Finally, the third possibility to adapt the system is to introduce new role components. The user agent is only the vehicle that carries and manages the roles. The roles provide the basic features to the triggers. Different users may need different basic features and therefore need different types of roles. For example, staff members belong to a different role than students (see the previous example in Sect. 3.2).

We do not allow users to assign or program roles themselves. The main reason is security. The role concept allows us to run a strict security policy (e.g., developers can access anything, staff services have read-write database access, student services have read-only database access) that the underlying agent platform can enforce. Otherwise we risk running into security problems as agent systems are not yet secure enough [13]. Therefore, if a user needs a special role, then she must ask the system designer and developer to assign or implement the new one.

5   Conclusion

Mobile Shadow is an infrastructure that provides support for proactive location-aware services. The main aims of the research project are to provide a sound software architecture that supports modifyability/adaptivity, scalability, and fault tolerance. The paper presents details about the concepts used in Mobile Shadow, its design and implementation that allow us to meet the requirements. In contrast to the related work, Mobile Shadow does not concentrate on technical solutions of locating an object but it concentrates on the software architectural issues of service platforms for location-aware services.

As future work, the researchers plan to open the agent system codebase so any user can program his own user agent given that it does not compromise the security of the system. Due to the success and the novelty of the approach, the researchers plan to evolve the Mobile Shadow prototype to a product.

References

1. Grasshopper. WWW Site. http://www.grasshopper.de.
2. Mobile Shadow. WWW Site. http://www.mobileshadow.net/.
3. G. D. Abowd, C. G. Atkeson, J. Hong, S. Long, R. Kooper, and M. Pinkerton. Cyberguide: A Mobile Context-Aware Tour Guide. *Baltzer/ACM Wireless Networks*, 3(5):421 to 433, October 1997.
4. M. Addlesee, R. Curwen, S. Hodges., J. Newman, P. Steggles, A.Ward, and A. Hopper. Implementing a Sentient Computing System. *IEEE Computer*, 34(8):42 to 48, August 2001.
5. P. Brown. The stick-e document: a framework for creating context-aware applications. *Electronic Publishing*, 8(2&3):259 to 272, Jun & Sep 1995.
6. N. Davies, K. Cheverst, K. Mitchell, and A. Efrat. Using and Determining Location in a Context-Sensitive Tour Guide. *IEEE Computer*, 34(8):35 to 41, August 2001.
7. F. Espinoza, P. Persson, A. Sandin, A. Nyström, E. Cacciatore, and M. Bylund. Geonotes: Social filtering of position-based information. In *Proceedings of Ubicomp 2001*, Atlanta, GA, 2001.
8. S. Fischmeister. Cell-based Pervasive Networking Technologies for Location- Aware Services. Technical Report C46, Software Research Lab, University of Salzburg, Austria, March 2002. (journal tutorial submission, available at http://www.mobileshadow.net/reports).
9. S. Fischmeister. *Location Context in Mobile and Pervasive Computing Systems: Technologies, Architectures & Implementation*. PhD thesis, University of Salzburg, December 2002.
10. S. Fischmeister. Mobile Software Agents for Location-based Systems. In *Net.Object Days*, number 2, page 234 to 247. IGT, Leipzig, October 2002.
11. S. Fischmeister and G. Menkhaus. L^2: A Novel Concept for Cell-based Location-Aware Services. Technical Report C45, Software Research Lab, University of Salzburg, Austria, February 2002. (concept sketch, available at http://www.mobileshadow.net/reports).
12. S. Fischmeister, G. Menkhaus, and W. Pree. MUSA-Shadow: Concepts, Implementation, and Sample Applications; A Location-Based Service Supporting Multiple Devices. In Proc. of TOOLS Pacific. Australian Computer Society Press, February 2002.
13. S. Fischmeister, G. Vigna, and R. Kemmerer. Evaluating the Security Of Three Java-Based Mobile Agent Systems. In G. P. Picco, editor, *Mobile Agents*, volume 2240 of *Lecture Notes in Computer Sciences*, page 31 to 41. IEEE, Springer-Verlag Heidelberg, December 2001.
14. A. Fuggetta, G. Picco, and G. Vigna. Understanding Code Mobility. *IEEE Transactions on Software Engineering*, 24(5):342 to 361, May 1998.
15. J. Hightower and G. Borriello. Location Systems for Ubiquitous Computing. *IEEE Computer*, 34(8):50 to 56, August 2001.
16. E. Horvitz. Principles of Mixed-Initiative User Interfaces. In *Proceedings of CHI'99*, ACM SIGCHI Conference on Human Factors in Computing Systems, May 1999.

17. IKV++. Grasshopper. WWW Site, 2000.
    `http://www.ikv.de/products/grasshopper/`.
18. J. F. McCarthy and T. D. Anagnost. EventManager: Support for the Peripheral
    Awareness of Events. In P. Thomas and H. W. Gellersen, editors, *Handheld and
    Ubiquitous Computing*, number 1927 in Lecture Notes in Computer Sciences, page
    227 to 236. Springer Verlag, Germany, September 2000.
19. B. O'Hara and A. Petrick. *The IEEE 802.11 Handbook: A Designer's Companion.*
    Standards Information Network IEEE Press, 1999.
20. N. B. Priyantha, A. Chakraborty, and H. Balakrishnan. The Cricket locations-
    upport system. In *Proc. of the sixth annual international conference on Mobile
    computing and networking (MobiCom 2000)*, pp. 32–43. ACM Press, 2000.
21. IKV++ GmbH Informations und Kommunikationssysteme. *Grasshopper Develop-
    ment System, Release 1.2, Basics and Concepts.* IKV++ GmbH Informations- und
    Kommunikationssysteme, February 1999.
22. IKV++ GmbH Informations und Kommunikationssysteme. *Grasshopper: The
    Agent Platform – Technical Overview.* IKV++ GmbH Informations- und Kom-
    munikationssysteme, February 1999.
23. R. Want, A. Hopper, V. Falcao, and J. Gibbons. The Active Badge Location Sy-
    stem. *ACM Transactions on Information Systems*, 10(1):91–102, January 1992.

Partner Detection and Selection in Emergent Holonic Enterprises

Mihaela Ulieru[1] and Rainer Unland[2]

[1]Electrical & Computer Engineering Department
The University of Calgary
Calgary, T2N 1N4, Alberta, Canada
ulieru@ucalgary.ca
http://isg.enme.ucalgary.ca

[2]Institute for Computer Science
University of Essen
45117 Essen, Germany
Unlandr@cs.uni-essen.de
http://www.cs.uni-essen.de/dawis/

Abstract. Web-centric virtual enterprises, to be successful, need to be highly innovative and competitive. Both features can only be achieved if the underlying structure is ever emerging and dynamic. Unsatisfactory member organizations need to be replaced by better ones and/or the conglomerate is to be extended if organizations are detected that can contribute substantially to the success of the virtual enterprise. This paper presents the key concepts of an emergence model and discusses how the best possible partners for a given (bunch of) task(s) can be found and evaluated to guarantee that always the best choice is made. Moreover, the overall objectives of a virtual enterprise and the individual objectives of involved agents need to be harmoniously integrated in order to guarantee the overall success of the conglomerate as well as the success of the individual agents. To achieve this intentional problem solving is introduced. It allows agents to cluster and collaborate in a way that fits best to the individual goals of each agent (e.g. maximum profit) as well as considering the overall objectives of the next higher level (e.g., virtual enterprise) as far-reaching as possible.

1 Introduction

Complex information systems in traditional, i.e. functionally structured, enterprises have in principle been constructed with a clear purpose in mind, therefore, top-down. Consequently, the requirements and aims of the underlying enterprise have both explicitly and implicitly influenced the design of these systems respectively of their components. As long as these requirements are relatively stable this is a feasible approach to system design, since changes occur seldom and thus can be accounted for by reengineering processes.

R. Kowalczyk et al. (Eds.): Agent Technology Workshops 2002, LNAI 2592, pp. 240-262, 2003.

A first important step to dilute this stability phenomenon has been the shift towards *process orientation* and *virtual enterprises*. They promise high flexibility, adaptability, dynamicity and fault-tolerance. A virtual enterprise is a *temporary alliance* of enterprises that *cooperatively* work together to *share skills* or *core competencies* and *resources* in order to better respond to *business opportunities*, and whose cooperation relies on *computer networks* and a *cooperative*, yet *distributed information systems structure. Emergent* virtual enterprises go even one step further by constantly monitoring their performance and the market in order to improve their overall performance and efficiency, i.e., they permanently check whether there are (more) suitable possible partners available on the market that may either replace existing ones or add to the overall business objectives of the virtual enterprise in a positive way. This feature may result in frequent organizational changes that have to be reflected by corresponding changes in the underlying information system (architecture). Another highly flexible organizational structure that has been discussed more recently are *supply webs* (cf., e.g., [Lase98], [FiFR02]). Nowadays, a supply chain is less a chain than a complex web of intersecting supply chains. Even from the perspective of one big company, a highly impressive number of nodes will make up this supply web, since a huge number of companies may deliver parts for a product, with many products being produced by the company in question. Partnerships between autonomous business entities in these supply webs can be flexibly contracted or withdrawn and are predominantly short-dated. This may cause complex coordination problems since the resulting many-to-many interactions and instantiated supply paths are not stable but may dynamically change. In this paper we will address such kinds of enterprises and will show how the features of emergence, dynamicity, flexibility, stability, robustness, and adaptability can be dealt with.

On the implementation level the above characteristics strongly imply the use of a multi-agent system (MAS) architecture in which enterprises are represented/cloned by one or more agent(s) at their interface. In a MAS environment it is assumed that every agent is autonomous and that the underlying agents architecture is in principle flat, i.e., agents are neither sub- nor superordinated. This is in congruence with the formation process of a virtual enterprise on the inter-enterprise level (*horizontal* integration). This is the only level on which it is usually discussed in literature. However, in contrast to this single level view we argue that the above demands can only be achieved if the involved member enterprises are deeply integrated and not only on the surface (*vertical* integration). Since a virtual enterprise is only a temporary conglomerate that is established to quickly react to complex demands of the market it is mandatory that every decision on the inter-enterprise level is propagated and reflected in each member enterprise through all levels down to the deepest level, the machine respectively autonomous atomic system level. This view extends the single level view to a hierarchical view. In order to integrate such a hierarchical structure in a society of autonomous agents, we advocate the concept of *holonic MAS*. It consists of several layers each of which represented by recursively nested self-similar structures (so called holons) which dynamically adapt/restructure themselves to achieve the design goals of the system. In holonic MAS these holons are groups of agents. In contrast to a normal MAS, agents that form a holon, need to accept a (partial) loss of their autonomy. However, they do not need to waive it completely. To a certain extent they can leave a holon and act autonomously, join other existing holons or rearrange themselves as new holons.

In this paper we will discuss how emergent Web-centric virtual enterprises can be created and organized. We will especially concentrate on the questions

- how possible partners/collaborators can be found,

- how it can be decided which partners are the better ones, and

- how an agent can be equipped with the ability to deal and consider its own goals (goals of the unit it represents) as well as the goals of the unit in which it is integrated (the higher level unit).

In the next section (Sect. 2) we will first introduce the basics, namely virtual enterprises, the demands they place on appropriate information systems, their extension in the direction of holonic enterprises and the notion of emergence. Section 3 constitutes the main part of this paper. It will discuss how the right (best) partners can be found, on the level of the holonic enterprise, as well as on the intra-enterprise level. Since especially in am emergent environment the number of possible partners may be high and my change continuously we need to lay down how the best partners can selected from a set of possible partners. Agents act on behalf of their enterprise. Therefore, they need to represent and follow the objectives of the enterprise they represent on the one hand and the objectives and aims of the holonic enterprise on the other hand. We will discuss these different kinds of goals and how they can be embedded into an agents belief system in a coherent and harmonious way. Finally, Sect. 4 will conclude the paper.

2   Foundations and State-of-the-Art

In this section we will introduce the basic concepts and discuss the state-of-the-art. Starting with an introduction of the concept of *virtual enterprises* we will extend this "single-layer" architecture to a multi-layered architecture by introducing the term *holonic enterprise*. Holonic enterprises form the basis for our proposal of *emergence* in *holonic enterprises* that we will introduce afterwards.

2.1   Basic Characteristics of Virtual Enterprises

Today's enterprises – to be successful – need to be highly flexible and adaptable. This insight has let to the introduction and discussion of *virtual enterprise*. Up to now a commonly agreed on more precise definition is still missing. However, the following four features are commonly accepted as mandatory characteristics:

1. Purpose-Driven
Virtual enterprises are formed for specific reasons and with clear overall common goals and objectives in mind. All members have agreed on and accepted these goals and objectives.

2. Flexible Organizational Structure

The organizational structure of a virtual enterprise can be quite dynamic; i.e. can change numerous times during the life time of the virtual enterprise. Davidow and Malone (cf. [DaMa92]) distinguish between the *inside* and the *outside* view of a virtual enterprise. From the outside the shape of a virtual enterprise may change continuously. It may extend if additional tasks are identified which cannot be covered by existing members or if an external company seems to be a useful supplement for the virtual enterprise. Companies may leave because their internal goals do no longer meet the goals of the virtual enterprise. Finally, companies may have to leave if their value for the virtual enterprise does not meet the expectations. In the inside we have to deal with a permanent demand driven restructuring process with respect to functional units like groups or departments.

3. Autonomy of Members

Members join a virtual enterprise intentionally. This implies that they are and will remain autonomous and independent. This, however, may result in that individual enterprises may be driven by diverging or even conflicting aims. With respect to the underlying IS it means that each member keeps the responsibility for and the control of information concerning its part in the virtual enterprise. This can only be realized respectively ensured by a decentralized design for the information system of a virtual enterprise. It is not possible to integrate the individual system components of its members into one monolithic whole but they need to be kept separate and under the respective control of the underlying member enterprise.

4. Temporal Membership

A virtual enterprise is a temporary network of independent enterprises. Temporal means that the lifetime of a virtual enterprise is normally either explicitly or implicitly restricted; the virtual enterprise is dissolved when its overall goal has been achieved (cf., e.g., [ByBP93]).

The life cycle of a virtual enterprise is defined by

- the *creation/configuration phase*, in which the partners are selected and the possible relationships and contracts are negotiated and agreed on,

- the *operation phase*, in which the necessary information needs to be exchanged between and within the member enterprises, unforeseen events and exceptions need to be handled with, and the operation processes need to be coordinated

- the *dissolution phase*, in which all necessary actions for the dissolution of the virtual enterprise are taken

In this paper we will mainly concentrate on the creation/formation phase. We especially will discuss how partners can be found and why and how the structure of a virtual enterprise may change constantly during the lifetime of a virtual enterprise.

2.2 Holonic Enterprises and Emergence

Since a virtual enterprise is only a temporary conglomerate that is established to quickly react to complex demands of the market it is mandatory that every decision on the inter-enterprise level (*horizontal* integration), is propagated through and re flected in all levels down to the lowest level, the machine level (*vertical* integration). This leads to the concept of a holonic enterprise. In [TiBU02], [TiUn02], [TiUn99a] and [TiUn99b] first approaches can be found that extend the vertical integration to some kind of mixture of vertical and horizontal integration. However, from our point of view, these approaches are still not going far enough to deal with today's dynamics on the market in an appropriate way. Instead we believe that full integration and exploitation of both axes is necessary, which leads to the concept of holonic enterprises.

About twenty-five years ago Arthur Koestler, a Hungarian philosopher, introduced the word *holon* to describe a basic unit of organization in social and biological systems (cf. [Koes67]). *Holon* is an artificial word, derived from the Greek word *holos*, meaning *whole,* and the suffix *on* meaning *particle* or *part.* Koestler observed that entirely self-supporting, non-interacting entities do not exist as such in living organisms and in social organizations. Instead, every identifiable unit of organization, such as a single cell in an animal or a family unit in a society, comprises more basic units (plasma and nucleus, parents and siblings) while at the same time forming a part of a larger unit of organization (a muscle tissue or a community). Also implicit is a recognition that systems (natural or organizational) cannot evolve from one level of complexity to significantly higher levels without the existence of stable intermediate forms to act as intermediate steps toward the goal state. A holon represents an autonomous and cooperative building block of such a system. Since holons may be made up of sub-ordinate parts and in turn may be part of a larger whole they form a tree-like hierarchy, called *holarchy.* If we would zoom into an inner node we see that it (recursively) consists of a set of (sub-)holons.

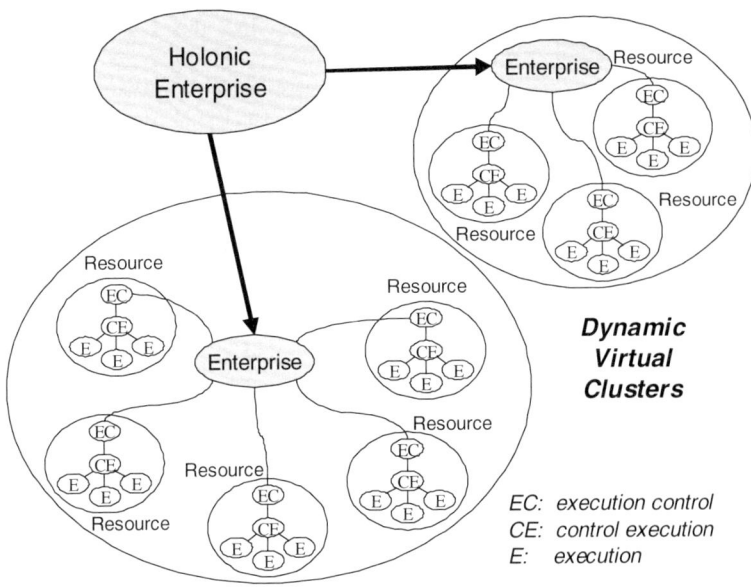

Fig. 1. Holonic enterprise

A holarchy that is not part of an encompassing holarchy is a *holonic enterprise* (cf. Fig. 1). In other words, a holonic enterprise is a holarchy of collaborative enterprises, where each unit is (recursively) regarded as a holon and is modeled by an agent with *holonic* properties, so that the software agent may be composed of other agents that behave in a similar way but perform different functions at lower levels of resolution (cf. [UIBW02], [UIWB01]).

The goal of a holonic enterprise is to attain the benefits that a holonic organization provides to living organisms and societies. The concept of holons enables the construction of very complex systems that are nonetheless efficient in the use of resources, highly resilient to internal and external disturbances, and adaptable and flexible in the face of changes in the environment in which they exist. It combines the best features of hierarchical (top down) and *heterarchical* (bottom up, cooperative) organizational structures. They are created in a way that fits best to a given specific situation. This proceeding can preserve the stability of hierarchies while providing the dynamic flexibility of *heterarchies*.

The stability of holons and holarchies stems from the fact that holons are self-reliant units, which have a degree of independence and handle circumstances and problems on their particular level of existence without needing to ask higher level holons for assistance. The other way round, holons may receive instruction from and, to a certain extent, be controlled by higher level holons. This self-reliant characteristic ensures that holons are stable and capable to survive disturbances. The subordination to higher level holons ensures the effective operation of the larger whole.

In contrast to the traditional, static enterprise structure the structure of a holonic enterprise is highly flexible and dynamic. On each level new groups/holons may be formed according to new and emerging needs or existing holons may reconfigure themselves in order to be more efficient or to deal with failures or unsatisfying performance of a member holon. In other words a holon forms a *dynamic virtual cluster*. Dynamic because it first forms and then dissolves as tasks are initiated and then completed, and virtual because it represents a logical or functional decomposition of the system interactions, and not a structural grouping of system components.

Emergence

On the basis of this notion of a *holonic enterprise* we want to develop and implement a model for highly efficient and competitive holonic enterprises in cyberspace that, in addition to addressing the above requirements and features, exhibits *emergent (self-organizing* and *evolutionary)* behavior.

As result of the process of evolution driven by the law of synergy, *emergence* endows the dynamics of composite systems with properties unidentifiable in their individual parts. The phenomenon of emergence involves

- *self-organization* of the dynamical systems such that the synergetic effects can occur;

- interaction with other systems from which the synergetic properties can *evolve* in a new context.

2.3   How Can a Holonic Enterprise Description Be Mapped onto a Multi-agent System Architecture

Multi-agent systems enable cloning of real-life systems into autonomous software entities with a 'life' of their own in the dynamic information environment offered by today's cyberspace. The *holonic enterprise* has emerged as a business paradigm from the need for flexible open reconfigurable models capable to emulate the market dynamics in the networked economy (cf. [MHWW95]), which necessitates that strategies and relationships evolve over time, changing with the dynamic business environment. Building on the MAS-Internet-Soft Computing triad to create a web-centric model endowing virtual communities/societies (generically coined as 'enterprises') with proactive self-organizing properties in an open environment connected via the dynamic Web, the holonic enterprise paradigm provides a framework for information and resource management in global virtual enterprises (cf. [Ulie02a]). According to this paradigm of holonic (virtual) enterprises, which - from our point of view - is the most sophisticated approach to virtual enterprises, the following three principle layers of collaboration/interaction need to be considered by a MAS in order to emulate a holonic enterprise.

Inter-enterprise Level
The highest level, the *Global Inter-Enterprise Collaborative Level* or *Inter-Enterprise Level* for short, is the level on which the holonic enterprise is formed. Each collaborative partner is modeled as an agent that encapsulates those abstractions relevant to the particular cooperation. By this, a dynamic virtual cluster emerges that – on the one hand – is supposed to satisfy the complex overall goal at hand as good as possible and – on the other hand – considers the individual goals of each enterprise involved.

Forces to Be Balanced
The following are the requirements that need to be balanced on this level:

- *Efficiency and Cost Minimization* - achieved via maximum synergy (obtained by clustering the 'best[1]' partners). Efficiency is obtained by an openness to continuously sense the market's pulse and rapid (re)configuration to respond quickly to changes, as well as by the ability to respond to errors in a timely fashion.

- *Fault-Tolerance* - On demand order tracking, on-line order error reporting, ability to quickly replace a collaborative partner if it does not fulfill its commitments in an appropriate way.

- *Competitiveness on the Global Market.* The collaborative cluster can achieve competitiveness only through continuous optimization of the collaborative cluster with maximum synergy as criteria. If a partner does not perform according to expectations (e.g. does not honor commitments, does not deliver on time, does not bid good enough to compete with its outside competitors) it will be replaced with a more suitable partner.

---

[1] Of course, finding the best partners is an NP-hard problem and can therefore not be achieved. Here and in the following when we use the term best we mean the best possible solution.

- *Autonomy* - Need to balance the autonomy of each individual partner with the co-operative demands of the collaborative cluster – through negotiation that can range from simple bidding (proposal and counter-proposal) to complex argumentation and persuasion strategies. An example may be the following: the cluster sets a deadline and requirements to coordinate among the partners while partners need to argue their position and integrate the deadline with their other priorities. The cluster sets the 'rules of the game' through component protocols [VuJe00]. Preferences can be captured via a utility function such that clustering best partners can be achieved via cost minimization (e.g. via fuzzy entropy minimization [UlNo00]).

- *Security* - An extremely important issue related to inter-platform accessibility at this level are security standards that would enable a fair balance of the autonomy and cooperative forces by enabling enough access to the collaborative cluster's entities to each-other services while keeping secrets safe.

Intra-enterprise Level

Before an enterprise can undertake responsibility for some subtask, it has to find out about its own internal resources to ensure that it can deliver on time according to the coordination requirements of the collaborative cluster and with the required quality. Since this is done within the enterprise at hand we call this level the *Intra-Enterprise Level*. This level covers all layers within an enterprise with the exception of the lowest level, the atomic autonomous system level.

More specifically the intra-enterprise level can be decomposed into the following well-established layers of abstraction in the control of a flexible (holonic) manufacturing system:

- *Production Planning Level*
  On this level the planning process for a production process is initiated, established, and controlled. It deals with
  ☞ the decomposition of a task into a sequence of production operations together with an appropriate workflow and with
  ☞ the nominal allocation of operations to resource types (but not specific resources or times)

- *Production Control (Scheduling) Level*
  This level mainly comprises the concrete scheduling task by
  ☞ allocating a production operation to a specific (complex) resource
  ☞ controlling the workflow of the production operations
  ☞ laying down the timing (start, duration, completion) for each operation

- *Flexible Cell/Shop Floor Level*
  This level comprises the holons that are formed by the physical systems that belong to the flexible cell/shop floor. More specifically this level
  ☞ initiates tasks (production, transport, etc) involving the actual start times and production settings
  ☞ controls the execution of tasks
  ☞ monitors the task status
  ☞ terminates the task
- *Autonomous System/Machine Level*

This layer is described by the physical body of an autonomous system (like a machine, a tool or an autonomous robot) together with its controlling unit/agent. It does not belong to the intra-enterprise level (instead forms the third and last level) and is mentioned here only for reasons of integrity.

Each level is represented by one or a number of holons each of which representing a unit on this level. Such a unit recursively consists of a set of holons each of which being responsible for a subtask of the overall task of the holon/holarchy at hand. These "sub-"holons do not need to be from lower levels only, but can as well be holons from the same level. For example, a flexible cell holon may comprise holons representing atomic autonomous systems as well as flexible cells which perform tasks that are necessary for the flexible cell holon at hand to be effective.

Planning and dynamic scheduling of resources on all these levels enable functional reconfiguration and flexibility via (re)selecting functional units, (re)assigning their locations, and (re)defining their interconnections (e.g., rerouting around a broken machine, changing the functions of a multi-functional machine). This is achieved through a replication of the dynamic virtual clustering mechanism having now each resource within the enterprise cloned as an agent that abstracts those functional characteristics relevant to the specific task assigned by the collaborative conglomerate to the partner. Re-configuration of schedules to cope with new orders or unexpected disturbances (e.g. when a machine breaks) is enabled through re-clustering of the agents representing the actual resources of the enterprise. The main criteria for resource (re)allocation when (re)configuring the schedules are related to cost minimization achieved via multi-criteria optimization.

Forces to Be Balanced

The following are the requirements that need to be balanced on the intra-enterprise level:

- *Trustworthiness* - Need to keep one's position within the collaborative cluster, by respecting deadlines. Here, planning and scheduling— with ability to reconfigure both plans and schedules—are required. Also relevant is to enforce trust-mechanisms, and the ability to accommodate new orders on-the-fly and prioritize the work flow via reconfigurable scheduling.

- *Autonomy* - Need to keep autonomy and to stay competitive in the market. Security policies; Advertising; Bidding; Interfacing with the cooperative cluster (what do we abstract for each cluster to enforce cooperation and keep position privileged within the cluster while keeping open for new deals and competitive on the market such that I will be preferred when new clusters look for services that I provide?); etc. How does one define 'look-out' ontologies to interact with the cooperative clusters that are 'out there' such as to be able to make the best choices from the offers available, given the limited resources one has and the commitments already made to partners that one does not want to lose.

Atomic Autonomous Systems or Machine Level

The lowest level is the *atomic autonomous systems* or *machine level*. It is concerned with the distributed control of the physical machines that actually perform the work. To enable agile manufacturing through the deployment of self-reconfiguring, intelligent distributed automation elements each machine is cloned as an agent that abstracts those parameters needed for the configuration of the virtual control system managing the distributed production.

Forces to Be Balanced

The following are the requirements that need to be balanced on the atomic autonomous systems or machine level:

- *Timing and precedence relationships*; need to be managed while executing e.g. distributed function blocks.

- *Monitoring and fault recovery*. The purpose of monitoring is to ensure that the control system performs as intended, or, in other words, that no latent faults occur. When monitoring for faults, the control system should watch for failures (events occurring at specific times) and errors (inherent characteristics of the system). The types of responsibilities that our control system will have in this area are: diagnosis of program execution, monitoring for exceptions that are thrown by function block code during execution, and monitoring the system state for inconsistencies (e.g., deadline control).

- *Safety*. To achieve a safe system, typically two general concepts are used. First, safety channels (i.e., fault monitoring and recovery code) are separated from non-safety channels (i.e., control code). This decomposition technique is typically referred to as the "firewall concept". Second, redundancy is applied in the system in the form of homogeneous redundancy where clones or exact replicas of code are used (only protects against random failures), or in the form of diverse redundancy where different means are used to perform the same function (this protects against random and systematic failures).

- *Run-time reconfiguration*, if changes are required unexpectedly. This may involve simply replacing portions of the running application at the granularity level of an individual function block or, the removal of a function block and the addition of a different function block or group of function blocks.

Mediator Holon/Agent

The basic condition for holonic systems is that a holon is simultaneously a "whole" and a "part" of some other whole/holon. Thus the system exhibits a tree-like structure. Each inner node of this tree in fact comprises a set of (sub-)holons that cooperatively work together to solve the specific task that is assigned to this node. In order to communicate with such a set of (sub-)holons one need to define a representative for it. In a MAS environment this representative can be a software agent that undertakes the task of a mediator. It fulfills two main functions. First, seen from the *outside*, it acts as the interface between the agents inside the holon and those outside it; conceptually, it constitutes the agent that represents the holon. Second, seen from the *inside*, it may initiate and supervise the interactions between the group of sub-holons/agents within the holon at hand; this also allows the system architect to implement (and later update) a variety of forms of interaction easily and effectively, thereby fulfilling the need for flexibility and reconfigurability. The mediator encapsulates the mechanism that clusters the holons into collaborative groups. The architectural structure in such holarchies follows the design principles for metamorphic architectures.

From the point of view of the mediator agent the first case describes its *outside view* while the second case describes its *inside view*. Both views are significantly different with the outside view usually being much more demanding than the inside view.

Since the mediator is the common representative for the outside world of the holon it represents, as well as for the inside world, it needs to understand both worlds. For this, two kinds of ontologies are necessary, namely for 'peer-to-peer' communication at each level (that is 'inter-agent' communication among entities that form a cluster); and for 'inter-level' communication that enables deployment of tasks assigned at higher levels (by the mediator) on lower level clusters of resources.

One main obstacle to the meaningful interoperation and mediation of services is the syntactic and semantic heterogeneity of data and knowledge the mediator agent does access and receive from multiple heterogeneous agents (cf. [Syca02]). The functional capability of the agent to resolve such heterogeneities refers to the knowledge-based process of *semantic brokering*. Most methods to resolve semantic heterogeneities rely on using partial or global ontological knowledge[2] that is to be shared among the agents. This requires a mediator agent to provide some kind of ontology services for statically or dynamically creating, loading, managing, and appropriately using given domain-specific or common-sense ontologies as well as inter-ontology relations when it wants to communicate and negotiate with different agents or understand complex tasks.

In fact a mediator agent that is looking for a specific task (service provider) has to parse, understand and validate the offers/service descriptions it gets. This is in order to efficiently determine which of the services and capabilities of other enterprise/holon agents are most appropriate. Typically for this ontology services are used as well. Depending on the complexity and required quality of the service a matching process may rely on simple keyword and value matching, use of data structure and type inferences, and/or the use of rather complex reasoning mechanisms such as concept subsumption and finite constraint matching. Semantically meaningful matching requires the matching service to be strongly interrelated particularly with the class of services enabling semantic interoperation between agents. We will come back to this issue when we will discuss the Semantic Web.

3   Search and Selection Process of Partners

One of the toughest problems with respect to the emergence of holonic enterprises is to find/set up the right mixture of component enterprises in order to maximize the overall outcome/impact. In principle several possibilities exist for the formation of a holonic enterprise. We assume that there exists one initiating agent which starts and is in charge of the formation process (mediator agent of some initiating enterprise).

Especially, when a new business opportunity arises, the initiator of the holonic enterprise has to find in the creation phase (the most) suitable partners. Also, during the normal operation of a holonic enterprise it may be useful to find further members or

---

[2] An *ontology* is a computer-readable representation of the real world in form of objects, concepts, and relationships.

even replace an existing member that does not fulfill its commitments in an appropriate way. With respect to this three questions are to be answered:

1. How to decide what partners are the best ones?

2. How can partners be found?

3. How can the individual goals/beliefs of an agent and the goals/beliefs of the outside world that wants to cooperate with this agent harmoniously be integrated?

3.1   How Can Partners Be Found

A holonic enterprise consists of collaborative entities, where each entity is regarded as a holon and is modeled by a software agent with *holonic* properties. As we have discussed already each such holon is represented by a mediator agent that may has to fulfill (a subset of) the following tasks:

- to understand and interpret the complex task at hand (overall objective of the holonic enterprise)

- decompose it into sub-tasks that can be dealt with by individual enterprises

- searching for and finding possible partners

- organizing the negotiation process and dealing with possible contractors/participants

- controlling and supervising the execution of the (sub-)tasks

- reacting to and solving emerging (unexpected) problems

- storing and maintaining a knowledge base in which experiences of the performance and efficiency of past holonic enterprises and its members are stored

These are very special abilities by which the mediator agent may be equipped. In the following we will first discuss the principle ways in which a mediator may search for/find the appropriate partners. In principle a mediator may engage a broker/facilitator, a matchmaker, may make use of a yellow page service or may rely (totally) on its acquaintances knowledge base.

Broker or Facilitator
A broker agent actively intermediates requested services by taking charge of all communication between requester and provider agents. It typically contacts (a set of) the most relevant provider agent, negotiates for, executes and controls appropriate transactions, and returns the result of the services to the requester agent. Broker agents typically do not provide a global, semantically integrated, consistent information model to their clients but store collected information together with associated ontological annotations in some standardized data (structure) format in one (or multiple) appropriate repositories like in a database.

Matchmaker
The task of a matchmaker agent is much simpler than the one of a broker. It simply pairs requester with provider agents by means of matching given requests with appro-

priate advertised services of registered/known provider agents. The matchmaker returns to the requester a ranked list of advertisements and the contact information of the agents whose advertisement matches the request. The requester chooses the most suitable agent for its needs and interacts/negotiates directly with it to get the requested service. Matchmaking gives a requester agent the full choice of selecting a (set of) provider agents a-posteriori out of the result of service matching. When a service-providing agent registers itself with a matchmaker together with a description of its capabilities, it is stored as an advertisement and added to the matchmaker's database. Thus, when an agent inputs a request for services, the matchmaker searches its database of advertisements for a service-providing agent that can fill such a request. Requests are filled when the provider's advertisement is sufficiently similar to the description of the requested service (service matching).

Yellow Pages
While a matchmaker is an active process since it provides a list of possible service providers yellow pages are a passive concept in that they are just a source of information about service providers. Every agent that wants to offer a service can register itself in the yellow pages and every agent that wants to get some service can read the yellow pages. Yellow pages are more primitive than matchmaker since they do not contain any further information about a service, only possible agents that may perform the service and their contact addresses.

Database of Acquaintances
If a (mediator) agent, which is looking for some service, uses a broker service it gives up control about the selection process and about the knowledge in what way which enterprise has contributed to the solution process. It fully depends on the quality and sincerity of the broker service. Matchmaking already allows the mediator agent to keep much more control about the choice of the participating agents since the mediator agent only gets a list of enterprises that are in principle capable to provide some service for the solution process. However, it is left to the mediator agent to decide on what service/enterprise is to be chosen. The most independent approach is that an agent maintains its own database of acquaintances in which all possible cooperation partners along with their capabilities and past experiences with them are stored. The agent will always first try to find appropriate partner agents in this database. Only if this is not successful it will make use of mediator services/yellow pages. While this approach promises the most independence of an agent from its environment it also requires a pretty sophisticated agent architecture since the agent needs to fully understand the overall goals and requirements of the requests it is supposed to find a solution for.

Types of Mediator Agents in a Holonic Enterprise
The most important level is the inter-enterprise level since here the foundation for a successful holonic enterprise is laid. It seems to be mandatory that the initiator agent of a holonic enterprise has the ability to be and is in full charge of the creation and operation process of the holonic enterprise. This is not only necessary to guarantee that the potential members of the holonic enterprise fully agree on the objectives of the holonic enterprise but it is also necessary to profit in the future from experiences

made during the time of existence of the holonic enterprise. Which member performed in what way? How flexible/motivated/dedicated/passionate/efficient/effecttive/competitive/ trustworthy were they? All these experiences can help to form even better holonic enterprises in the future. This requires the mediator agent to be capable to understand and dissolve complex tasks, to negotiate with possible partners, to control the operation of the holonic enterprise and, especially, to judge its success and the role which each member of the holonic enterprise has played in achieving the results. To meet these demands the mediator agent of a holonic enterprise has to maintain its own database of acquaintances and experiences/history. Additionally in order to improve the overall performance of the holonic enterprise and to react to ever changing demands of the market it may make use of matchmakers respectively yellow pages. Since the overall objectives of a holonic enterprise may be far beyond the expertise, knowledge and capabilities of the holonic enterprise the mediator is a representative of it is necessary that the mediator can understand and interpret semantics in a broad area (possesses a broad world model).

In contrast to that the scope/knowledge/capabilities of a mediator agent on the *intra*-enterprise level do(es) not need to be that brilliant. The area of expertise is limited, namely constraint to the area of expertise of the holonic enterprise at most. The set of possible partners is restricted as well because these are first of all other units within the enterprise. More specifically, a holon deals in its outside view with holons on the next higher level of the architecture while it deals in its inside view with the (sub-)holons that constitute the holon at hand.

Of course, in case of emergencies (machine break-down) or simply because a needed capacity/service/capability is not available even on the intra-enterprise level the necessity may arise to publicly advertise a task/service in the outside world. However, in this case the mediator may make use of a broker in order to settle and process the transaction at hand. The broker can, of course, as well be the mediator agent representing the enterprise at hand.

As we move down the layers of a holonic enterprise we can observe a decrease with respect to the following features:

Time Scale
From top to bottom time scales become shorter and real-time constraints change from soft to hard real-time. While, e.g., a formation process of a new holonic enterprise on the inter-enterprise level may take a longer time and time pressures are only soft (more deliberative style of agent) an agent that represents a tool needs to usually react extremely fast (more reactive style of agent that reacts directly to patterns ("prefabricated" reaction)).

Complexity/Sophistication
From top to bottom the degree of agency decreases. The higher level agents are more sophisticated but slower, while lower agents are fast and light-weight. For example, the mediator agent that represents an enterprise needs to have a very broad knowledge and a huge number of sophisticated skills. On the contrary, the agent on the level of an atomic system just needs to have an understanding of this system together with some basic skills to communicate with the next higher level agents.

Autonomy

Since each agent will act according to its own goals conflicting situations may occur when a contacted agent does not want to contribute (fully) to the task at hand (due to overload, preference/priority given to more lucrative offers, etc.). In this case the superordinated (higher-level) agent has to decide on how to deal with such obstacles and, if possible, to develop an alternative solution on the basis of all the information obtained from all its subordinated agents (using its knowledge base and goals). Especially in emergency cases it may be necessary to compromise the autonomy of agents completely. Again, agents that represent an enterprise will be fully autonomous and, therefore, cannot be forced from the outside to do something they do not want to do. On the intra-enterprise level the autonomy will decline with decreasing level. If, e.g., a task was accepted for which a very high contract penalty is to be paid in case it cannot be executed on time higher levels may decide that lower levels need to execute some task in order not to risk a delay of the high-priority task, even if the lower level holons are already fully booked.

The above observation requires designing mediators on different levels differently. We will come back to this topic in section 3.2 when we will discuss how outside goals (that are imposed on an agent from the outside world) and inside goals (the goals of an agent) can be harmoniously integrated.

3.2   Calculation of and Determination on the Best Partners

In the open environment created by the dynamic Web opportunities for improvement of an existing holonic enterprise arise continuously. New partners and customers alike come into the virtual game bidding their capabilities and money to get the best deal. Staying competitive in this high dynamics requires openness and ability to accommodate changes rapidly through a flexible strategy enabling re-configuration of the organization. In response to this need we have designed an evolutionary search strategy that enables the holonic enterprise to continuously find better partners fitting the dynamics of its goals as they change according to the market dynamics.

We regard 'the living Web' as a genetic evolutionary system. The selection of the agents (partners) that best fit the holarchy's objective is done in a similar way to the natural selection by 'survival of the fittest' through which the agents/partners that best suit the holonic enterprise with respect to the goal accomplishment are chosen from the offers available on the Web. In this search model the mutation and crossover operators represent probabilities of finding 'keywords' (describing the attributes required from the new partners searched for) inside the search domain considered. Our construction is based on the observation that the search process on an agent domain[3] (cf. [UlCN01]) containing information about a set of agents that 'live' on the Web (e.g. a directory 'look-up'-like table of 'yellow page' agents describing the services that the possible partners offer (cf. [UNKS00])) is analogous to the genetic selection of the most suitable ones in a population of agents meant to 'fit' the holonic enterprise's goals.

The main idea is to express the fitness function (measuring how well the new agent fits the holarchy's goal) in terms of fuzzy entropy. With this, minimizing the entropy

---

[3] See the FIPA architecture standard at www.fipa.org

across the extended MAS (which includes the agents from the search domain) according to holonic enterprise goals equates optimizing the fitness function which naturally selects the best agents fitting the optimal organizational structure of the holonic enterprise (cf. [UlRa99]). Once a goal is assigned for the MAS agents may cluster in various ways to work cooperatively towards the goal's accomplishment. We define a source-plan as a collection of different clustering configurations in which the agents team-up to accomplish the holarchy goal.

Optimal knowledge at the holarchy's highest level of resolution (inter-enterprise level) corresponds to an optimal level of information organization and distribution among the agents within all levels of the holarchy. We consider the entropy as a measure of the degree of order in the information spread across the multi-agent system modeling the holarchy. One can envision the agents in the MAS as being under the influence of an information "field" that drives the agent interactions towards achieving "equilibrium" with other agents with respect to this entropy[4]. This information is usually uncertain, requiring several ways of modeling to cope with the different aspects of the uncertainty. Fuzzy set theory offers an adequate framework for dealing with this uncertainty (cf. [KlFo88]).

The generalized fuzzy entropy is the measure of the "potential" of this information field and *equilibrium* for the agents under this influence corresponds to an optimal organization of the information across the MAS with respect to the assigned goal's achievement ([UlRa99]). When the circumstances change across the holarchy (due to unexpected events) the equilibrium point changes as well, inducing a new redistribution of information among the agents with new emerging agent interactions.

The essence of this evolutionary search process stems from the recursive modification of the chromosomes in the genotype in each generation while monitoring the fitness function. At each iteration (that is whenever a new agent domain is searched) all members of the current generation (that is the existing agents in the holarchy and the new ones searched for) are compared with each other in terms of the preference measures. The ones with highest preferences are placed at the top and the worst are replaced with the new agents. The subsequent iteration resumes this process on the partially renewed population. In this way the openness to new opportunities for continuous improvement in the holonic enterprise constituency is achieved and with this the emergence of an optimal structure for the holarchy. Embedding this strategy in the holon endows the holonic enterprise with the capability to continuously evolve towards a better and better structure by bringing to the table better and better partners as they are found. A detailed description of this fuzzy-evolutionary approach can be found in [Ulie02a] and [Ulie02b].

3.3 Intentional Problem Solving

While we have discussed in the previous sections what methods/concepts can be used to find and compare the offers of different provider agents we will now discuss how this ability can be implanted into an agent. More specifically, we will concentrate on

---

[4] The information 'field' acts upon the agents much in the same manner as the gravitational and electromagnetic fields act upon physical and electrical entities respectively.

- what kind of goals need to be deeply embedded in a (mediator) agent's belief system in order for it to be capable to truly represent its underlying holon on the one hand and consider the goals of the environment it is embedded in on the other hand and how this can be realized.

We will now concentrate on a single organizational unit and its (problem solving) behavior. Such a unit can be a profit-center (on the level of production planning, shop floor or flexible cell) or an individual corporation. In order to provide complex products or service bundles first of all a comprehensive planning process is to be performed. To achieve flexibility and dynamicity the process needs to rely on the concept of modularized capabilities. Thus planning enables the identification of those elementary products or services that can be combined for product or service bundles. As is illustrated by Fig. 2 planning can be performed top-down, that is by problem decomposition[5], and bottom-up, that is by service or product aggregation.

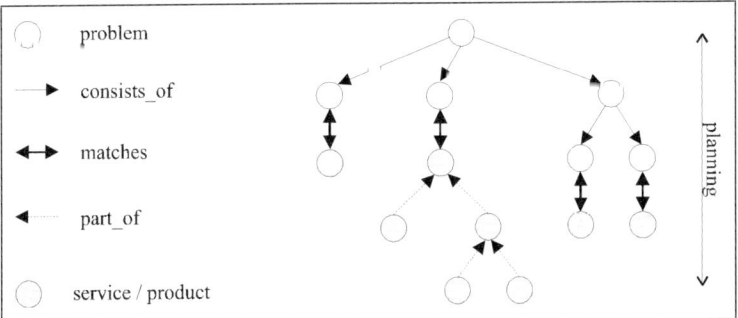

Fig. 2. Planning

To find the optimal solution for such a kind of planning in general is impossible since it is NP-hard. To nevertheless provide a solution simplifications or heuristics have to be applied which are supposed to be good enough to provide a solution that is at least better than the average. From the perspective of the organizational unit that alternative has to be chosen that promises the highest profit[6]. Therefore, the organizational unit has to open up its planning for this profit dimension in order to behave economically reasonable. Consequently, two different kinds of goals can be distinguished and described: *output* goal and *system* goal. According to [Perr70] the former stands for the plan goal that is to be achieved, that is the problem that has to be solved, and the latter describes the real purpose of the organizational unit, which often means profit maximization.

An agent that represents such a unit must consider both goals. Systems showing a behavior that is dependent on certain goals in a reasonable way are characterized as intentional [Denn87]. Intentionality is a key characteristic of agent-based technology.

---

[5] It is important to understand that the decompositions have to completely cover the original problem. If, as a result of the decomposition process, a subsequent synthesis is required, it has to be encoded in one of the sub-problems.

[6] For reasons of simplicity we concentrate exclusively on profit. Of course, many other aspects can also be taken into consideration.

In addition agent methodology genuinely addresses planning respectively deliberative behavior [WoJe95] and cooperation.

The coexistence of two kinds of goals that have both to be ensured by the generated plan necessitates a two-staged planning process. Due to the overriding importance of the system goals in comparison to output goals the traditional planning process that mainly concentrates on the realization of the output goal has to be revised in a way that it reflects the system goals in an appropriate way. However, system goals, to become effective, must strongly influence the planning process as such. Planning processes can best be influenced if they are controlled by meta-planning processes, that is, processes on a more abstract level. In such an architecture the basic problem solving behavior of an agent can be illustrated as in Fig. 3 which visualizes the interdependencies between the most important entities involved.

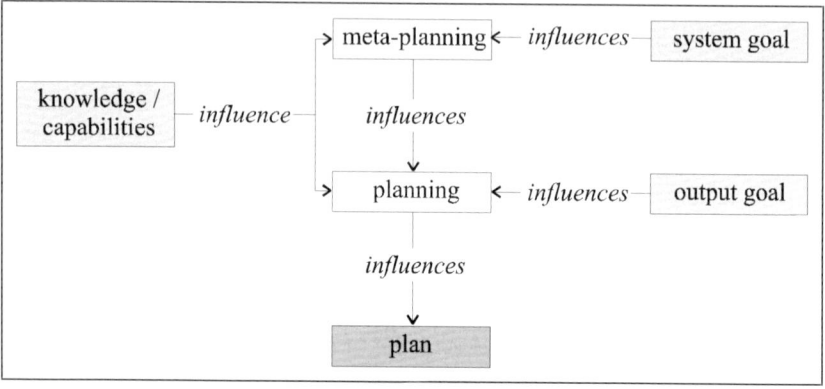

Fig. 3. Meta-panning vs. panning

An in-depth discussion of both planning and meta-planning procedures is beyond the scope of this paper, especially since traditional planning has already been intensively studied in literature. Instead, we will only concentrate on meta-planning. Meta-Planning has to ensure that, based on the system goals, a rational choice can be made among planning actions that constitute alternatives[7] with respect to the given output goal. Consequently meta-planning has to control the planning process either in a dynamic or in a static manner:

Dynamic meta-planning stands for the direct control of the planning process. This means that whenever the planning process has to make a choice among alternative decompositions or, correspondingly, among alternative products or services the meta-planning process will determine the one from which it assumes that it is the most appropriate one with respect to the system goals.

---

[7] This choice among alternatives does not comprise all choices that are available to the planning procedure. Alternative in this context means that the respective products or services are likewise applicable within a given planning stage. Without meta-planning the planning procedure would use heuristics to decide on one of them, because from a planning perspective they equally serve the output goal. Otherwise meta-planning would actually comprise planning.

Static meta-planning describes the procedure where first of all a number of relevant alternative plans are generated. Subsequently that plan is chosen that reflects the system goals best. This corresponds to an extensive, however, non exhaustive[8] search through the set of all possible plans. This set represents the overall search space and is obviously relatively complex.

Both alternatives operate analogously on two subsequent levels of aggregation: Depending on the respective system goal both select one element from a set of alternatives. Dynamic meta-planning does so for a set of decompositions respectively services or products whereas its static counterpart works on a set of plans that again constitute a set of decompositions and products or services. Since the system goal has to enable a total ordering of the alternative elements it has to be defined by referencing some of their characteristics (e.g., profit contribution of each product). Consequently, system goals can be distinguished according to the kind of element they refer to. Thus dynamic meta-planning requires a system goal that is located at a decomposition / product or service level and static meta-planning necessitates system goals that are situated at plan level. By continuing analogously the aggregation by combining alternative plans we get to a system goal at a plan-set level.

Within the scenario of this paper such system goals can be formulated as follows:

Decomposition / Product or Service Level:
i. "Make as much profit as possible by preferring certain products/services."

Plan Level:
ii. "Make as much profit as possible from each task respectively problem."

Plan-Set Level:
iii."Make as much profit as possible within a certain period of time."

Achieving a system goal at the plan-set level would constitute an "ex-post" meta-planning and can consequently not be implemented directly, because a runtime control of the planning procedure can only be realized by applying either dynamic or static meta-planning. However, both require system goals located at "lower" levels, i.e., plan or decomposition / product or service level. This motivates a top-down transformation of system goals that in general cannot be performed unambiguously[9] and consequently requires the use of heuristics[10].

ad i.   Dynamic meta-planning can be applied, e.g. by prioritizing the decompositions respectively products or services so that the planning procedure obey the system goal as far as possible. This means it will apply the specific decomposition or use the specific product whenever possible for a given problem.

ad ii. Either static or dynamic meta-planning can be applied. Static meta-planning simply has to evaluate all plans that have been generated for the problem at hand and has to select the most profitable one. Dynamic meta-planning would have to rely on heuristics, since the system goal cannot be implemented directly at the product level. Such a heuristic could be to prefer always the most profitable product among a set of alternative ones.

---

[8] An exhaustive search would be NP-hard.
[9] In general it is not possible to derive the most profitable plan by simply choosing the most profitable decomposition or activity among emerging alternatives.
[10] Algorithm that can not guarantee an optimal performance in any case but an improvement for the average-case performance [RuNo95, p. 94].

ad iii. Either static or dynamic meta-planning can be applied but both have to rely on heuristics. Such a heuristic could be to prefer always the cheapest plan for the customer based on the assumption that most profit can be made if as much customers as possible can be attracted by extremely attractive offers. Static meta-planning would consequently select the cheapest plan out of the set of alternatives. Dynamic meta-planning would have to rely on a further heuristic, since the former cannot be implemented directly at the decomposition / product or service level. This second heuristic could analogously prefer always the cheapest product among a set of alternative ones.

Figure 4 summarizes and visualizes the respective kinds of system goals.

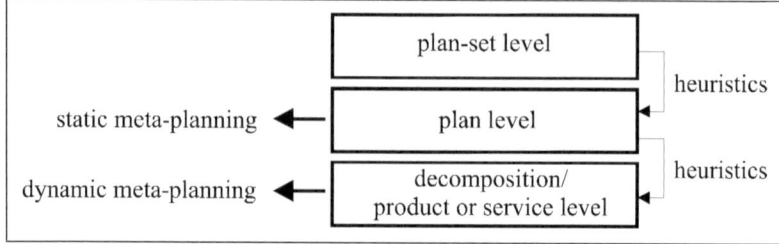

Fig. 4. System goal classification

Up to now we have (implicitly) concentrated on the highest level, the inter-enterprise level. If we want to apply the discussed concept of goal implantation on lower level mediator agents as well we need to consider the ranking/importance of the goals for these kinds of agents. For example, the mediator agent representing an enterprise needs to be highly autonomous, which means that is strictly acts only according to the objectives of its enterprise and cannot be prevented by some other agent to do that. This clearly indicates that goals of the agent need to be modeled as system goals. On the other hand, the agent representing an atomic system/machine has low autonomy. Only if the superordinated levels do not intervene it can react only according to its own goals. Therefore, here the goals of the agent need to modeled as the output goals. A more detailed discussion of intentional problem solving can be found in [Wank97].

4   Conclusion

In the context of a global, web-driven economy, the capability of enterprises to flexibly join groups on the inter-enterprise level and form highly efficient groups on intra-enterprise is of high importance for the possible survival of enterprises. This paper presented the key concepts of a new e-Business model that enable the development of global collaborative e-Commerce/e-Business applications. With intentional problem solving an approach has been presented that allows agents to cluster and collaborate in a way that fits best to the individual goals of each agent (e.g. maximum profit) as well does consider the objectives of its encompassing environment.

Currently we are working on a smoother integration of agent technology, Web Services and the specification of workflows on the Internet (Web Services flow). We believe that this will lead to highly competitive enterprises which can flexibly deal with all kinds of problems in a timely way.

References

[ByBP93]   Byrne, J.; Brandt, R.; Port, O.: *The Virtual Corporation.* In: Business Week, 8. Feb. 1993, pp. 36-40.

[CaAf01]   Camarinha, L.; Afsarmanesh, H.: *Virtual Enterprise Modeling and Support Infrastructures: Appyling Multi-agent Sytsem Approaches;* in: Multi-Agent Systems and Applications, 9th ECCAI Advanced Course; ACAI 2001 and Agent Link's 3rd European Agent Systems Summer School, EASSS 2001, Prague, Czech Republic, July 2-13, 2001, Selected Tutorial PapersLNAI 2086, p. 335 ff.

[CaAf99a]   Camarinha, L.; Afsarmanesh, H.: *Infrastructures for Virtual Enterprises: Networking Industrial Enterprises*; IFIP 27; Kluwer Academic Publishers; ISBN: 0-7923-8639-6; 1999

[CaAf99b]   Camarinha, L.; Afsarmanesh, H.: *The Virtual Enterprise Concept*; in [CaAf99a]; 1999

[CaAf99c]   Camarinha, L.; Afsarmanesh, H.: *Tendencies and General Requirements for Virtual Enterprises*; in [CaAf99a]; 1999

[Cast95]   Castelfranchi, Ch.: *Guarantees for Autonomy in Cognitive Agent Architecture.* In: Wooldridge, M. and Jennings, N. (eds.): Intelligent Agents. Lecture Notes in AI 890, Springer-Verlag, 1995, pp. 56 - 70.

[DaMa92]   Davidow, W.; Malone, M.: *The virtual corporation: structuring and revitalizing the corporation for the 21st century.* Harper-Collins Publishing Comp., New York, 1992.

[Denn87]   Dennett, D. C.: *The Intentional Stance.* The MIT Press, 1987.

[FaSJ00]   Faratin, P., Sierra, C., N. Jennings (2000) *"Using similarity criteria to make negotiation trade-offs"*, Proc. 4th Int. Conf. On Multi-Agent Systems, Boston, pp. 119-126.

[FiFR02]   Klaus Fischer, Petra Funk, and Christian Ruß; *Specialized Agent Applications*; in [LMST02]; 2002

[Ioni97]   Ionita, S.: *Genetic Algorithms for Control Engineering Applications* (in Romanian), ECIT-97, Pitesti, Romania, 21-22 Nov. 1997, p. 77-84.

[JFNO00]   Jennings, N, R., Faratin, P., Norman, T.J., O'Brien, P. and Odgers, B. (2000) *"Autonomous Agents for Business Process Management"*, Int. J. of Applied Artificial Intelligence 14(2), 145-190.

[KlFo88]   Klir, G; Folger, T: *Fuzzy sets, Uncertainty, and Information*, Prentice Hall Publishing Company; 1988

[Koes67]   Koestler, A. (1967), The Ghost in the Machine, Hutchinson & Co, 1967

[Lase98]   T.M. Laseter. *Balanced Sourcing: Cooperation and Competition in Supplier Relationships.* Jossey-Bass, October 1998. ISBN: 0787944432.L

[LaSw99]   Ora Lassila and Ralph R. Swick: *Resource Description Framework (RDF) Model and Syntax Specification*; W3C recommendation, Feb. 1999.

[LiYe01]   Liu, J.; Ye, Y (Eds.): *E-Commerce Agents: Marketplace Solutions, Security Issues, and Supply and Demand*; Springer Publishing Company, Lecture Notes in Artificial Intelligence LNAI 2033; 2001

[LMST02]   Michael Luck; Vladimír Mařík; Olga Stěpánková; Robert Trappl (Editors): *Proc. Advanced Course on Artificial Intelligence (ACAI 2001) "Multi-Agent Systems and Their Applications (MASA)"*, Prague, Czech Republic; July 2001; Joint event of ECCAI (European Coordinating Committee for Artificial Intelligence) and AgentLink (European Network of Excellence for Agent-Based Computing); Springer Lecture Notes in Artificial Intelligence (LNAI) 2086; 2002

[LSDD00]   E. Lupu, M Sloman, N. Dulay, N. Damianou: *Ponder: Realising Enterprise View-point Concepts* 4th International Enterprise Distributed Object Computing Conference (EDOC2000), Makuhari, Japan, 25-28 Sept. 2000, pp.66-75.

[LuSl99]   E. Lupu and M. Sloman "*Conflicts in Policy-based Distributed Systems Management*" IEEE Transactions on Software Engineering - Special Issue on Inconsistency Management, Vol 25, No.6 Nov. 1999, pp. 852-869.

[MaNo96]   Maturana, F. and D.H. Norrie (1996) *Multi-agent Mediator Architecture for Distributed Manufacturing.* Journal of Intelligent Manufacturing, 7, pp.257-270

[MHWW95] McHugh, P., Wheeler, W., Merli, G. (1995). *Beyond Business Process Reengineering: Towards the Holonic Enterprise*, John Wiley & Sons, Inc., Toronto.

[Perr70]   Perrow, Ch.: *Organizational analysis: A sociological view*. Wadsworth Pub Co; ASIN: 0818502878; (June 1970)

[RuNo95]   Russell, S.; Norvig, P.: *Artificial intelligence: a modern approach*. Prentice-Hall, Upper Saddle River, NJ, 1995.

[ShSN99]   Shu, Sudong and D.H. Norrie (1999) "*Patterns for Adaptive Multi-Agent Systems in Intelligent Manufacturing*", Proc. of the 2nd International Workshop on Intelligent Manufacturing Systems (IMS'99), Leuven, Belgium, pp. 67-74, September 22-24, 1999.

[SmDa81]   R. Smith, R. Davis: *Frameworks for Cooperation in Distributed Problem Solving*, IEEE Transactions on Systems, Man and Cybernetics, Vol. SMC-11 1981, pp. 61-70.

[Syca02]   K. Sycara: *Multi-agent Infrastructure, Agent Discovery, Middle Agents for Web Services and Interoperation;* in [LMST02]; 2002

[TiBU02]   Tianfield, H.; Binder, Z.; Unland, R.: *Multi-Agent Interactions under Organizational Hierarchy*. 2nd IFAC/IFIP/IEEE Conference on Management and Control of Production and Logistics (MCPL'00), Grenoble, France, July 5-8, 2000

[TiUn02]   Tianfield, H.; Unland, R. (guest eds.) Special Issue "*Virtual Organisations and E-Commerce*"; International Journal of Information Technology and Decision Making; World Scientific Publishing Co.; Vol. 1, Issue 3, Sept. 2002

[TiUn99a] Tianfield, H.; Unland, R.: *On the Hierarchical Structure of Multi-Agent Systems;* 1st International IFAC/IFIP/IFORS Workshop on Multi-Agent Systems in Production (MAS'99), Vienna, Austria, December 2-4, 1999

[TiUn99b] Tianfield, H.; Unland, R.: *Formulating Enterprise Federation Into Multi-Agent Systems*; 1st International IFAC/IFIP/ IFORS Workshop on Multi-Agent Systems in Production (MAS'99), Vienna, Austria, December 2-4, 1999

[UlBW02]   Mihaela Ulieru, Robert Brennan and Scott Walker, "*The Holonic Enterprise – A Model for Internet-Enabled Global Supply Chain and Workflow Management*", International Journal of Integrated Manufacturing Systems, No 13/8, 2002, ISSN 0957-6061.

[UlCN01]   Ulieru, M, Cobzaru, M. and Norrie, D. *A FIPA-OS Based Multi-Agent Architecture for Global Supply-Chain Applications*, IPMM 2001 International Conference on Intelligent Processing and Manufacturing of Materials, July 29-August 3, 2001, Vancouver, BC.

[Ulie02a] Ulieru, M. (2002). "Emergence of Holonic Enterprises from Multi-Agent Systems: A Fuzzy-Evolutionary Approach", Invited Chapter in Soft Computing Agents, (V. Loia; Ed.), IOS Press , 2002 (in print).

[Ulie02b] Ulieru, M.; *Internet-Enabled Soft Computing Holarchies for e-Health Applications-Soft Computing Enhancing the Internet and the Internet Enhancing Soft Computing-*, Invited Chapter in Fuzzy Logic and the Internet, (Zadeh, L.A. and M. Nikravesh, Eds.), Physica Verlag; 2002

[UlNo00]   Ulieru, M. and D.H. Norrie (2000), "*Fault Recovery in Distributed Manufacturing Systems by Emergent Virtual Re-Configuration: A Fuzzy Multi-Agent Modeling Approach*", Information Science, 7669, ISSN # 0020-0255, September 2000.

[UlRa99]   Ulieru M, Ramakhrishnan S (1999): *"An Approach to the Modelling of Multi-Agent Systems as Fuzzy Dynamical Systems"*, Advances in Artificial Intelligence and Engineering Cybernetics, Vol. V: Multi-Agent Systems/Space-Time Logic/Neural Networks (George Lasker, Ed.), IIAS-68-99, ISBN 0921836619

[UlWB01]   Mihaela Ulieru, Scott Walker and Robert Brennan, *"Virtual Enterprise as a Collaborative Information Ecosystem"*, Proc. Workshop: Holons, Autonomous and Cooperative Agents for the Industry, Montreal, Canada, May 20, 2001 (AA 2001).

[UNKS00]   Mihaela Ulieru, Douglas Norrie, Rob Kremer and Weiming Shen: *A Multi-Resolution Collaborative Architecture for web-Centric Global Manufacturing*, Information Science, Volume 127, Journal no.: 7669, ISSN # 0020-0255, August 2000.

[VuJe00]   Vulkan, N. and N.R. Jennings (2000), *"Efficient mechanisms for the supply of services in multi-agent environments"*, Int. Journal of Decision Support Systems 28(1-2), pp. 5-19.

[Wank97]   Wanka, U.: *Multiagent Problem Solving*; PhD-Thesis; University of Essen; 1997

[WoJe95]   Wooldridge, M.; Jennings, N.: *Agent Theories, Architectures, and Languages: A Survey*. In: Wooldridge, M. and Jennings, N. (eds.): Intelligent Agents. Lecture Notes in AI 890, Springer-Verlag, 1995, pp. 1-39

A Multi-agent System for E-insurance Brokering

Luís Nogueira[1] and Eugénio Oliveira[2]

[1] Instituto Superior de Engenharia do Porto
Instituto Politécnico do Porto
luis@dei.isep.ipp.pt
[2] Faculdade de Engenharia
Universidade do Porto, LIACC
eco@fe.up.pt

Abstract. Agent-based systems suitable for dealing with applications where the environment is both dynamic and populated with competitors demand for sophisticated characteristics including adaptation, negotiation and coordination. In this paper we propose an agent-mediated insurance brokering system using a flexible negotiation model that includes multi-attribute bidding as well as some kind of learning capabilities. Moreover, in the core of the provided brokering facility, we are using conceptual clustering procedures as an approach to better match customers and insurance product offers providing a valuable add-on to both customer's and sellers' sides. Intelligent agents engage themselves in a negotiation process by exchanging proposals and counter-proposals trying to convince opponents to modify their bidding values. We are now developing a Java based multi-agent infrastructure specifically dedicated to the insurance e-commerce domain, exploiting Toshiba's Bee-gent framework. For both acceptability and generalisation purposes, XML (including appropriate ontology-based messages) has been chosen as our agent communication format.

1   Introduction

In most current e-commerce applications, the buyers are generally humans who typically browse through a catalogue of well defined products (books, computer components, CDs) and make fixed price purchases. However, there are important differences between selling this type of goods and selling insurance and other financial products over the Internet. This paper explores some of those differences from both the insurers' and customers' perspectives and describes the development of a multi agent system through which products (and services) offered by insurance companies could be better evaluated and selected. From the customers side point of view, more interesting information can be found, even things that the customer did not think of before. On the other hand, the electronic market system, with its intimate knowledge of who the user is and what he wants, can shorten the time needed for finding an appropriate insurance product. Insurers can then use information automatically collected during negotiation to develop

R. Kowalczyk et al. (Eds.): Agent Technology Workshops 2002, LNAI 2592, pp. 263–282, 2003.

a more customer-directed kind of marketing strategy. Information about the customer can be used to find out what he is interested in and, therefore, a more personalised product could be offered.

2   Electronic Markets and Insurance

Despite e-commerce's huge impact on business in general, the insurance industry has yet to fully embrace it. The insurance industry faces several obstacles to e-commerce, including customer attitudes, complex insurance policies, state regulations and the traditional agency distribution system.

The insurance business is based on selling a service at a given risk. Insurers have to make premiums high enough to cover the forecast level of claims but also to keep them low enough to be attractive in an increasingly competitive market. The balance between profit and risk is fundamental to the success of any insurance business.

Insurance companies traditionally have segmented the policyholders into separated lines of business such as auto, life and business markets. Each segment would carry its own underwriting, claims and marketing strategy. This made it impossible to get a complete view of multiple relationships a customer might have with one company.

E-insurance involves the advertisement, recommendation, negotiation, purchase and claim settlement of insurance policies through the Internet. At present, most of these processes are not automated. Some insurance sites offering web-based policies are little more than passive catalogues of alternatives available to customers.

The success of the sale of an insurance policy depends on how good the requirements of the insurer have been matched with the terms of the policy. In the conventional insurance industry, the insurance company initially informs its customers through advertisements. Advertisements are made either through passive channels like newspapers, magazines, billboards, radio and television, or, through active channels like human insurance agents. E-insurance employs the Internet to reach customers through advertisements more effectively since it integrates the traditional passive and active channels of advertisement into one. Advertisement banners, e-mail notifications and coupons are used to replace passive media, while software agents replace their active human counterparts.

However, it's a much difficult task to match the insurance requirements of a customer with appropriate financial complex products than, for example, finding those shops that sell a specific book. An insurance policy has benefits, conditions and exclusions that add detail to the high level coverage features advertised by the insurer.

The availability of an insurance product may be determined according to the risk profile of the customer. Customers will be asked to supply enough information from which insurers have the ability to make a proposal. An electronic market will therefore need to provide a facility, capable of exploiting this bidirectional exchange of information.

3   Existing Online Insurance

Despite the increasingly rapid advances made in computer technology, companies in the insurance industry are making the most use of the new technical possibilities offered today in their internal operations only, while hardly employing them at all when negotiating with their customers. A recent comparative study of 25 web-based Internet sites offering comparative term life insurance information from Consumer Federation of America has showed us that not all sites are useful for getting quotes. Some of them are too difficult to use and others little more than referral services where you are put in touch with an agent, something you, most of the time, do not need the Internet to achieve. Several of the quote services do not include no commission insurance companies, because many of this sites make money through commissions on sales and do not show this companies in their service because it would reduce their incomes [9].

Most web sites offering online quotation and purchase of insurance products are implemented by the insurers and sell directly to the customer, excluding the broker. Brokers, however, provide a valuable service and are widely used by customers. Yet, online brokerage is rare.

Those sites, which do offer a brokerage service, do so by drastically simplifying the problem: they standardise offered products. The broker's role is then reduced to collect a standard set of information from the customer and negotiate standard coverages. This gives no advantage for any of the players because:

- customers are provided with a more limited choice of products, which may not meet all their requirements.
- insurers have limited flexibility in product design, targeting and pricing.
- brokers lose their traditional role.

4   Insurance Brokerage

Insurance brokerage is a process involving three types of players:

4.1   Customers

The customer wishes to buy an insurance to cover certain risks. This requirement will usually be incomplete and uncertain and, possibly, the customer will not be aware of all the options available and may be prepared to compromise on certain aspects. The customer expects the broker to help and advise in defining his needs, to select an insurance product and to appropriately deal with insurance companies always taking customers' preferences into consideration.

4.2   Insurers

Each insurance company offers a number of insurance products, covering different risks and aimed at different groups of customers. Each product is only available to customers who satisfy a complicated set of rules, designed to minimise the risk of the insurer. The product is usually configurable, allowing the customer to select the amount of each type of cover and a number of other different optional extras. The premium charged by the insurer takes into account the characteristics of the customer and the risk insured, as well as the configuration of the product.

4.3   Broker

The broker mediates between customers and insurers, attempting to supply each customer with a product appropriate to his needs at an acceptable price to both parties. To accomplish this role, a broker needs to execute four different, although related tasks:

- To gather and filter information about available products.
- To discuss customers' requirements and provide adequate advice.
- To negotiate product details with insurers on behalf of the customer.
- To establish the final contract involving both the customer and the selected insurance product provider.

5   Agent Mediated Insurance Brokerage

5.1   Our Proposal

It is our belief that intelligent agents are well suited to deal with the insurance brokering problem in a distributed manner. By configuring a society of intelligent agents, each one charged with autonomously carrying out different specific functionalities, the insurance broker system will not only be able to analyse products being offered, but will also deduce useful information regarding the current state of the market.

We here present a distributed, intelligent agent-based system making it possible the electronic commerce of insurance products. Our approach for an agent-based insurance products assisted electronic market includes an agent representing each of the insurers, an agent representing the customer and a broker agent for intelligent brokering services. Each insurer has full ownership of its agent, ensuring that all strategic information remains confidential. The goal is to support distribution of a full range of insurance products from several different insurers without the need to modify or constrain them in standard rigid formats for electronic commerce purposes.

By increasing the degree and the sophistication of the automation process, commerce becomes much more dynamic, personalised and context sensitive. These changes can be of benefit of both customers and insurers. From the customers' perspective, it is desirable to have software that could search all the available offers to find the most suitable one and that could then go through the process of actually purchasing the product.

From the insurer's perspective it is desirable to have software that could vary its own offering depending on the customer it is dealing with, what its competitors are doing and the current state of its own business.

We use a common descriptive language for describing the terms, conditions and relationships of insurance products. These terms can define any aspect of the market. Relationships can be used to define dependencies between terms and values of these terms. A product is then defined as an encapsulation of an attribute set to define an individual item or service for sale by an individual insurance company in the market. A requirement is the specification of what a customer wishes to buy. It stores attributes about the customer and the details of his insurance requirements.

Given a customer's details and requirements, the broker must find a match with the list of insurance products proposed by insurers. The customer will have specific requirement constraints and preferences as to his ideal product. Insurance products will have a set of eligibility criteria which may exclude some customers or require further information to be determined. In our system both products and requirements have attributes, constraints and preferences.

Customers need to supply enough information from which insurers have the ability to make a proposal. Different insurers will require different information in order to make a decision. The simplest solution would be to display all the required questions to answer. However, this may prove to be too extensive list to be acceptable to most customers and eligibility often involves dependencies rather than straightforward constraints.

Communities of users can be used to improve the negotiation of insurance products. In this paper we examine *unsupervised learning* for the acquisition of user communities. The question is whether there is any meaning in the generated communities, that is if they associate users with a limited set of common interests. For this reason we use a metric to decide which preferences are most representative for each community. This approach allows the insurance company to target product configurations at specific market segments, and avoids the need to ask all customers the same typically large number of product specific questions.

In order to reach an agreement about a particular insurance product a negotiation process is started by the broker. This negotiation process comprises several rounds, starting when the broker sends and announcement for all the insurer agents in the market. The negotiation ends when a deadline is reached or a satisfactory proposal is received. At each negotiation round bids are evaluated. Bids evaluation is done through a multi-issue function that encodes the customer's preferences.

A Customer Agent coordinates the dialogue between the customer and the Broker Agent, passing on information as appropriate. It offers the customer a flexible navigation tool that allows the exploration of the received proposals. This is particularly useful because the customer wish to express product feature preferences and view the corresponding proximity of each offered product. The result is a ranking of products, which can be tuned by the customer by varying the preferences and viewing the consequent effect on the ranked list [18]. Such a navigation tool encourages the user to consider non-price related features and helps the customer to explore the trade-off between product features and price. This is not just of benefit to the customer, because insurers have also the means of drawing attention to their products' distinguishing features other than price [7]. This helps the customer to make an informed purchase decision.

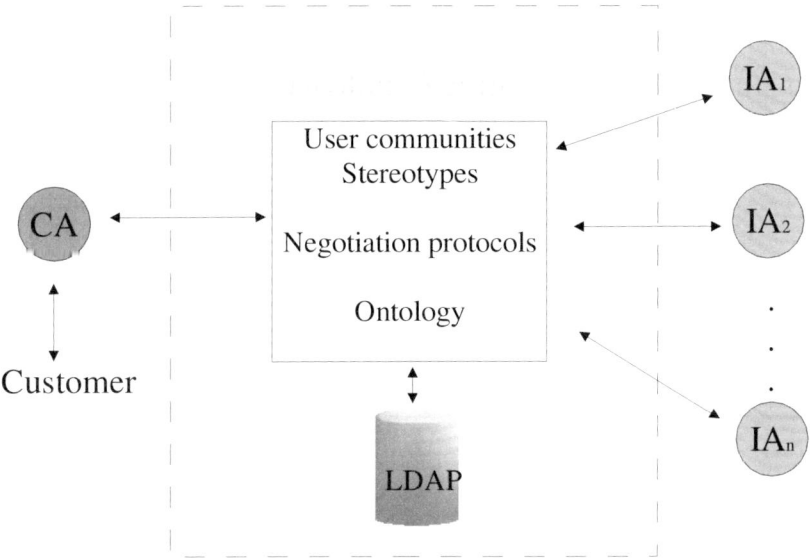

Fig. 1. Agent based architecture for the Insurance Brokering System

5.2   Stages of the Brokering Problem

Figure 1 briefly depicts the overall architecture including needed agents as well as agent interaction links in our agent-based insurance brokering system. Notice that, besides relevant services like negotiation protocols and ontology-based services, BA, the Broker Agent, provides a facility to build up, memorise and exploit User's Stereotypes. Although this concept will be elaborated later on Sect. 5.2, the reader can see a stereotype as a classification of a set of features and characteristics associated with a specific sub-set of customers.

Our model for the insurance brokering activities divides the interaction process aiming at solving the problem into five stages as follows:

In **Stage 1**, the user defines the allowed range for the insurance product attributes' values. This definition includes attaching a degree of importance (weight value) for each one of the product's attributes (in a range from *low* to *high*) and the increasing order of preference for the attributes' values.

Stage 2 , starts when the Customer Agent (CA), instructed by the user, sends a request to the Broker Agent (BA). This request is constructed by selecting the most important value for each one of the user-defined attributes. When it is possible to obtain, according to the defined values, the direction in which the user is willing to relax the value's constraint, this direction is also sent in the initial request. The purpose of all that information is helping the BA evaluate the received proposals.

The BA replies with a stereotype to CA. Based on this received stereotype, the user can refine its initial request. CA re-sends a, possibly improved, request to BA. Now, BA also asks user's preferences to CA.

In **Stage 3**, BA sends an announcement to each Insurer Agent (IA), starting a negotiation process. Each IA replies with bids to BA, which are then evaluated according to CA preferences, extracting relevant features from these bids. At each negotiation round bids are evaluated, and the one with the greatest evaluation value is considered the winner in current round.

When this negotiation process ends, BA starts **Stage 4**, by starting a new interaction with CA. This interaction will ultimately direct the system to a suitable solution through a constraint satisfaction process.

If this conversation has produced a valid number of alternatives, BA initiates **Stage 5** by ranking selected proposals according to user utility function results, and send them plus relevant information to CA. The user either rejects or agrees with one of the received proposals.

If the user has selected one of the proposed insurance products, BA resumed to **Stage 6**, establishing a contract with the winning IA.

6   Learning about User Communities and Stereotypes

6.1   User Communities

Machine learning methods have been applied to user modelling problems mainly for acquiring models of individual users interacting with a system, e.g. [1,2,16]. Recently, other authors like [14] have approached this subject including a higher level of generalisation of the users' interests and leading to the identification of different user communities in a population of users.

The choice of a learning method or algorithm largely depends on the kind of training data that is available. The main distinction in machine learning algorithm paradigm is between *supervised* and *unsupervised* learning. Supervised learning requires the training data to be preclassified. This means that each example is assigned a unique label, signifying the class to which the item belongs. Given these data, the learning algorithm builds a characteristic description of each class, covering the examples of this class. The important feature of this approach is that the class descriptions are built conditional to the preclassification of the examples in the training set. In contrast, unsupervised learning methods do not require preclassification of the training examples. Through these latter methods clusters of examples are built up, which share common characteristics.

When the cohesion of a cluster is high, i.e, the examples described in it are similar, such cluster defines a new class.

User communities can be automatically constructed using an unsupervised learning method. Unsupervised learning tasks have been approached by a variety of methods, raging from statistical clustering techniques to neural networks and symbolic machine learning. The branch of symbolic machine learning that deals with this kind of unsupervised learning is called *conceptual clustering* and a popular representative of this approach is the COBWEB algorithm [3]. Conceptual clustering is a type of learning by observation that is particularly suitable for summarising and explaining data. Summarisation is achieved through the discovery of appropriate clusters, which involve determining useful subsets of an object set. In unsupervised learning each example is an object. Explanation involves determining a useful concept description for each cluster.

COBWEB is an incremental clustering algorithm that employs the concept of *category utility* [5] to create a clustering that maximises inter-cluster dissimilarity and intra-cluster similarity. The category utility of a partition is measured by the following equation:

$$CU = \frac{\sum_k \left(P(C_k) \left[\sum_i \sum_j P(A_i = V_{ij}|C_k)^2 - \sum\sum P(A_i = V_{ij})^2 \right] \right)}{k} . \quad (1)$$

where k is the number of categories or classes, C_k is a particular class, A_i refers to one of the I attributes and V_{ij} is one of the J values for attribute A_i.

COBWEB performs its hill-climbing search of the space of possible taxonomies and uses the expression above for category utility to evaluate and select possible categorisations. It initialises the taxonomy to a single category whose features are those of the first instance. For each subsequent instance, the algorithm begins with the root category and moves through the tree. At each level it uses category utility expression to evaluate the taxonomies resulting from the following four steps algorithm:

1. Classifying the object with respect to an existing class.
2. Creating a new class.
3. Combining two classes into a single class (merging).
4. Dividing a class into several classes (splitting).

Using COBWEB on a small data set generated a concept hierarchy presented in Fig. 2. An important property of the hierarchy is the balanced split of objects in different branches. Therefore the underlying concepts are of similar strength.

6.2   Stereotypes of Customer's Communities

Our insurance brokering system is then using COBWEB as a tool for grouping potential customers in meaningful classes we call stereotypes.

The clusters generated by COBWEB, should represent well-defined customer communities. Besides the customer data used for incrementally build up customer's communities, a customer is characterised by its own preferences and by the

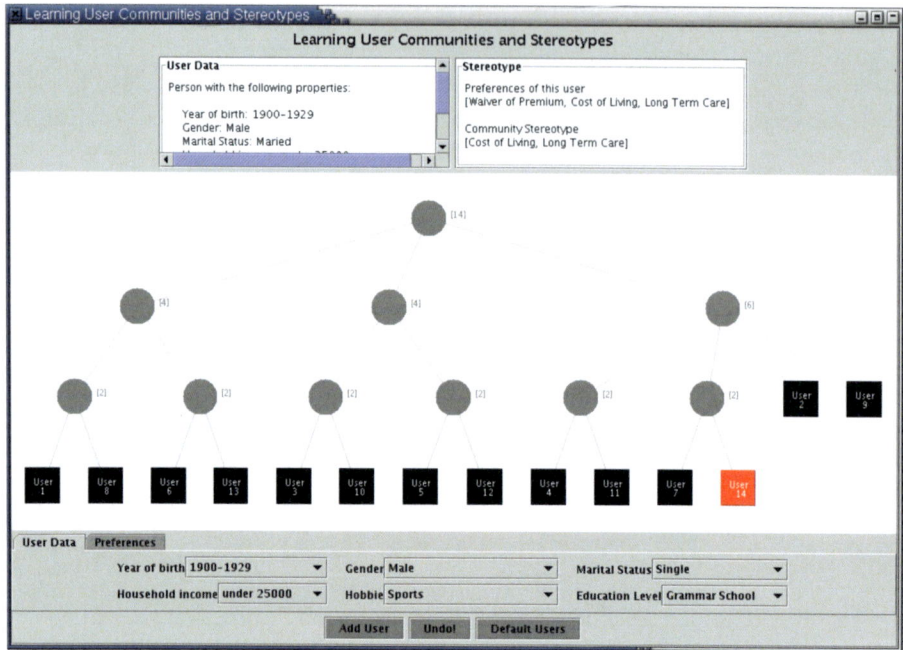

Fig. 2. Learning user communities and stereotypes

insurance configuration chosen in the negotiation process. Thus, the natural way to define meaningful stereotypes associated to customer's communities is by trying to identify patterns that are representative of the participating users. We try to construct a prototypical model for each community, which is representative of its users and significantly different from other communities of users.

In order to build appropriate stereotypes, our system is using a specific metric to measure the increase in the frequency of a specific preference or negotiation result within a given community, as compared to the default frequency in the whole number of available observations [15]. In [14] and [4] the increase in frequency was used as an indication of the increase in the predictability of the feature (a given preference, for example) within the community. Given a component c (a user preference), with the default frequency f_c, if the frequency of this component within a community i is f_i, the frequency increase is defined as a simple difference of the squares of the two frequencies:

$$FI_c = f_i^2 - f_c^2 . \qquad (2)$$

When FI_c is negative there is a decrease in frequency and the corresponding component is not representative of the community. A community's representative characteristic is found through $FI_c > \alpha$, where α is pre-established as the required threshold for considering that frequency increase enough relevant.

The hierarchy generated in Fig. 2 has showed us that the insurance attributes that are chosen by either to few or to many users do not appear in the constructed community stereotype. In the former case the algorithm ignores them during learning and in the latter case, they correspond to such general interests, that they cannot be attributed to particular communities. Filtering out this to types of preferences is a positive feature of the used metric [15].

7   Negotiation

7.1   Negotiation Protocol

Most of the well-known negotiation protocols deal with a single dimension (usually the price). Moreover, [20] also says that the preferences' reduction to the price single figure is what characterises the market. However, this seems to be a bit simplistic and unrealistic for the insurance domain.

Some authors [19,10] advocate the use of auctions protocols for agent mediated electronic commerce, arguing that they are widely recognised by economists as the most efficient way of resolving one-to-many negotiations.

Other authors point out the limitations of auctions protocols and look for more flexible negotiation models [7]. They stress out that online auctions are in fact less efficient and more hostile than it would be desired. For example, the winner's curse, i.e. "the winner pays more than the real value of the product", seems to be a consequence of auctions and, therefore, they propose more cooperative multi-attribute decision analysis tools and negotiation protocols using distributed constraint satisfaction policies to support it.

However, both lines of research are aware of the need of enhancing simple price-based auction mechanisms, transforming them into new protocols encompassing multi-dimensional issues to be negotiated among the market participants.

In order to reach an agreement about a particular insurance product, customers and brokers usually engage themselves in a sequential negotiation process composed of multiple rounds for exchanging proposals and counter-proposals. A negotiation protocol should then be defined in order to select the participants in the electronic market that, based on its capabilities and availability, will be able to make the optimal deal according to its own goals.

Real negotiation in the insurance domain implies taking into consideration not only one, but multiple attributes for defining the insurance under discussion. For instance, although the policy premium is an important attribute, the length of the coverage, the renewability and convertibility of the policy are complementary issues to include in the decision about whether to buy or not a specific insurance product.

Attaching utility values to different issues, helps to solve the problem of multi-issue evaluation. However, in some cases, it can be a very difficult task to attach absolute values to issues' utilities. A more natural and realistic situation is to simply impose a multi-issue evaluation based on a qualitative, not quantitative, measure.

One way of enhancing agent's autonomy in dynamic environments, is to endow its architecture with learning capabilities. Agent learning process may include several different facets, from simple, step by step, adaptation to the changes in a dynamic environment, to heavier and more sophisticated processes of gathering new knowledge about the environment and other agents based on the past history.

This capability is included, in our negotiation protocol, through a Reinforcement Learning algorithm. Reinforcement learning algorithms support continuous, on-line learning during the negotiation process itself by making decisions according to the environment reactions in the past. The history of a negotiation is a crucial information to be considered when to decide what to do in the next negotiation round. Q-Learning also includes not only exploitation but also exploration facilities. In dynamic environments or in the presence of incomplete information, exploration, i.e., trying out different possibilities different from the obvious ones, becomes a powerful technique.

We also have adapted the Q-Negotiation algorithm [11,12] to the insurance brokering problem. This algorithm uses a reinforcement learning strategy based in Q-learning for the formulation of new proposals. The Q-learning algorithm is a well-known reinforcement learning algorithm that maps values (Q-values) to state/action pairs.

The Q-Negotiation algorithm has the ability to maintain information private to individual agents, and at the same time, includes the capability to evaluate multi-attribute proposals, to learn how to make better proposals during the negotiation process and to resolve attributes' inter dependencies.

Intelligent agents technology is a flexible paradigm suitable for dynamic and open environments, since agents can effectively cope with the complexity and large amount of information. Intelligent agents technology seems to be an appropriate paradigm to use in our case, since electronic commerce environments are both very dynamic and complex.

7.2 Negotiating with Insurers

Generally, an evaluation of a received proposal is a linear combination of the attributes' values weighted by their corresponding importances. In this way, a multi-attribute negotiation is simply converted in a single attribute negotiation, once there is a single result of the combined evaluation.

The multi-attribute function presented in the following formula encodes the attributes' and attributes values' preferences in a qualitative way and, at the same time, accommodates attributes intra-dependencies.

$$Evaluation = \frac{1}{Deviation} \ . \tag{3}$$

$$Deviation = \frac{1}{n} * \sum_{i=1}^{n} \frac{i}{n} * dif(PrefV_i, V_i) \ . \tag{4}$$

$$dif(PrefV_i, V_i) = \begin{cases} \dfrac{V_i - PrefV_i}{max_i - min_i} & \text{if continuous domain .} \\[2em] \dfrac{Pos(V_i) - Pos(PrefV_i)}{nvalues} & \text{if discrete domain .} \end{cases} \qquad (5)$$

where n is the number of attributes that defines a specific insurance product component.

A proposal's evaluation value is calculated by the Broker Agent, as the inverse of the weighted sum of the differences between the optimal $PrefV_i$ and the proposed value V_i of each one of the attributes. The proposal with the highest evaluation value so far is the winner, since it is the one that contains the attributes' values more closely related to the preferred ones from the customer point of view.

The function $dif(PrefV_i, V_i)$ quantifies for an issue i, the degree of acceptability of the current value V_i proposed by a specific Insurer Agent when compared to its preferable value $PrefV_i$.

If the insurance product component's values domain is of continuous values, this quantification simply is a normalised difference between the two values V_i and $PrefV_i$.

If the insurance product component's values domain is of discrete values, the difference is now calculated as a normalised difference between the preference attached to V_i and $PrefV_i$. This difference can be calculated considering the relative position of the two values in the ontology's enumerated domain values specification.

The negotiation process is considered as a set of rounds where Insurer Agents concede, from round to round, a little bit more trying to approach the customer preferences, in order to be selected as the winning insurance company. The winner bid in the current round is selected as the one that presents the highest evaluation value, since it is the solution that contains attributes values the closest to the preferable ones. The winner bid in the current round is compared with the bid in all past rounds, and the best on is selected. The Broker Agent helps Insurer Agents in their task of formulating new proposals by giving them some hints about the direction they should follow in their negotiation space. The Broker Agent gives this hints as comments about attributes' values include in current proposals.

This qualitative feedback reflects the distance between the values indicated in a specific proposal and the best one received so far, and is formulated as a qualitative comment on each of the proposal' attributes values, which can be classified in one of three categories: *sufficient, bad* or *very_bad*. The Broker Agent compares a particular proposal with, not its optimal, but the best one received so far because its more convincing to say to an Insurance Agent that there is a better proposal in the market than saying that its proposal is not the optimal one.

Insurer Agents will use this feedback information about its past proposals, in order to formulate, in the next negotiation rounds, new proposals trying to follow the hints included in the feedback comments. The negotiation process ends when:

- Broker Agent receives a bid which has a satisfactory evaluation value. This is the winner bid.
- A deadline is reached. The winner bid is the one that presents the highest evaluation value among all bids received until then.

7.3 Interacting with the Customer Agent

When the negotiation process ends it is possible that the received bids do not satisfy all the constraints imposed by the customer Agent. In that case a conversation with customer Agent is initiated. This interaction takes the form of a sequence of questions whose aim is to reduce alternatives rather than simply sort them.

We use the schematic format (*performative sender_agent receiver_agent conversation_operator content*). The outermost performative, or message type, represents the general class of a message. In our system, we make use of the classes *request*, *query* and *inform*.

According to [8], we may view a request as having an associated level of commitment. A request performative is tightly coupled with advancing the task of selecting an insurance product, as it always concern constraints. When a request is made the Broker Agent is making a pre-commitment to how the progress on the selection of an insurance product might be accomplished and it prompts the Costumer Agent for information in order to do this. The Costumer Agent must respond with an appropriate *inform* message.

A query performative is about exchanging information. Its objects of discourse are the domain ontology and relevant information to the current state of the interaction. When a query is sent by the Costumer Agent, the Broker Agent must respond with an *inform* followed by an appropriate conversation operator.

The first operator available to the Broker Agent, CONSTRAIN-ATTRIBUTE, involves asking the Customer Agent to provide a value for an attribute that does not yet have one. In some cases, the process of introducing a constraint can produce a situation in which no insurance products are satisfactory. When this occurs, the Broker Agent applies RELAX-ATTRIBUTE, which asks whether the user wants to drop a particular constraint.

Another operator, SUGGEST-VALUES, informs the Customer Agent of possible values for an attribute. In this case, the Broker Agent sends only the most adequate options rather than all possible choices. A similar operator, SUGGEST-ATTRIBUTES, informs the Customer Agent about the possible attributes for an insurance product.

Once the conversation has produced a valid number of alternatives, the Broker Agent invokes RECOMMEND-INSURANCE, an operator that proposes a complete insurance product to the Customer Agent.

Now let us discuss the operators available to customer Agent. The action, PROVIDE-CONSTRAINT, involves specifying the value of some attribute. This

Table 1. Broker Agent Conversation Operators

Performative	Conversation Operator	Description
Request	CONSTRAIN-ATTRIBUTE	Asks a question to obtain a value for an attribute
	RELAX-ATTRIBUTE	Asks a question to modify a value for an attribute
Inform	SUGGEST-VALUES	Suggests a set of possible values for an attribute
	SUGGEST-ATTRIBUTES	Suggests a set of unconstrained attributes
	RECOMMEND-INSURANCE	Recommends an insurance product that satisfies the constraints

can be a value for the attribute just asked by the Broker Agent, a value for a different attribute, or a replacement for a previously specified value. If the proposed value for an attribute is found to be inappropriate by the customer Agent or less relevant than some other factor, it can reject the attribute or even replace it with another. The REJECT-SUGGESTION is used for explicit rejection.

In addition, the customer Agent can explicitly accept or reject other proposals that the Broker Agent makes, for relaxing a certain attribute (ACCEPT-RELAX or REJECT-RELAX), or for a complete insurance product (ACCEPT-INSURANCE or REJECT-INSURANCE). The customer Agent can also query the Broker Agent about available attributes (QUERY-ATTRIBUTES) or about possible values of an attribute (QUERY-VALUES).

Table 2. Customer Agent Conversation Operators

Performative	Conversation Operator	Description
Inform	PROVIDE-CONSTRAINT	Provides a value for an attribute
	REJECT-SUGGESTION	Rejects the proposed attribute
	ACCEPT-RELAX	Accepts the new value of an attribute
	REJECT-RELAX	Rejects the new value of an attribute
	ACCEPT-INSURANCE	Accepts proposed insurance product
	REJECT-INSURANCE	Rejects proposed insurance product
Query	QUERY-ATTRIBUTES	Asks broker for information about possible attributes
	QUERY-VALUES	Asks broker for information about possible values for an attribute

8   Agent Communication Using XML

8.1   Agent Communication

Autonomous agents cooperate by sending messages including concepts from an appropriate domain ontology. A standard message format with meaningful structure and semantics has become a key issue in agents understanding each other. Furthermore, the message format should be accepted not only by the agent research community, but also by all information providers. As XML is fast becoming the standard for data interchange on the Web, we choose XML as the message format for agent communication.

Agents send and receive information through XML encoded messages. We use a FIPA ACL-like format, encoded in XML. XML tags markup the information and break up the data into parts, with meaningful structure and commonly agreed semantics.

The power of XML, its role in e-commerce, and even the use of XML for agent communication, have been recognised. However, although XML is well structured for encoding semantically meaningful information, it must be based on an appropriate ontology.

Generally speaking, a domain ontology provides a set of concepts that can be asked for, advertised and used to control the agent cooperation behaviour. These concepts can be marked using XML tags, and then a set of commonly agreed tags, underlie message interpretation. The structure and semantics of the message used in a particular problem domain are represented by the corresponding DTDs.

8.2   XML Benefits in Insurance

Standardisation in information representation and transfer is crucial to both B2B and B2C e-commerce. XML is platform and application independent, and vendor-neutral mechanism. XML relies on existing technologies, in particular, SGML for syntax, URIs for name identifiers, EBNF for grammar, and Unicode for character encoding, which are all standards.

The advantage of data being independent of any particular platform, application or vendor, is that it can be transformed to produce different types of outputs for different media devices (Web browser, paper, CD-ROM) without the need to modify the original content. When modifications are required, only the original version of the content need to be edited before republishing to the various target media. This leads to efficiency and ease-of-maintainability, without the inherent problems of version control and the effort required in making modifications in medium-specific document versions.

Conducting e-commerce requires communicating with other companies and often poses a challenge. XML simplifies business-to-business communication, particularly in vertical industries for the following reasons: (1) The only thing that is to be mutually agreed upon is the XML vocabulary that will be used to represent data. (2) Neither company has to know how the other's back-end systems (platforms, operating systems, programming languages) are organised, which does not put any extra technical burden while keeping the privacy. All that is required is that each company develop the mapping to transform XML documents into the internal format used by the back-end systems. (3) XML-based solution is scalable: If there is an addition of another partner, there is no need by the host company to interact with the systems of the new company. All that is required is that they follow the protocol (the XML vocabulary).

A major advantage of conducting business on the Web is that it broaden the customer-base towards globalisation, without the necessity of having physical office locations. However, in order to communicate, a business must still "speak" the language of the region in context. With the Unicode support in XML, Web sites can be multi-lingual. XML also includes a method to signal what language and encoding is being used.

XML provides companies opportunities for customer services that did not exist previously. Corporate data that was previously stored in disparate sources and considered to be non-integrable, can be transformed in an XML format. By consolidating different data sources, opens the doors for the companies to make a variety of such data available to be explored. It gives insurance companies a powerful way to transact, manage and share data over the Web.

XML is far less expensive than other data exchange alternatives, such as Electronic Data Interchange (EDI), opening the door for small, low-tech companies to participate in online data exchange.

9   Knowledge Representation

Sharing common understanding of the structure of information among people or software agents is one of the more useful and common goals in developing ontologies [6]. Adopting a common ontology guarantees information consistency and compatibility for a community of agents. The information consistency is satisfied when each specific expression has the same meaning for every agent in the market. The information compatibility is verified when any concept is described by the same expression, for all the agents. This is why knowledge representation becomes an important issue in the context of agent-based insurance brokering as well.

Fundamentally we need a common descriptive language that all insurers can agree on. These include the terms, conditions, relationships and categorisations of insurance products. Such definition must be extensible in that it can support new terms and relationships being added later. Given a standard definition language, there needs to be agreement on how to define an insurance product so that it can be searched, displayed and its terms negotiated over. These include defining attributes, constraints, eligibility criteria, preferences and negotiable aspects of the product.

In our system a class is described by a set of slots, and slots are described by a set of attributes, which are instantiated with values. The schema we are using for an ontology is then the knowledge representation scheme suitable for properly identify classes, slots, attributes and values, together with relationships that map classes to slots, slots to attributes and attributes to values. Such ontology can be represented by the following structure:

$$Ont = \{Class, Slot, Attr, Val, CS_r, SA_r, AV_r, Deps\}$$

where $Class$ is the set of item's identifiers, $Slot$ is the set of component's identifiers, $Attr$ is the set of attributes' identifiers, Val is the set of attribute values' identifiers, IC_r is a relationship that assigns to each class in $Class$ a set of slots in $Slot$, AV_r is a relationship that assigns to each attribute in $Attr$ a specific value in Val and $Deps$ is a set of relationships defining the dependencies between attributes' values.

Each value is represented by the tuple:

$$
\begin{aligned}
Val_i &= \{Type, Domain\} \\
Type &= \{integer, real, string\} \\
Domain &= \{continuous, discrete\}
\end{aligned}
$$

Each relationship CS_r is represented by:

$$Class_i \rightarrow \{Slot\}, \forall Class_i \in Class$$

Each relationship SA_r is represented by:

$$Slot_i \rightarrow \{Attr\}, \forall Slot_i \in Slot$$

Each relationship AV_r is represented by:

$$Attr_i \rightarrow Val_k, \forall Attr_i \in Attr, \exists^1 Val_k \in Val$$

Each dependency in $Deps$ is represented by:

$$Dep_{ij} = f(Val_{ki}, Val_{mj}), \forall Attr_i, Attr_j \in Attr$$

10   Related Work

In [17] a system was proposed to support a market place for complex products and services such as insurance. Such an electronic market place is represented by a network of match-maker objects or traders, which perform a symmetric match-making between product features and customer's requirements. Each tier of traders is responsible for determining the suitability and availability of products following each successive stage of the dialogue between insurers and their customers.

This paper proposes a distributed multi-agent system where customers are grouped together, exploiting user modelling and machine learning techniques. Although clustering seems a computationally expensive task compared to the case-based approach, this is not a real problem since communities change far less often than individual user models. In the other hand communities can be very useful since they can be used to tailor products offering to the needs of customers. However, this can be done effectively, only when the generated communities are meaningful. That's why our system is using a specific metric to characterise the generated communities of users.

We propose a common format for describing products and requirements that can be agreed upon by all participants and that is rich enough to incorporate all the current and future functionality of the system. Thus, if the product definition ontology is agreed, then it becomes relatively easy for submit new products at any time or to make changes in existing product terms and conditions and make these changes automatically and instantly.

Several other researchers have also addressed learning algorithms for agents in electronic commerce negotiation. Reinforcement learning algorithms have been used in the work reported in [13], where simulation results showed that learning agents performed better than their non-learning competitors agents in different market situations. However the algorithm only addresses single-issue type of negotiations. Learning in multi-issue negotiation is addressed in [21], where negotiation occurs in a one-to-one basis. Our work combines both one-to-many and multi-issue negotiation characteristics. Learning is an important characteristic that should be available in automated e-commerce negotiation. Past proposals can, and should, constraint the value of the next insurer proposal. Moreover, in open systems, as it is the case of e-commerce, learning becomes a powerful capability for deal with the environment dynamics. Therefore, we consider both multi-issue and adaptive negotiation features as being of great importance for agent-mediated electronic commerce.

11   Conclusions

The system described in this paper presents a new approach to insurance products brokering and has the potential to improve the quality of customer service by ensuring that individual customer needs are reflected in the products offered.

Different communities of users can be identified and used to improve the exploitation of an insurance brokering service. The construction of those communities is achieved using an unsupervised learning technique. We also use a specific metric to decide which are the representative preferences of a user's community. The proposed ontology includes multi-attribute definition as well as attributes intra and inter dependencies. This scheme is suitable for properly identify items, components, attributes and values of a insurance product, together with relations that map items to components, components to attributes and attributes to values.

We also have adapted an advanced negotiation protocol, suitable for multi-issue negotiation in electronic commerce activity. A learning capability was also included enabling agents to become more effective in a dynamic market by learning with past experience through the qualitative feedback received from their opponents. Our platform for automatic insurance brokering services is now ready to be exploited for future experiments in several realistic scenarios.

References

1. Marko Balabanovic and Yoav Shoham. Learning information retrieval agents: Experiments with automated web browsing. In *Proceedings of the AAAI Spring Symposium on Information Gathering from Heterogeneous, Distributed Resources*, pages 13–18, 1995.
2. P. Chiu. Using c4.5 as an induction engine for agent modelling: An experiment for optimisation. In *Proceedings of the User Modelling Conference UM'97*, 1997.
3. D. H. Fisher. Knowledge acquisition via incremental conceptual clustering. *Machine Learning*, 2:139–172, 1987.
4. J.B. Weinberg G. Biswas and D. Fisher. Iterate: A conceptual clustering algorithm for data mining. *IEEE Transactions on Systems, Man and Cybernetics*, 28:100–111, 1998.
5. M.A. Gluck and J.E. Corter. Information, uncertainty and the utility of categories. In *Proceedings of the 7th Conference of the Cognitive Science Society*, pages 283–287, 1985.
6. T.R. Gubber. A translation approach to portable ontology specifications. *Knowledge Acquisition*, 5(2):199–220, 1993.
7. R. Guttman, A. Moukas, and P. Maes. Agent-mediated electronic commerce: A survey. *Knowledge Engineering Review*, 13(2):147–159, 1998.
8. A. Haddadi. Communication and cooperation in agent systems: A pragmatic theory. *Lecture Notes in Computer Science*, 1056, 1996.
9. J. Robert Hunter and James H. Hunt. *Term Life Insurance on the Internet: An Evaluation on On-line Quotes*. Consumer Federation of America, http://consumerfed.org/termlifeinsurance.pdf, 2001.
10. J. Moore. Implementation, contracts and negotiation in environments with complete information. *Advances in Economic Theory*, 1, 1992.
11. Eugénio Oliveira and Ana Paula Rocha. Agents advanced features for negotiation in electronic commerce and virtual organisations formation process. *Lecture Notes in Computer Science*, 1991:77–96, 2001.
12. Eugénio Oliveira and Ana Paula Rocha. Electronic institutions as a framework for agent's negotiation and mutual commitment. *Lecture Notes in Artificial Intelligence*, 2258:232–245, 2001.
13. Eugénio Oliveira, Jos M. Fonseca, and Nick Jennings. Learning to be competitive in the market. In *American Association of Artificial Intelligence Workshop on Negotiation*, Orlando, USA, July 1999.
14. G. Paliouras, V. Karkaletsis, C. Papatheodorou, and C. Spyropoulos. Exploiting learning techniques for the acquisition of user stereotypes and communities. In *Proceedings of the International Conference on User Modelling (UM '99).*, 1999.
15. G. Paliouras, C. Papatheodorou, V. Kakaletsis, C. Spryropoulos, and V. Malaveta. Learning user communities for improving the services of information providers. *Lecture Notes in Computer Science*, 1513:367–384, 1998.
16. B. Raskutti and A. Beitz. Acquiring user preferences for information filtering in interactive multimedia services. In *Proceedings of the Pacific Rim International Conference on Artificial Intelligence*, pages 47–58, 1996.
17. Y. Hoffner S. Field. Vimp - a virtual market place for insurance products. *EM - Electronic Markets*, 8(4):3–7, 1998.
18. M. Stolze. Soft navigation in electronic product catalogs. *International Journal in Digital Libraries*, 3(1):60–66, 2000.

19. Nir Vulkan and Nick Jennings. Efficient mechanisms for the supply of services in multi-agent environments. In *Proceedings of the First International Conference on Information and Computation Economics*, Charleston, South Carolina, October 1998.

20. M. Wellman and P. Wurman. Market-aware agents for a multiagent world. *Robotics and Autonomous Systems Journal*, 24:115–125, 1998.

21. Dajun Zeng and Katia Sycara. How can an agent learn to negotiate. In *Proceedings of the Third International Workshop on Agent Theories, Architectures and Languages*, Budapest, Hungary, August 1996.

An XML Multi-agent System for E-learning and Skill Management

Alfredo Garro[1] and Luigi Palopoli[2]

[1] D.E.I.S. – Università della Calabria, Via P. Bucci, 87030 Rende (CS), Italy
`garro@si.deis.unical.it`
[2] D.I.M.E.T. – Università di Reggio Calabria, 89060 Reggio Calabria, Italy
`palopoli@ing.unirc.it`

Abstract. E-learning is nowadays recognized as one of the key components of Enterprise Knowledge Management platforms. Given a project
specification, the platform should be able to suggest a project team,
to measure human resources competence gaps and to contribute to reduce them by creating personalized learning paths. In this paper we
propose an XML based Multi-Agent System to perform the following
tasks: *(i)* supporting Chief Learning Officers in defining roles, associated
competencies and knowledge level required; *(ii)* managing the *skill map*
of the organization; *(iii)* measuring human resources competence gaps;
(iv) supporting employees in filling their competence gaps as related to
their roles; *(v)* enriching a given courseware or creating personalized learning paths according to feedbacks user provides in order to optimize the
acquisition of needed competencies; *(vi)* assisting Chief Learning Officers
in choosing the most appropriate employee for a given role.

1 Introduction

Among the themes that are currently central in the market of the ICT (Information and Communication Technology) a special place is occupied by e-learning.
The increasing and generalized attention toward e-learning, especially in the
sector of business training, has various causes, that are briefly addressed below.
Whereas the advantages of e-learning has been for years characterized simply in
terms of gained cost effectiveness (both in terms of time and space), it is nowadays widely recognized that its potentialities go far beyond, involving issues like
diversification of learning paths and general business competitive advantage [7,
12,8,6]. Nowadays, it is quite agreed that organizations have not much to gain in
adopting e-learning platforms that only provide educational content. An advantageous e-learning platform should have instead the capability to help enriching,
sharing and circulating organization knowledge, thus being a tool able to make
the organization dynamic and flexible. Because of the dynamism of the market,
often organizations cannot program in the medium-long term, but need to work
in a project-shaped, short-to-medium term perspective. When an organization is
to carry out a project, new competencies are typically to be acquired, which are
frequently expensive and hard to find in the external market (the *skill shortage*

R. Kowalczyk et al. (Eds.): Agent Technology Workshops 2002, LNAI 2592, pp. 283–294, 2003.

problem). Therefore it is often the case that such competencies must be found (or constructed) inside. An appropriate and properly used e-learning platform thus becomes an important component of the Enterprise Knowledge Management [4, 9]. Given a project specification, the platform should be able to suggest a project team, to measure human resources competence gaps and to contribute to reduce them by creating personalized learning paths. Moreover, the platform should be able to dynamically readapt learning paths according to feedbacks users provide in order to optimize the acquisition of needed competencies. In this scenario, it appears meaningful to look at educational contents as organized in relatively small independent units (Learning Objects), which can be combined in order to create personalized learning paths. In order to achieve an appropriate management of a Learning Object database, it is necessary to describe its content in an efficient and effective way. In other words, we need a meta-knowledge that allows us to classify learning objects (documents, slides, simulations, role plays, questionnaires, pre-recorded lessons, classroom lessons, ...) and their relationships with respect to their objective, topic, used media, etc... (i.e. Learning Object Metadata). Furthermore, in order to allow exchanging, rousing and sharing of learning objects, we have to express the meta-knowledge using standard formats and protocols. And, in fact, standardization of learning objects description is one important goal for the scientific community that operates in the field of e-learning. The authoritative organization IMS [1] proposes to describe the Learning Objects through an XML document validated with respect to an *XML schema* established by the standard [2]. This standard has been defined and most of the commercial e-learning platforms support it. But the fundamental problem of managing Learning Objects is still to be solved. In particular, within the Skill Management context, the following main tasks are to be considered:

- Individuating the *student* learning objectives and evaluating his competence gaps.
- Building, starting from a database of Learning Objects, the courseware able to fill such competence gaps.
- Controlling the *student* improvements and (re)adapting and integrating the courseware content and presentation structure.
- Creating a bridge between single user learning objectives and general organization learning objectives.
- Managing the *skill map* [2] of the organization and updating it according to the *student* learning improvements.

All those tasks are currently carried out "manually". The idea that we propose here is trying to automatize, as far as possible, those activities by designing suitable Multi-Agent environments. Agents technologies seem to be well suited to

---

[1] Instructional Management System, http://www.imsproject.org
[2] A skill map stores, for each employee, information about his/her role in the organization, the competencies required for that role and the current level of employee competencies

carry out the main activities listed above. In fact, those activities require communications between distributed components, sensing and monitoring of the environment and autonomous operations; agents have the ability to reason, they can easily perform sequences of complex operations based on messages they receive, their own internal beliefs and their overall goals and objectives. Furthermore an e-learning agents platform is expected to be proactive, interactive, adaptive and cognitive.

In this paper we propose an XML based Multi-Agent System to perform the following tasks: *(i)* supporting Chief Learning Officers in defining roles, associated competencies and knowledge level required; *(ii)* managing the *skill map* of the organization; *(iii)* measuring human resources competence gaps; *(iv)* supporting employees in filling their competence gaps as related to their roles; *(v)* enriching a given courseware or creating personalized learning paths according to feedbacks user provides in order to optimize the acquisition of needed competencies; *(vi)* assisting Chief Learning Officers in choosing the most appropriate employee for a given role.

We call our system *Multi-Agent System for E-Learning e skill management (MASEL)*. The system we propose includes seven kinds of agents:

1. *CLO Assistant Agent*, that supports Chief Learning Officers in defining a learning strategy in terms of roles and competencies required for the organization.
2. *Skill Manager Agent*, that stores, for each employee, information about his/her role in the organization, the competencies required for that role and the current level of employee competencies (*Skill Map*);
3. *Student Assistant Agent*, that supports the student in improving his/her competencies;
4. *Learning Paths Agent*, that deals with creation of learning paths;
5. *Content Agent*, that is specialized in managing a Learning Object Database;
6. *CCO Assistant Agent*, that supports Chief Content Officer in managing a Learning Object Database.
7. *User Profile Agent*, that stores all the needed information about the users (e.g., login information, learning activities information and so on);

Such agents will have to be FIPA[3] compliant [1], and able to operate, at least, in e-learning contexts based on the LOM (Learning Object Metadata) paradigm according to the IMS standards [2]. It should be clear, from the description provided above, that information exchange plays a central role in MASEL. Actually, the efficient management of information exchange has become a challenging issue in most areas of computer science research, particularly for Web-based information systems. In this area, the most promising solution to this problem has been the definition of XML, a language for representing and exchanging data over the Internet. XML embodies both representation capabilities, typical of HTML, and data management features, typical of DBMS's; it is presently considered as a standard for the Web. XML capabilities make it particularly suited to be

---

[3] Foundation for Intelligent Physical Agents, `http://www.fipa.org`

exploited in the agent research; as a matter of fact, agents using XML for both representing and handling their own ontology have been proposed, especially in the e-commerce area [5,16]. MASEL is an attempt to extend this trend to *E-Learning and Skill Management agents*; in particular, in our multi-agent system, XML is exploited both for representing and handling agents' ontology as well as for managing data exchange among them.

The rest of this paper is organized as follows: Sect. 2 describes the MASEL architecture by pointing out both provided services and the ontology of the MASEL Agents. An overview of MASEL behaviour is presented in Sect. 3. Finally, in Sect. 4 we draw our conclusions.

2   MASEL Architecture

As previously pointed out, MASEL consists of seven typologies of agents, namely CLO Assistant Agent (CLO), Skill Manager Agent (SMA), Student Assistant Agent (SAA), Learning Paths Agent (LPA), Content Agent (COA), CCO Assistant Agent (CCO), User Profile Agent (UPA).

2.1   The CLO Assistant Agent

Provided Services. A generic *CLO* supports a Chief Learning Officer in managing the *skill map* of an organization and defining a learning strategy in terms of roles and required competencies. *CLO* provides the following services:

- managing roles and competencies: CLO supports the user in defining roles, associated competencies and knowledge level required;
- managing employees with relative roles and competence levels;
- suggesting the most suitable employees for taking a specific role;
- defining priorities or/and temporal constrains relative to competencies that have to be acquired by employees;
- showing, for each employee, the learning activities history.

Ontology. The ontology of a generic *CLO* consists of an XML document storing information about the *skill map SM* of the organization and, for each employee Em_i, her/his *learning history* lh_i. The ontology of a MASEL *CLO* is described by the DTD shown in Fig. 1.

```
<!ELEMENT CLOAssistantOntology (SkillMap,LearningHistory*)>

<!-- SkillMap is defined in SMA Ontology -->
<!-- LearningHistory is defined in SAA Ontology -->
```

Fig. 1. The DTD describing the ontology of a CLO assistant agent

2.2   The Skill Manager Agent

Provided Services. A SMA manages the *skill map SM* of the organization. SM stores the information, previously inserted by a CLO, about roles and associated competencies. Moreover, for each employee Em_i, SMA manages the information about Em_i role and Em_i current competence levels. SMA provides the following services:

- inserting, deleting and updating organization roles and competencies;
- inserting, deleting and updating employees' roles and competencies;
- querying the *skill map*. For example: searching the most suitable employees for a specific role; evaluating an employee competence gaps.

Ontology. The ontology of a generic SMA consists of an XML document storing information about the *skill map SM* of the organization. The skill map SM consists of a set of roles, competencies and employees. A *role* is described by an identifier, its name and its required competencies. Each *required competence* specifies, for a particular role, the minimum level of knowledge required for the competence to fit that role. A *competence* is described by an identifier, its name and its description. Finally, an *employee* is characterized by an identifier, its personal data, its role and its competencies. In such a case SM stores, for each *employee competence*, the current level of employee knowledge. The ontology of a MASEL SMA is described by the DTD shown in Fig. 2.

```
<!ELEMENT SkillManagerOntology (SkillMap)>     <!ELEMENT EmployeeCompetence (Competence)>
<!ELEMENT SkillMap (Role*, Competence*,          <!ATTLIST EmployeeCompetence
Employee*)>                                          CurrentLevel CDATA #REQUIRED
<!ELEMENT Role (RequiredCompetence+)>            >
    <!ATTLIST Role                             <!ELEMENT PersonalData (firstName, lastName,
       Identifier ID #REQUIRED                 gender, birthDate, birthPlace, nationality+,
       Name CDATA #REQUIRED                    languageUnderstood*, address+, phoneNumber+,
    >                                          emailAddress+,note?)>
<!ELEMENT RequiredCompetence (Competence) >    <!ELEMENT firstName (CDATA)>
    <!ATTLIST RequiredCompetence               <!ELEMENT LastName (CDATA)>
       RequiredLevel CDATA #REQUIRED           <!ELEMENT gender (Male|Female)>
    >                                          <!ELEMENT birthDate (CDATA)>
<!ELEMENT Competence EMPTY>                     <!ELEMENT birthPlace (CDATA)>
    <!ATTLIST Competence                       <!ELEMENT nationality (CDATA)>
       Identifier ID #REQUIRED                 <!ELEMENT languageUnderstood (CDATA)>
       Name CDATA #REQUIRED                    <!ELEMENT address (CDATA)>
       Description CDATA #IMPLIED               <!ELEMENT phoneNumber (CDATA)>
    >                                          <!ELEMENT e-mailAddress (CDATA)>
<!ELEMENT Employee (PersonalData, Role,         <!ELEMENT note (CDATA)>
EmployeeCompetence*)>
    <!ATTLIST Employee
       Identifier ID #REQUIRED
    >
```

Fig. 2. The DTD describing the ontology of a skill manager agent

2.3   The Student Assistant Agent

Provided Services. A generic Student Assistant Agent SAA_i is associated with a student (employee) Em_i and assist her/him in filling her/his competence gaps related to her/his roles. SAA_i provides Em_i with the following services:

- showing the gaps between her/his current competencies and the competencies required for her/his role in the organization, possibly associated with priorities or/and temporal constrains. This task is carried out in cooperation with the Skill Manager Agent;
- evaluating a specific competence gap by *pre-assessment* tests and finding or creating courseware allowing for filling that competence gap. This task is carried out in cooperation with the Learning Paths Agent;
- enriching and modifying a courseware according to feedbacks user provides in order to optimize the acquisition of needed competencies. This task is carried out in cooperation with the Learning Paths Agent;
- managing information about current learning activities of the student.

Ontology. The ontology of a generic SAA, associated with a student Em_i, consists of an XML document storing information about the *profile* up_i of Em_i and the Learning Objects which are currently used by Em_i. A *user profile* up_i consists of a *learning history*, a *login information* and a set of *session log*. A *learning history* consists of a set of selected Learning Objects. For Each selected Learning Object, SAA stores information about its use like download status, start learning timestamp, utilization rate and so on. A *session log* consists of a set of *user actions*. The ontology of a MASEL SAA is described by the DTD shown in Fig. 3.

```
<!ELEMENT StudentAssistantOntology        <!ELEMENT UserAction EMPTY>
(UserProfile,lom*)>                           <!ATTLIST UserAction
<!ELEMENT UserProfile (Employee, LearningHistory,   ActionDescription CDATA #REQUIRED
LoginInformation, SessionLog*)>            >
<!ELEMENT LoginInformation EMPTY>         <!ELEMENT SelectedLO
    <!ATTLIST LoginInformation            (lom,LearningObjectStatus)>
    Username CDATA #REQUIRED              <!ELEMENT LearningObjectStatus EMPTY>
    Password CDATA #REQUIRED                  <!ATTLIST LearningObjectStatus
    >                                         Downloaded (true|false) #REQUIRED
<!ELEMENT LearningHistory (SelectedLO*)>      StartLearningTimestamp CDATA #IMPLIED
    <!ATTLIST LearningHistory                 UtilizationRate CDATA #IMPLIED
    Employee IDREF #REQUIRED                  >
    >
<!ELEMENT SessionLog (UserAction*)>       <!-- lom is defined in IMS Specification -->
    <!ATTLIST SessionLog
    SessionId ID #REQUIRED
    >
```

Fig. 3. The DTD describing the ontology of a student assistant agent

2.4   The Learning Paths Agent

Provided Services. A Learning Paths Agent deals with creation of learning paths able to fill the competence gaps of an employee Em_i supported by a Student Assistant Agent SAA_i. LPA provides SAA_i with the following services:

- creating a *pre-assessment* test allowing to evaluate a specific competence gap;
- selecting and composing the Learning Objects suitable to build a learning path;
- enriching and modifying a learning path according to feedbacks user provides for example by answering tests and assessments;

Moreover LPA possibly informs CCO Assistant Agent about the lack of some Learning Objects needed to build specific learning paths.

Ontology. The ontology of a generic LPA consists of an XML document storing information about the *skill map* SM of the organization, the Learning Objects available for building learning paths and, for each employee Em_i, her/his *learning history* lh_i. The ontology of a MASEL LPA is described by the DTD shown in Fig. 4.

```
<!ELEMENT LearningPathOntology (lom*,SkillMap,
LearningHistory*)

<!-- lom is defined in IMS Specification -->
<!-- SkillMap is defined in SMA Ontology -->
<!-- LearningHistory is defined in SAA Ontology -->
```

Fig. 4. The DTD describing the ontology of a learning paths agent

2.5   The Content Agent

Provided Services. A Content Agent COA is specialized in managing a Learning Object Database. COA assists a generic Learning Paths Agent LPA in building a courseware and a generic CCO (Chief Content Officer assistant agent) in managing learning contents providing the following services:

- inserting, deleting and updating Learning Objects and relative descriptions;
- querying the Learning Object Database by accessing Learning Object Metadata (LOM);

Ontology. The ontology of a generic COA consists of an XML document storing information about the Learning Objects currently stored in the Learning Object Database. Each Learning Object is described by an XML document(*lom*) validated with respect to an XML Schema established by the IMS standard [2]. The ontology of a MASEL COA is described by the DTD shown in Fig. 5.

```
<!ELEMENT ContentAgentOntology (lom*)>

<!-- lom is defined in IMS Specification -->
```

Fig. 5. The DTD describing the ontology of a content agent

2.6   CCO Assistant Agent

Provided Services. A generic CCO supports a Chief Content Officer in managing a Learning Object Database providing the following services:

– inserting, deleting and updating Learning Objects and relative descriptions. This task is carried out in cooperation with the Content Agent;
– showing, for each employee, the learning activities history. This task is carried out in cooperation with the User Profile Agent.

Ontology. The ontology of a generic CCO consists of an XML document storing information about the *skill map SM* of the organization, the Learning Objects available for building learning paths and, for each employee Em_i, her/his *learning history lh_i*. The ontology of a MASEL CCO is described by the DTD shown in Fig. 6.

```
<!ELEMENT CCOAssistantOntology (lom*,SkillMap,
LearningHistory*)>

<!-- lom is defined in IMS Specification -->
<!-- SkillMap is defined in SMA Ontology -->
<!-- LearningHistory is defined in SAA Ontology -->
```

Fig. 6. The DTD describing the ontology of a CCO assistant agent

2.7   The User Profile Agent

Provided Services. A User Profile Agent stores useful information about all the registered users. For instance: personal data, log-in data, current and past learning activities. A UPA provides the following services:

– managing users log-in to the system;
– managing users profile information;
– updating the competence levels of a employee according to her/his learning activities. This task is carried out in cooperation with the Student Assistant Agent, associated with the employee, and with the Skill Manager Agent.

Ontology. The ontology of a generic UPA consists of an XML document storing the *profiles* of the registered users. The ontology of a MASEL SAA is described by the DTD shown in Fig. 7.

```
<!ELEMENT UserProfileOntology (UserProfile*)>

<!-- UserProfile is defined in SAA Ontology -->
```

Fig. 7. The DTD describing the ontology of a user profile agent

3   An Overview of MASEL Behaviour

As previously pointed out, a MASEL multi-agent system consists, at least, of one CLO Assistant Agent, one Skill Manager Agent, one Content Agent, one Learning Paths Agent, one CCO Assistant Agent, one User Profile Agent and n Student Assistant Agents SAA_1, $SAA_2,\ldots,SAA_n$. Figure 8 shows the architecture of a generic MASEL multi-agent system. Each box represents a MASEL Agent and each arrow between two boxes indicates cooperation between the two represented agents. A CLO Assistant Agent supports Chief Learning Officer in managing the *skill map* and in defining a learning strategy in terms of roles and competencies required for the organization. This task is carried out in cooperation with the Skill Manager Agent. A SMA stores the information, previously inserted by a CLO, about roles, associated competencies and employees. Moreover, for each employee Em_i, SMA manages the information about Em_i role and Em_i current competence levels. A Chief Learning Officer can require

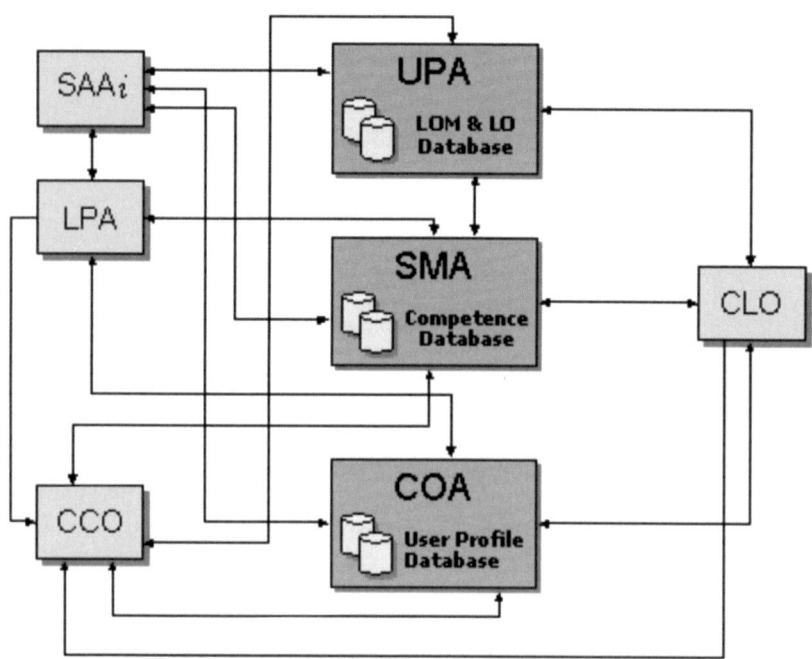

Fig. 8. A diagram showing MASEL architecture

her/his CLO to suggest the most suitable employees for taking a specific role and can assign a specific role to an employee, in such a case SMA evaluates the employee competence gaps according to the assigned role.

When Em_i logs in, her/his Student Assistant Agent SAA_i shows the gaps between her/his current competencies and the competencies required for her/his role in the organization; these are specified along with priorities or/and temporal constrains related to the competencies that have to be acquired. This task is carried out in cooperation with the Skill Manager Agent. When Em_i decides to fill a specific competence gap, say $Cgap_k$, SAA_i asks the LPA to perform three tasks:

- creating a *pre-assessment* test by which to evaluate $Cgap_k$;
- selecting and composing some Learning Objects $\{LO_{k_1},\ldots,LO_{k_n}\}$ in order to build a learning path Lp_k;
- enriching and modifying Lp_k according to feedbacks user provides (for instance by answering tests and assessments).

These tasks are carried out by LPA in cooperation with the Content Agent COA that provides LPA with the suitable Learning Objects. Moreover LPA may inform CCO Assistant Agent about the lack of some Learning Objects needed to build specific learning paths.

SAA stores all the information about current learning activities of the student. Periodically these information are sent to the User Profile Agents. Thus, eventually, UPA, in cooperation with SMA, updates the current competence level Ccl_k of Em_i according to her/his learning activities.

4   Conclusions

In this paper we have presented MASEL, a Multi-Agent System for E-Learning and Skill Management. MASEL is strongly based on a continuous interaction among involved agents; such an activity is facilitated by the choice of XML for both representing agent ontologies and handling data exchange. MASEL has been thought as an attempt to exploit agents and their benefits in E-Learning research. Agents in MASEL have been designed to perform the following tasks: *(i)* supporting Chief Learning Officers in defining roles, associated competencies and knowledge level required; *(ii)* managing the *skill map* of the organization; *(iii)* measuring human resources competence gaps; *(iv)* supporting employees in filling their competence gaps as related to their roles; *(v)* enriching a given courseware or creating personalized learning paths according to feedbacks user provides in order to optimize the acquisition of needed competencies; *(vi)* assisting Chief Learning Officers in choosing the most appropriate employee for a given role.

Presently we are working towards the construction of a prototype tool implementing MASEL using $JADE$ (*Java Agent DEvelopment Framework*). $JADE$ [3] is a software framework fully implemented in Java language. It simplifies the implementation of multi-agent systems through a middle-ware that complies with FIPA specifications and through a set of tools that support debugging and deployment phases. As far as future work is concerned, we plan to extensively test our system on real application cases in order to validate the effectiveness of the approach. In addition, we plan to improve the cooperation among the agents of the system by extending it to support other tasks like *optimal team building* and definition of *career path* for employees taking into account gaps between their possible future roles and current competencies.

References

1. Foundation for Intelligent Physical Agents (FIPA) Specifications. [http://www.fipa.org].
2. Instructional Management System (IMS) Global Learning Consortium Specifications. [http://www.imsproject.org].
3. Java Agent DEevelopment Framework Project. [http://jade.cselt.it].
4. F. Soliman G. Duncan, R. Beckett. A comprehensive organizational model for Enterprise Knowledge Management. In *Proc. of the 1st International Conference on Enterprise Information Systems*, page 783, Setubal, Portugal, March 1999.
5. R.J. Glushko, J.M. Tenenbaum, and B. Meltzer. An XML framework for Agent-Based E-commerce. *Communications of the ACM*, 42(3):106–114, 1999.
6. H. Takeuci I. Nonaka. In *The Knowledge Creating Company*, New York, USA, 1995. Oxford University Press.
7. P. Kaipa. Design organizations that learn: An executive guide to learning. In *Chinmaya Management Review*, July 1998.
8. Y. Malhotra. Organizational learning and learning organizations: An overview. http://www.brint.com/papers/orglrng.htm.
9. D. E. O'Leary. Enterprise Knowledge Management. *IEEE Computer Journal*, 31(3):54–61, 1998.
10. G.A. Papadopoulos. Models and technologies for the coordination of Internet Agents: A survey. In Andrea Omicini, Franco Zambonelli, Matthias Klusch, and Robert Tolksdorf, editors, *Coordination of Internet Agents: Models, Technologies, and Applications*, chapter 2, pages 25–56. Springer-Verlag, March 2001.
11. A. Patel and Kinshuk. Adaptive educational environments for cognitive skills acquisition. In *Proc. of International Conference on Advanced Learning Technology*, pages 502–504, Madison, Wisconsin, USA, August 2001. IEEE Computer Society Press.
12. R. Lassleben R. Klimecki. What causes organizations to learn? In L. Araujo M. Easterby-Smith and J. Burgoyne, editors, *Organizational Learning (Proceedings of the 3rd In-ternational Conference on Organizational Learning)*, volume 2, pages 551–577, Lancaster, England, June 1999.
13. T. Okamoto S. Belkada, A. I. Cristea. Measuring knowledge transfer skills by using constrained-student modeler autonomous agent. In *Proc. of International Conference on Advanced Learning Technology*, pages 375–378, Madison, Wisconsin, USA, August 2001. IEEE Computer Society Press.
14. A. Seffah and P. Grogono. Learner-centered software engineering education: From resources to skills and pedagogical patterns. In *Proc. of 15th Conference on Software Engineering Education and Training*, pages 14–21, Covington, Kentucky, USA, February 2002. IEEE Computer Society Press.

15. P. Stone and M. Veloso. Multiagent systems: A survey from a machine learning perspective. *J. of Autonumous Robotics*, 8(3):345–383, 2000.
16. J.M. Tenenbaum, T.S. Chowdhry, and K.Hughes. Eco system: An internet commerce architecture. *IEEE Computer Journal*, 30(5):48–55, 1997.

Integrating Mobile and Intelligent Agents in Advanced E-commerce: A Survey

Ryszard Kowalczyk[1], Mihaela Ulieru[2], and Rainer Unland[3]

[1]CSIRO
Private Bag 10, South Clayton 3169, Australia
ryszard.kowalczyk@cmis.csiro.au

[2]University of Calgary, Electrical Engineering, T2N 1N4 Canada
ulieru@ucalgary.ca

[3]University of Essen, Schützenbahn 70, 45117 Essen, Germany
UnlandR@cs.uni-essen.de

Abstract. The paper attempts to survey the existing research and development efforts involving the use of mobile agents and intelligent agents for advanced e-commerce solutions. In particular it aims at providing a representative view of the current research trends in developing intelligent and mobile agent-mediated e-commerce including location-aware, mobile and networked comparison-shopping, auction bidding and contract negotiation. A number of selected agent systems are presented with short overviews and then discussed in the context of agent-mediated e-commerce including the scope and specific solutions they address, provided support for agents' migration and users' mobility, and deployment approaches. Finally, a summary of the considered systems is presented together with some concluding remarks including the current state and trends in developing mobile and intelligent agents in advanced agent-mediated e-commerce.

1 Introduction

Future global information networks are often described as rich collaborative environments in which intelligent automated services (such as advertising, negotiation, financial transactions, order placement and delivery) enable enterprises to dynamically discover, interact and do business with each other. To be competitive in the new "e-economy" it is vital that researchers and industry alike are able to exploit emerging technologies that will form the basis of tomorrow's global information networks and act as the worldwide infrastructure for automated commerce. There are clearly many challenges in realizing such environments and today's technology already provides many of the tools for creating such dynamic service environments:

- Agent technology as a paradigm for structuring, designing and building systems that require complex interactions between autonomous distributed components. The Foundation for Intelligent Physical Agents, FIPA (www.fipa.org), is the lead-

R. Kowalczyk et al. (Eds.): Agent Technology Workshops 2002, LNAI 2592, pp. 295–313, 2003.

ing international organization in research and development of standards for interactive software systems based on the agent-oriented software engineering paradigm.

- Novel business-to-business (B2B) technologies such as market places, smart directories (Universal Description, Discovery and Integration - UDDI), trading standards (electronic business eXtensible Markup Language - ebXML), novel interaction paradigms (such as peer-to-peer) and web based integration (e.g., Simple Open Access Protocol - SOAP).

- New web and semantic web standards such as eXtensible Markup Language (XML), that enrich the semantic content of information published and form the basis of new interoperability mechanisms.

Following the web-centric paradigm shift [Ulieru, et.al. 2002] enabling the sharing of information, services and applications among suppliers, employees, partners and customers, an international effort referred to as the Global Dynamic Service Environment (Global DSE), is being conducted to provide an integrated information framework for inter-enterprise collaboration. This framework creates an environment by which enterprises can deploy and exchange services as well as create new, compound services. The value of this web-enabling environment extends across many domains, encompassing e-Business and e-Commerce; Internet-enabled services to society (such as e-Health [Ulieru and Geras 2002], e-Education, e-Collaboration); global communications; distributed manufacturing automation and robotics.

The Global DSE approach is to build upon the FIPA Agent standard as the underlying model for services – identifying the concept of a dynamic service with that of an agent as an autonomous system, capable of interacting with other agents in order to satisfy its design objectives. Other technologies then build on top of or into the agent framework to provide a rich/flexible environment. Although the technologies involved are mature enough to create and use an initial functional test bed and begin exploiting some of its benefits, extensive research is required to address the issues related to dynamic service composition. Such issues include: architectural requirements for interactive software systems, ontologies for multi-agent systems, security norms for multi-agent collaborative environments, market and other economic models in agent-systems engineering, gateways technologies for telecommunications networks, and coalition formation in multi-agent enabled organizations.

Electronic commerce offers new channels and business models for buyers and sellers to effectively and efficiently trade goods and services over the Internet. Agent-mediated e-commerce is concerned with providing agent-based solutions for different stages of trading processes in e-commerce including need identification, product brokering, merchant brokering, contract negotiation and agreement, payment and delivery, and service and evaluation (e.g. [Bailey and Bakos, 1997], [Chavez et al, 1997], [Guttman and Maes, 1998], [Gutman et al, 1998]). As the market quickly evolves, new advanced dynamic e-commerce (called also negotiated e-commerce or e-negotiation) solutions emerge to enable mapping more sophisticated and efficient negotiation models in business transactions to e-commerce, in particular in the contract negotiation and agreement stage of the trading process. It involves the development of e-commerce agents with more intelligent decision-making and learning capabilities in the context of automated contracting that can include comparison shopping, bidding in auctions and contract negotiations. At the same time the e-commerce environment also becomes more complex and dynamic due to the business trends to trade in several inter-connected marketplaces and use new wireless communication channels and portable computing devices (e.g. PDAs, mobile phones) in emerging location-aware mobile e-commerce (m-commerce). Here the mobility aspects of agent technology are predicted to play a significant enabling role.

The creation of the DSE involves a whole range of technologies ranging from emerging business-to-business facilitators to web standards and Agent and Artificial Intelligence technologies. In particular however the Global DSE uses the Foundation for Intelligent Physical Agents (FIPA) standard as the basis for intelligent interoperability in the network. Other technologies are then considered to augment this.

The growing number of research efforts in the areas of mobile e-commerce agents (e.g. [Kotz and Gray, 1999], [Papaioannou, 2000]) and agent-mediated e-commerce (e.g. [Bailey and Bakos, 1997], [Chavez et al, 1997], [Faratin et al, 1999], [Guttman and Maes, 1998], [Gutman et al, 1998], [Kowalczyk and Bui, 2000], [Lomuscio et al, 2000], [Maes et al, 1999], [Sandholm and Lesser, 1995] testify the potential benefits of integrating mobile agents in more advanced e-commerce applications. Some research has also been directed towards the use of mobile agents in intelligent agent-mediated e-commerce (e.g. [Dasgupta et al, 1999], [Griffel at al, 1997], [Sandholm, 2000], [Tu at al, 1999]).

The paper attempts to survey the existing research and development efforts involving the use of mobile agents and intelligent agents for advanced e-commerce solutions. It should be stressed that due to the practical and technical reasons the paper is not intended to be an exhaustive survey of all related research in this area. However it aims at providing a representative view of the current research trends in developing intelligent and mobile agent-mediated e-commerce. Some stationary agent-agent-mediated e-commerce systems are presented in Sect. 2. A number of selected intelligent and mobile agent systems in agent-mediated e-commerce including location-aware, mobile and networked comparison-shopping, auction bidding and contract negotiation are presented in Sect. 3. They are discussed and summarized in Sect. 4 in the context of agent-mediated e-commerce including the scope and specific solution they address, provided support for agents' migration and users' mobility, and used deployment approaches. Section 5 presents concluding remarks including summary of the current state and trends in research and development of integrated intelligent mobile agents for advanced e-commerce.

2 Intelligent Agents in Dynamic E-commerce

There are many challenges in realizing the Global DSE. Highly interdisciplinary research (e.g., industrial engineering and control systems, distributed artificial intelligence and logic programming, information systems and communication technologies) is required to develop and implement dynamic services for a Networked Economy. The *Agent paradigm* for structuring, designing, and building software systems that have complex interactions among autonomous distributed components has proven to be an excellent tool for modeling large-scale distributed information environments (e.g., factory control and global supply chain coordination).

To enable the deployment of dynamic e-commerce environments the European and the US Agentcities initiatives (www.agentcities.org), combined with international initiatives in FIPA (www.fipa.org), aim to create a global open information-exchange environment where dynamic services from geographically distributed organizations can be deployed, tested, interconnected and composed. Examples of such dynamic services are: B2B (Business to Business) dynamic value chain creation; dynamic pricing through trading exchange; automatic discovery of business partners; advertising and marketing. At this time there are large-scale initiatives in Japan, Europe, USA, Australia and New Zealand. In particular, Europe has two major Projects involving more than 50 organizations that already have been funded through the European Union's Fifth Framework Program.

Research on intelligent agents for dynamic e-commerce has already resulted in a number of interesting developments within the scope of agent-mediated e-commerce in particular for contract negotiation and agreement (e.g. [Bailey and Bakos, 1997], [Beam and Segev, 1997], [Gutman et al, 1998], [Maes et al, 1999], [Jennings et al, 2001]). A survey on agent-mediated e-commerce including well known early agent systems for e-commerce developed prior to 1998 such as Kasbah [Chavez et al, 1997] and Tête-à-Tête [Guttman and Maes, 1998] was presented in [Beam and Segev, 1997] and [Gutman et al, 1998]. For completeness, in this section we overview some selected new systems that have been proposed or further advanced since the above survey was conducted. Latest results regarding the implementation of e-Commerce strategies founded on the Global DSI infrastructure can be found in: [Sadeh et. Al, 2002] and [Dale and Ceccaroni, 2002]

AuctionBot [AuctionBot] is a well-known experimental Internet auction server developed at the University of Michigan. Its users can create new auctions by choosing from a selection of auction types and then specifying its parameters (such as clearing times, method for resolving tie bids, and number of sellers permitted). Buyers and sellers can then bid according to the auction's multilateral distributive negotiation protocols. In a typical scenario, a seller bids a reservation price after creating the auction and lets AuctionBot manage and enforce buyer bidding according to the auction's protocols and parameters. AuctionBot also provides an application-programming interface for users to create their own bidding agents to autonomously participate in the AuctionBot auctions. In particular AuctionBot has successfully been used to host Trading Agent Competitions [TAC] designed to promote, evaluate and showcase trading agents competing in a challenging market game consisting of several simultaneous auctions. The TAC competitions have resulted in a number of interesting systems of bidding agents developed by different research groups (e.g. [Greenwald and Boyan, 2001; Stone et al, 2001])

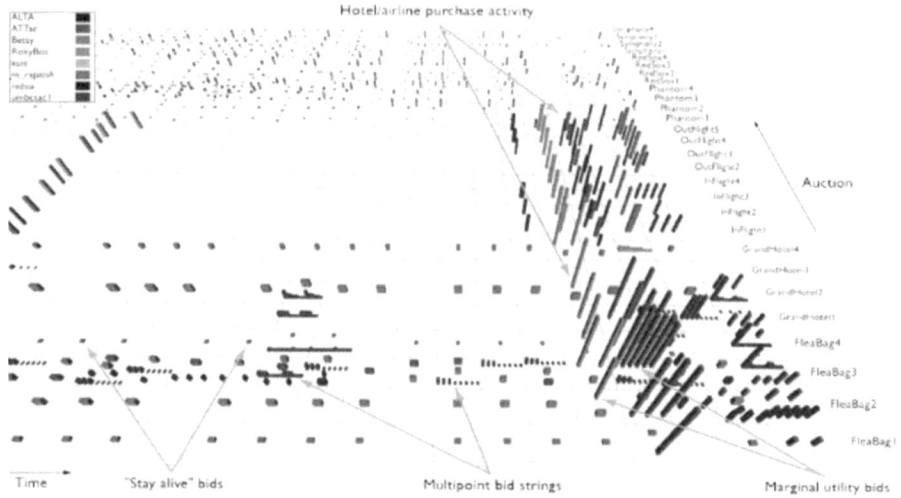

Fig. 1. Visualization of TAC games [TAC]

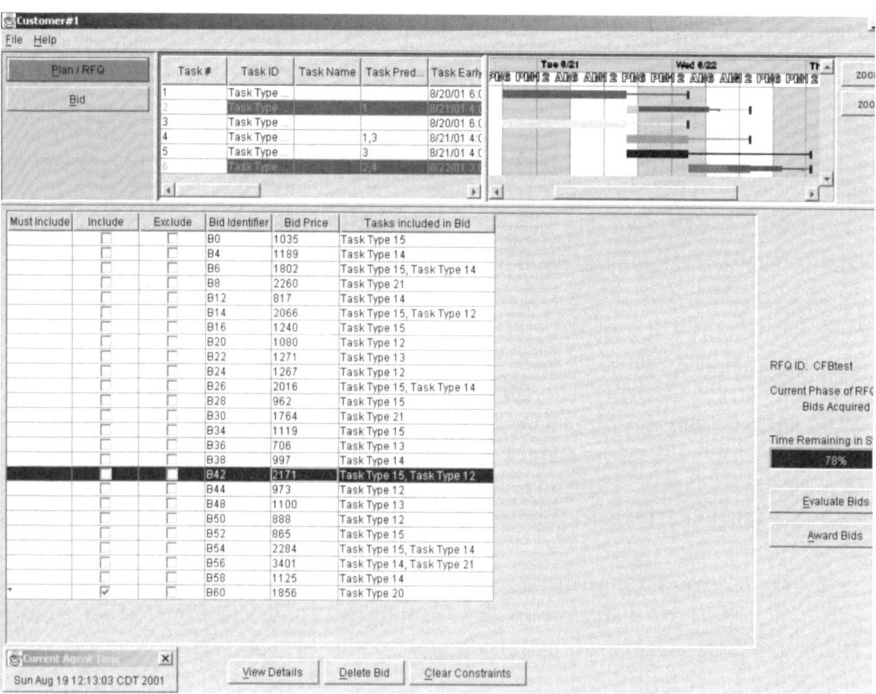

Fig. 2. Bid evaluation in MAGNET

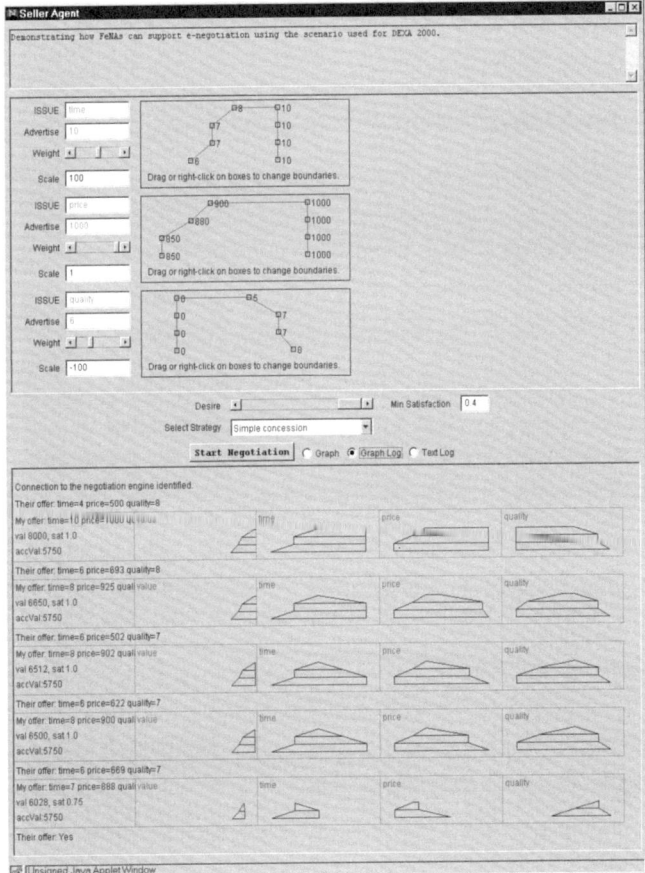

Fig. 3. FeNA agent displaying propagation of fuzzy constraints during negotiation

MAGNET (Multi AGent Negotiation Testbed) [MAGNET] is an experimental architecture developed at University of Minnesota to provide support for complex agent interactions, such as in automated multi-agent contracting, as well as other types of negotiation protocols. Agents in MAGNET negotiate and monitor the execution of contracts among multiple suppliers. A customer agent issues a Request for Quotes for resources or services it requires. In response some supplier agents may offer to provide the requested resources or services, for specified prices, over specified time periods. Once the customer agent receives bids, it evaluates them based on cost, risk, and time constraints, and selects the optimal set of bids that can satisfy its goals. An example of the bid evaluation screen shot is shown in Fig. 2. Suppliers are then notified of their commitments, and the Execution Manager is called to oversee completion of the plan. Plan maintenance includes re-negotiating existing commitments, re-bidding portions of the plan and re-planning [Collins and Gini, 2000].

eNAs (e-Negotiation Agents) [Kowalczyk and Bui, 2000a] and *FeNAs* (Fuzzy eNAs) [Kowalczyk, 2001] are prototypical intelligent trading agents developed at CSIRO [ITA] to autonomously negotiate multiple terms of transactions in e-commerce trading. The agents can engage in integrative negotiations in the presence of limited common knowledge about other agents' preferences, constraints and objectives through an iterative exchange of multi-attribute offers and counter-offers.

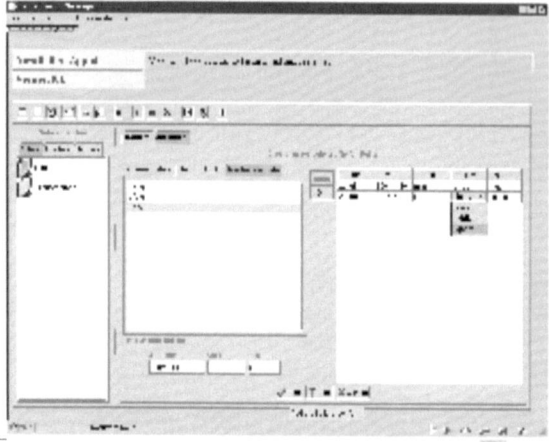

Fig. 4. Casba Agent Interface

Fuzzy eNAs can also flexibly negotiate with fuzzy constraints and preferences (see Fig. 3). The F/eNAs environment can consist of many autonomous trading agents representing buyers and sellers that can engage in concurrent bi-lateral negotiations according to a number of user-selected negotiation strategies. The eNAs and FeNAs agents have been demonstrated with a number of testbeds of e-commerce trading [Kowalczyk and Bui, 2000a; Kowalczyk and Bui, 2000b].

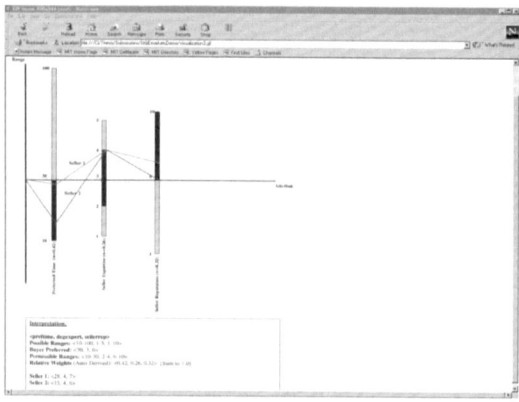

Fig. 5. MARI Visualization

Casba (Competitive Agents for Secure Business Applications) [Casba] is an Esprit funded project to develop a framework for an electronic marketplace using multi-agent technology. Casba aims at providing a set of tools for setting up, administering and managing an electronic agent market on the Internet.

Several auction types are supported and automated using agent technology. The libraries include flexible negotiation strategies that can be adapted to various product categories. Casba's toolset incorporates objects and interfaces for interacting with existing Web servers, database management systems and e-commerce platforms.

MARI (Multi-Attribute Resource Intermediaries) [MARI] is a project initiated at MIT Media Lab to develop an intermediary architecture intended as a generalized platform for the specification and brokering of heterogeneous goods and services in e-commerce. It aims at allowing both buyers and sellers to specify relative preferences for the transaction partners and different product attributes. MARI facilitates automation in the brokering process with the agents acting as proxies for buyer/ seller interests. Figure 5 presents an example of MARI visualization of changing preferences during that process.

It should be noted that there also are a growing number of other important on-going research efforts that have already provided several significant contributions to intelligent agents in e-commerce. For example AgentLinks' SIG on Agent-Mediated e-Commerce [AMEC] and ACM SIG on e-commerce [SIGecom], which bringing together active researchers and research groups from academia and industry working on different aspects of intelligent negotiation agents, are good reference sites for the latest developments in that dynamic R&D area.

3   The Mobility Aspect in Dynamic and Mobile E-commerce

Mobile agents have been recognized as a very prospective technology for both dynamic and mobile e-commerce applications (e.g. [Sandholm, 2000], [Griffel at al, 1997]) but the research in that area is still in very early stages. Although most of the related research considers mobile communication and location-aware computing, there is also growing research on deploying mobile and intelligent agents in advanced e-commerce including location-aware, mobile and networked comparison shopping, mobile auction bidding and mobile contract negotiation.

3.1   Location-Aware Shopping

Agora [Fonseca et al, 2001] is a project conducted at HP Labs to develop a test-bed for applications of agent technology to a mobile shopping mall. A scenario involves mobile shoppers with personal digital assistants (PDA) interacting with store services while in the mall, on the way to the store, or in the store itself. Mall-wide services,

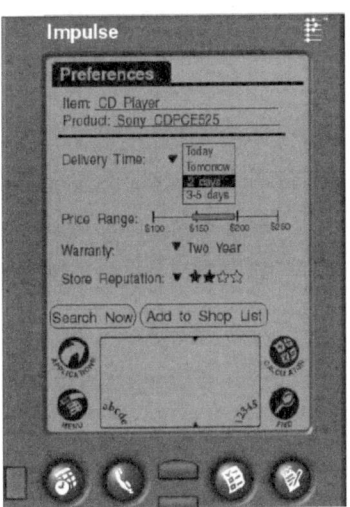

Fig. 6. Impulse displaying user's preferences

such as directories and locators are available through the PDA connected to the wireless network that provides an URL-based access to the virtual presence of the mall and its services on the Web. Intelligent agents represent both shoppers and the store, and participate in on-line auctions to bid for desired products based on shopper's preferences. A lightweight version of a scenario involving mall infrastructure agent, store agents, a shopper personal assistant and bidding agent, and an English auction agent has been implemented with a multi-agent system Zeus from BT Labs [Zeus] and additional Java-based support software.

Impulse [Impulse] is an on-going research project at MIT Media Lab that explores a scenario in which the buying and selling agents can run on wireless mobile devices and engage in multi-parameter negotiation for comparison-shopping at the point of purchase. The buyer agent resides on a PDA equipped with a GPS receiver and a wireless Internet connection that enables the URL access and communication with the seller (provider) agents (see Fig. 6). The agents have been implemented with a Java-based mobile agent system called Hive [Taylor, 2000] also developed at MIT Media Lab.

3.2 Mobile Comparison Shopping

An agent-based framework for mobile commerce has been proposed by [Mihailescu and Binder, 2001] (referred as MB later in this paper). It provides three types of agents, i.e. device agents, service agents and courier agents. The device agent is a stationary agent that resides on a mobile device and provides access to wireless services such as a location-based comparison-shopping. The service agents are owned by service providers and handle service requests from the users. They are heavy-weighted mobile agents operating within the wired network. The courier agents are single-hop light-weighted mobile agents that can migrate from a service agent to a mobile device in order to establish communication with the user. Figure 7 shows a display of a mobile courier agents presenting product options available for the user. A test-bed has been developed for a shopping center scenario where consumers can access a web portal wirelessly via their PDA devices for services such as product location, product comparison and store location with the envisaged possibility of negotiation. The test-bed has been implemented with the use of Java-based tools including Aglets SDK for service agents and KVM SDK for the device agent and courier agents.

Fig. 7. MAgNET's Cart view: The quotes brought back by the shopping aglets [MAgNET]

3.3 Networked Comparison Shopping

MAgNET (Mobile Agents for Networked Electronic Trading) [MAgNET] is a mobile agent-based system prototype developed at University of California with Java and IBM Aglets SDK to enable buyers to comparison shop for items from different online

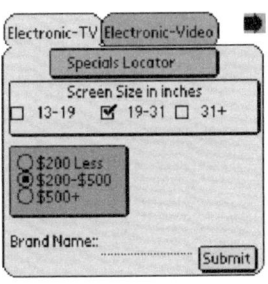

Fig. 8. Display of a mobile courier agent presenting product options

sellers. In MAgNET, a human buyer creates a mobile shopping aglet that compares quotes for an item from different online sellers by visiting those seller sites and returns to the buyer with the best offer that it obtains (see Fig. 8). It can also allow the buyer to send a mobile agent to various suppliers to purchase component parts needed to produce a complex product [Dasgupta et al, 1999]. It can also be possible for a supplier to create and dispatch a mobile agent to potential buyers to survey customer responses, determine market values of products, and sell products.

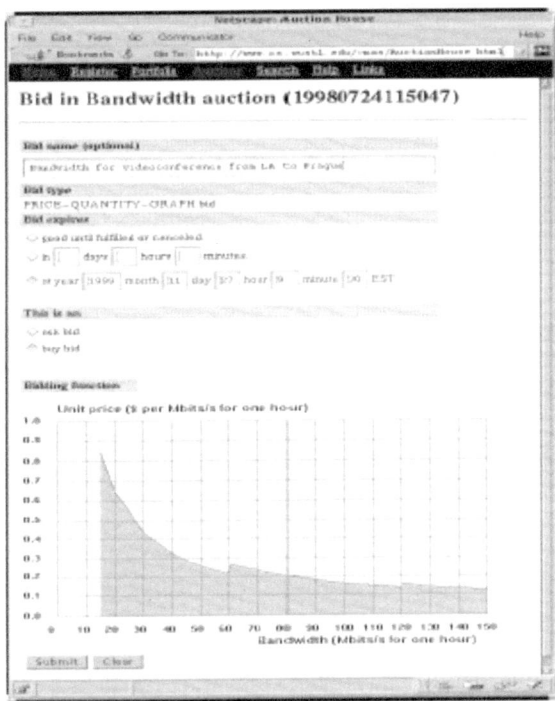

Fig. 9. eAuctionHouse bidding agent

3.4 Mobile Auction Bidding

eAuctionHouse [eAuction-Site] is a prototype of the Internet auction server developed as a component of a dynamic e-commerce platform eMediator [eMediator] at the University of Washington. It supports combinatorial auctions with bidding via quantity-price graphs through the Web browsers (see Fig. 9), and use the integrated mobile agent system called Nomad for automated bidding and auction monitoring in selected auctions [Sandholm, 2000]. Nomad is based on the Concordia mobile agents system [Concordia] and allows the users to generate mobile agents within the eAuction-House through the Web browser and launch them onto the agent dock within the eAuctionHouse site. The agents can then be executed locally to actively participate in two auction types on the user's behalf even when the user is disconnected from the network. In bidding the agents follow game-theoretical dominant strategies based on the user's reservation price (English auctions) and number of bidders (single-item, single-unit, sealed-bid first-price auctions).

BiddingBot [Fukuta el al, 2001] is a multi-agent system developed at Nagoya Institute of Technology, which can support attending, monitoring, and bidding in multiple auction sites. It consists of several cooperative bidding agents that have been implemented with a mobile Java-based agent framework called MiLog [Fukuta el al, 2001]. In BiddingBot, multiple bidding agents can attend different auctions and bid on behalf of users simultaneously in order to obtain the items at the best price. BiddingBot's bidding agents can bid according to autonomous and coordinated bidding mechanisms designed for the agents. Figure 10 shows an example of the user interface of BiddingBot

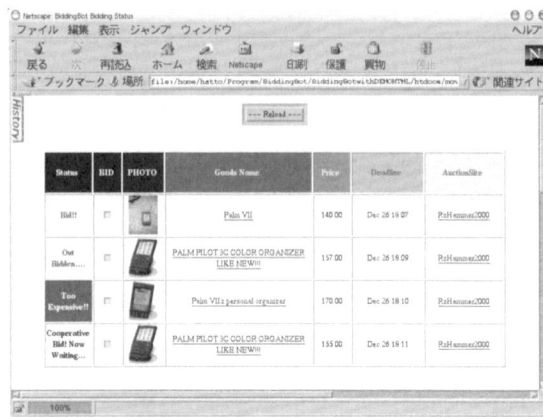

Fig. 10. An example of BiddingBot

3.5 Mobile Contract Negotiation

Electronic contract negotiation has been investigated as an application niche for mobile agents by [Griffel at al, 1997] in the scope of the OSM project [OSM] at University of Hamburg. In particular, a mechanism for contract document circulation has been developed to support the contracting parties engaged in the electronic contract negotiation with contract documents represented as mobile objects. More specifically, mobile agents are used to circulate the contract data objects between the negotiating participants who can review and alter the contract terms. A contract-carrying agent can also have a responsibility and a role for dealing with the contract while avoiding explicit locking mechanisms. An experimental prototype of a mobile agent system has been built to allow the users to get involved in the contract negotiation process through their Web browsers.

DynamiCS [DynamiCS] is an actor-based framework for mobile negotiation agents proposed by the same group at University of Hamburg [Tu et al, 2000]. It involves integration of intelligent decision-making capabilities into mobile agents based on plug-in mechanisms enabling dynamic composition of mobile negotiation agents [Tu et al, 1999]. In particular the DynamiCS framework aims at providing flexibility in integrating negotiation strategies into mobile agents dynamically. It also uses rule management mechanisms to manage actors and coordinate plug-ins' mobility. DynamiCS has been implemented with Java using Voyager system [Voyager] as the basic mechanism for distribution and mobility.

InterMarket [InterMarket] is a research project proposed to develop an Intelligent Mobile e-Marketplace System at Intershop Research and Fredrich Schiller University based on a mobile agent system Tracy [Tracy]. It aims at enabling mobile access and automated trading in e-marketplaces based on integration of mobile agents and intelligent decision-making agents offered as an add-on component to a commercial e-marketplace platform. InterMarket proposes stationary (or networked mobile) intelligent trading agents to automate the users' decision-making and negotiation tasks in e-marketplaces, and mobile agents to support deployment of the trading agents and provide mobile access and communication to e-marketplaces from mobile devices such as Personal Digital Assistants (PDA) or mobile phones.

4 Discussion

The systems presented in the previous section aim at providing new capabilities for advanced e-commerce solutions with the use of the approach based on agent technology. The mobility factor adds several aspects to the scope of agent-mediated e-commerce and specific solution they address, provided support for agents' migration and users' mobility, and their implementation and deployment approaches that are briefly discussed in the remainder of this section.

E-commerce Solutions
The e-commerce application areas mobile agents considered in this paper include comparison shopping, auction monitoring and bidding, and contract negotiation that are common application areas in agent-mediated e-commerce. They extend however the scope of the previous agent-mediated e-commerce applications to wireless m-commerce (Agora, Impulse, MB, InterMarket) and mobile agent-mediated networked e-commerce (MAgNET, eAuction-House, BiddingBot, OSM / DynamiCS). Figure 11 summarizes the extended scope of agent-mediated e-commerce with the agent systems considered in the paper including also stationary agents e-commerce systems from Sect. 2.

	Networked e-commerce	Wireless m-commerce
Stationary agents	*Kasbah* *T@T* *AuctionBot* *MAGNET* *FeNAs* *Casba* *MARI*	*Agora*
Mobile agents	*MAgNET* *eAuctionHouse* *BiddingBot* *OSM* *DynamiCS*	*Impulse* *MB* *InterMarket*

Fig. 11. The extended scope of AMEC with mobile agents and m-commerce

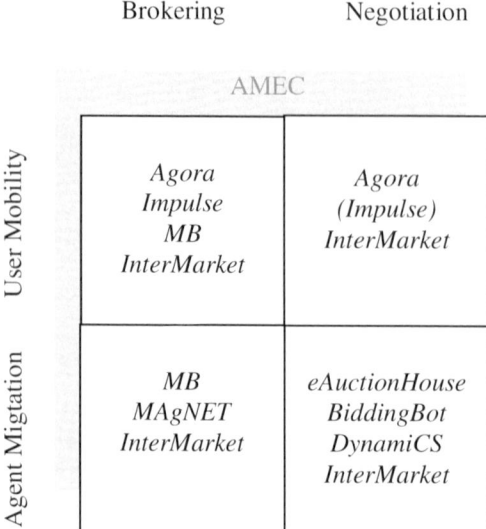

Fig. 12. Mobility/migration and decision support

Agent Support

he intelligent agents aim at providing automation support for decision-making tasks in e-commerce. The mobile agents extend that support by allowing for participation in several marketplaces in networked e-commerce and enabling users' mobility and wireless participation. The agents' mobility add ubiquity power to the participants to the e-Commerce game. In particular it allows the agents to quicker respond to local changes in marketplaces and make trading decisions faster than remote agents or human participants could. The agents can move across the network to reduce network traffic and communication latency and can also perform their trading tasks when the users are disconnected from the network. In addition some of them enable portable devices that support users' mobility, i.e. they allow the users to access, move and disconnect from the wireless network while the agents perform trading tasks on their behalf. Figure 12 summarizes the scope of decision support and mobility in the considered agent systems

Deployment Approach

Most presented systems have been deployed with general-purpose mobile agent systems that have been extended with decision-making (intelligent agent) capabilities. Typically they have used commercial agent development tools such as Concordia in eAuctioHouse, IBM Aglets in MB and MAgNET, and Voyager in OSM/Dyna-miCS. Some systems have also used prototypical mobile agent systems such as Hive in Impulse, Tracy in InterMarket and MiLog in BiddingBot. The use of third-party systems provides specific mobility and communication capabilities, and permits implementation of some high-level application specific functions. The use of proprietary systems

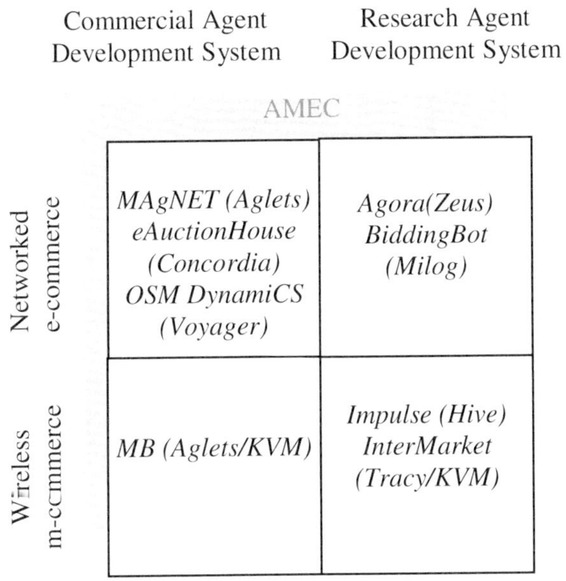

Commercial Agent   Research Agent
Development System   Development System

AMEC

	Commercial Agent Development System	Research Agent Development System
Networked e-commerce	MAgNET (Aglets) eAuctionHouse (Concordia) OSM DynamiCS (Voyager)	Agora(Zeus) BiddingBot (Milog)
Wireless m-commerce	MB (Aglets/KVM)	Impulse (Hive) InterMarket (Tracy/KVM)

Fig. 13. Deployment tools used in agent-mediated e-commerce

typically allows for more flexible implementation at the expense of the development costs. However in almost all cases an additional Java-based software component has been required even when commercial tools have been used. Figure 13 summarizes the used deployment tools.

Most efforts focus on developing intelligent trading agents to automate the users' decision-making and negotiation tasks, and mobile agents to support deployment of the trading agents and enable agents' mobility and users' access from mobile devices. However there is a trade-off between decision-making capabilities that make the agents "heavy" and mobility of agents that requires the agents to be "light weighted" due to the limited network bandwidth and device computational resources. For example eAuctionHouse deploys mobile agents that are created within the e-marketplace by the users through the Web browser for participation in the bidding processes on the users' behalf. InterMarket also deploys intelligent trading agents within the e-marketplace with a mobile agent system, but it enables mobile access with the use of mobile communication agents that can reside on the users' personal computers and mobile devices, and move to the e-marketplace to deliver the users' instructions to the trading agents. [MAgNET] integrates some decision-making capabilities into mobile agents that allow the shopping agents to compare quotes from different seller sites they can visit. DynamiCS proposes plug-in decision-making capabilities for mobile agents that permit flexible change of the decision-making capabilities within the agents keeping them reasonable small. To cope with the limited computational resources of the mobile devices and wireless network bandwidth, some systems like MB and InterMarket adopt "light weighted" mobile communication agents to deliver instructions to "heavier" decision-making agents operating within the networked environment.

Table 1 summarizes the intelligent mobile agent systems presented in this paper in the context of the all aspects considered in the paper.

Table 1. Summary of intelligent mobile agent e-commerce systems

	Application areas		Agent support		Implementation	
	Networked e-commerce	Wireless m-commerce	Decision Tasks	Mobility and Migration	Stationary Agent System	Mobile Agent System
Agora		Comparison shopping	Product and Merchant Brokering Bidding	User	Zeus Java	
Impulse		Comparison shopping	Product and Merchant Brokering (Negotiation)	User		
MB	Comparison shopping		Product and Merchant Brokering (Negotiation)	User Agents		Aglets SDK KVM SDK Java
MAgNET	Comparison shopping		Product and Merchant Brokering	Agents		Aglets SDK Java
eAuctionHouse	Auction Bidding		Bidding	Agents		Concordia Java
BiddingBot	Auction Bidding		Bidding	Agents		MiLog
OSM DynamiCS	Contract Negotiation		Negotiation	Agents		Voyager
InterMarket	Comparison shopping and negotiation		Product and Merchant Brokering Negotiation	User Agents		Tracy KVM SDK Java

5 Conclusion

This paper presents a survey of the existing research and development efforts involving the use of mobile agents and intelligent agents for advanced e-commerce solutions. A number of the selected agent systems have been overviewed with an aim to provide a representative view of the current research trends in developing intelligent and mobile agent-mediated e-commerce including location-aware, mobile and networked comparison shopping, auction bidding and contract negotiation.

In general the considered systems aim at providing new capabilities for advanced e-commerce solutions with the use of an approach based on agent technology, in particular focusing on integration of the complementary capabilities of the intelligent and mobile agents. The intelligent agents aim at providing automation support for decision-making tasks in e-commerce. The mobile agents extend that support by allowing for participation in several marketplaces in networked e-commerce and enabling users' mobility and wireless participation. It extends the scope of the agent-mediated e-commerce to wireless m-commerce and mobile agent-mediated networked e-commerce.

In most cases a considerable software development effort was required to implement decision-making capabilities and support software enhancing the agent systems used during the development. It is consistent with several views on the early stage of maturity of agent development tools available today. Therefore there is still a need for agent development tools that can support efficient deployment of both agents' mobility and decision-making in e-commerce applications.

Recently a number of standards and technologies have emerged in relation to electronic business-to-business trading processes. Dominant examples are ebXML, RosettaNet, eCo and WFMC standards. They cover the interoperability spectrum from the data exchange to business process collaboration, and are becoming widely used as business process infrastructures. Agent technology can augment these existing standards and frameworks to enable truly open and flexible trading processes with reduced negotiation and transaction costs. The end result of this fusion will be an intelligent infrastructure for cross-organisational business processes that enables e-Commerce service to interact dynamically to discover each-other to form new, compound, more complex services.

The focus of our future research and collaboration with the Global Agentcities Task Force will be to demonstrate the value added by agents in dynamic e-Commerce by focusing on service level agreements and trading contracts. We thus hope to stimulate work on agent-based prototypes which can demonstrate dynamic and open trading processes.

References

[Advice] http://www.advice.iao.fhg.de

[AMEC] http://www.iiia.csic.es/AMEC/

[AuctionBot] http://auction.eecs.umich.edu/

[Bailey and Bakos, 1997] J. Bailey and Y. Bakos. An Exploratory Study of the Emerging Role of Electronic Intermediaries. International Journal of Electronic Commerce, vol. 1, no. 3, Spring 1997.

[Baruceanu and Lo, 2000] M. Baruceanu and W. Lo. A multi-attribute utility theoretic architecture for electronic commerce. Proceedings of 4th Int. Conf. On Autonomous Agents, Barcelona, Spain, pp. 239-247.

[BargainFinder] http://bf.cstar.ac.com/bf

[Beam and Segev, 1997] C. Beam and A. Segev. Automated Negotiations: A Survey of the State of the Art. CMIT Working Paper 97-WP-1022. May, 1997, http://haas.berkeley.edu/~citm/wp-1022.pdf

[Braun et al, 2001] P. Braun, J. Eismann, C. Erfurth, W. Rossak. Tracy - A Prototype of an Architected Middleware to Support Mobile Agents. Proceedings of the 8th Annual IEEE Conference and Workshop on the Engineering of Computer Based Systems (ECBS), Washington D.C. (USA), April 2001, pp. 255-260.

[Casba] http://www.casba-market.org/

[Chaves and Maes, 1996] A.Chavez, P. Maes. Kasbah: An Agent Marketplace for Buying and Selling Goods, Proceedings of the First International Conference on the Practical Application of Intelligent Agents and Multi-Agent Technology, London, UK, April 1996..

[Chavez et al, 1997] Chavez, A., Dreilinger, D., Guttman, R., and Maes, P. A real-life experiment
in creating an agent marketplace. In Proceedings of the Second International Conference on the Practical Application of Intelligent Agents and Multi-Agent Technology PAAM'97 (London, U.K., Apr.). Practical Application Company, London, 1997.

[Collins and Gini, 2000] J. Collins and M. Gini, "Exploring decision processes in multi-agent automated contracting", Technical Report, TR 00-53, University of Minnesota, 2000. An edited version appeared in IEEE Internet Computing, pp 61-72, March/April 2001.

[Concordia] http://www.meitca.com/HSL/Projects/Concordia/
http://www.concordiaagents.com/

[Dale and Ceccaroni, 2002] J. Dale and Ceccaroni L. 'Pizza and a Movie: A Case Study in Advanced Web-Services', Proceedings of the First International Workshop on Agent-cities – Challenges in a Open Environment – AAMAS 2002, Bologna, Italy.

[Dasgupta et al, 1999] P. Dasgupta, N. Narasimhan, L. Moser, P.M. Melliar Smith, "MAgNET: Mobile Agents for Networked Electronic Trading", IEEE transactions on Knowledge and Data Engineering, Special Issue on Web Technologies, vol. 24, no. 6, July/August 1999, pp 509-525 [eAuctionSite]
http://ecommerce.cs.wustl.edu/

[Erfurth et al, 2001] C. Erfurth, P. Braun, W. Rossak. Migration Intelligence for Mobile Agents. Artificial Intelligence and the Simulation of Behaviour (AISB) Symposium on Software mobility and adaptive behaviour. University of York, United Kingdom, 21st - 24th March 2001. pp. 81-88

[Faratin et al, 1998] P. Faratin, C. Sierra, and N. Jennings. Negotiation decision functions for autonomous agents. International Journal of Robotics and Autonomous Systems 24 (3-4), pp. 159-182

[Faratin et al, 1999] P. Faratin, C. Sierra, N. Jennings and P. Buckle. Designing Flexible Automated Negotiators: Concessions, Trade-Offs and Issue Changes, 1999, Institut d'Investigacio en Intelligencia Artificial Technical Report, RR-99-03

[Fonseca et al, 2001] FonsecaS., Griss M., Letsinger R. An Agent-Mediated E-Commerce Environment for the Mobile Shopper. HP Technical Report HPL-2001-157, 2001.

[Foroughi, 1995] A. Foroughi. A Survey of the Use of Computer Support for Negotiation. Journal of Applied Business Research, Spring 1995, 121-134.

[Fukuta el al, 2001] N. Fukuta, T. Ito, T. Ozono, and T. Shintani. A Framework for Cooperative Mobile Agents and Its Case-Study on BiddingBot," In the Proceedings of the JSAI 2001 International Workshop on Agent-based Approaches in Economic and Social Complex Systems (AESCS 2001), pp.91—98.

[Greenwald and Boyan, 2001] A. Greenwald and J. Boyan. Bid Determination in Simultaneous Auctions-A Case Study. In the Proceedings of the Third ACM Conference on Electronic Commerce., Tampa, October, 2001.

[Griffel at al, 1997] F. Griffel, M. Tuan, M. Munke, M. da Silva. Electronic contract negotiation as an application niche for mobile agents. Proc. EDOC – International IEEE Workshop on Enterprise Distributed Object Computing, Australia 1997.

[Gutman et al, 1998] R. H. Guttman, A. G. Moukas, and P. Maes. Agent-mediated Electronic Commerce: A Survey. Knowledge Engineering Review, June 1998

[Guttman and Maes, 1998] Guttman, R., and Maes, P. Agent-mediated integrative negotiation for retail electronic commerce. In Proceedings of the Workshop on Agent-Mediated Electronic Trading AMET'98 (Minneapolis, May 1998).

[Impulse] agents.www.media.mit.edu/groups/agents/projects/impulse

[ITA] http://www.cmis.csiro.au/aai/ITA.htm

[Jango] jango.excite.com

[Jennings] http://www.ecs.soton.ac.uk/~nrj/neg-arg.html

[Jennings et al, 2001] N. Jennings, P. Faratin, A. Lomuscio, S. Parson, C. Sierra and M. Wooldridge. Automated Negotiation: Prospects, Methods and Challenges (to appear in Journal of Group Decision and Negotiation)

[Keeney and Raiffa, 1976] R. Keeney and H. Raiffa. Decisions with Multiple Objectives: Preferences and Value Trade-offs. John Willey and Sons, 1976.

[Kotz and Gray, 1999] D. Kotz and R. Gray. Mobile Agents and the Future of the Internet. ACM Operating Systems Review, August 1999, pp. 7-13.

[Kowalczyk and Bui, 2000a] Kowalczyk R. and Bui V. On Constraint-based Reasoning in e-Negotiation Agents. In F. Dignum and U. Cortés (Eds.) Agent Mediated Electronic Commerce III, LNAI (2000), Springer-Verlag, pp. 31 - 46.

[Kowalczyk and Bui, 2000b] R. Kowalczyk, V. Bui (2000). On Fuzzy e-Negotiation Agents: Autonomous negotiation with incomplete and imprecise information. DEXA Workshop on e-Negotiation, UK, 2000.

[Kowalczyk, 2001] R. Kowalczyk. Fuzzy e-Negotiation Agents (to appear)

[Maes et al, 1999] P. Maes, R. Guttman and A. Moukas. Agents That Buy and Sell. Communications of the ACM, March 1999/Vol. 42, No. 3, pp. 81-91.

[MAgNET] http://alpha.ece.ucsb.edu/~pdg/magnet/

[Milojicic, 1999] D. Milojicic. Trend Wars: Mobile agent applications. A review article in IEEE Concurrency, July-September 1999, pp. 80-90.

[MAGNET] http://www.cs.umn.edu/Research/airvl/magnet/

[Mihailescu and Binder, 2001] Mihailescu P. and Binder W. A Mobile Agent Framework for M-Commerce

[Lomuscio et al, 2000] A. Lomuscio, M. Wooldridge and N. Jennings. A classification scheme for negotiation in electronic commerce. In F. Dignum and C. Sierra (Eds.). Agent-Mediated Electronic Commerce: A European Perspective, Springer Verlag 2000, pp. 19-33.

[Oliveira and Rocha, 2000] E. Oliveira and A.-P. Rocha. Agents advanced features for negotiation in electronic commerce and virtual organisation formation process.

[OSM]. http://osm-www.informatik.uni-hamburg.de/

[Papaioannou, 2000] T. Papaioannou. Mobile Information Agents for Cyberspace – State of the Art and Visions. In Proc. of Cooperating Information Agents (CIA-2000)

[Rosenschein and Zlotkin, 1994] J. Rosenschein and G. Zlotkin. Rules of Encounter: Designing Conventions for Automated Negotiation among Computers. MIT Press, 1994

[Sandholm and Lesser, 1995] T. Sandholm and V. Lesser. Issues of Automated Negotiation and Electronic Commerce: Extending the Contract Net Framework. Proc. 1st International Conference on Multiagent Systems (ICMAS'95), San Francisco, 1995

[Sandholm, 2000] T. Sandholm and Q. Huai. Nomad: Mobile Agent System for an Internet-Based Auction House. IEEE Internet Computing, March-April 2000, pp. 80-86.

[SIGecom] http://www.acm.org/sigecom/

[Stone et al, 2001] P. Stone, M. Littman, S. Singh and M. Kearns. ATTac-2000: An adaptive autonomous bidding agent. In the Proceedings of the Fifth International Conference on Autonomous Agents (Agents-01).

[Sadeh, et.al., 2002] N. Sadeh, E. Chan and Van L. 'Open-Agent Environment for Context-Aware M-Commerce, Agentcities Workshop – at AAMAS 2002, Bologna, Italy, July 2002.

[TAC] http://auction2.eecs.umich.edu/

[Taylor, 2000] D. Taylor. Agents that move for things that think. IEEE Intelligent Systems, March/April 2000, pp. 4-6.

[Tete-a-Tete] http://ecommerce.media.mit.edu/Tete-a-Tete/

[Tu et al, 1999] M. Tu, F. Griffel, W. Lamersdorf. Integration of Intelligent and Mobile Agents for E-commerce – A Research Agenda. in: St. Kirn, M. Petsch. Workshop 'Intelligente Softwareagenten und betriebswirtschaftliche Anwendungsszenarien', TU Ilmenau, FG Wirtschaftsinformatik 2, Arbeitsbericht

[Tu et al, 2000] M. Tu, C. Seebode, F. Griffel and W. Lamersdorf. DynamiCS: An Actor-based Framework for Negotiating Mobile Agents.

[Ulieru, et.al. 2002] Mihaela Ulieru, Robert Brennan and Scott Walker, "The Holonic Enterprise – A Model for Internet-Enabled Global Supply Chain and Workflow Management", Int. Journal of Integrated Manufacturing Systems, No 13/8, 2002, ISSN 0957-6061.

[Ulieru and Geras 2002] Ulieru, M and Geras, A., Emergent Holarchies for e-Health Applications – A Case in Glaucoma Diagnosis, IECON 2002 – 28th Annula Conference of the IEEE Industrial Electronics Society, November 5-8, 2002, Sevilla, Spain (accepted).

[Wurman et al, 1998] PR Wurman, WE Walsh, MP Wellman Flexible double auctions for electronic commerce: Theory and implementation. Decision Support Systems 24:17-27, 1998.

[Zeus] Nwana Hyacinth, Divine Ndumu, et al., ZEUS: A Tool-Kit for Building Distributed Mutli-Agent Systems, Applied Artificial Intelligence Journal, Vol. 13(1), 1999, 129-186

An Agent-Oriented Approach to Industrial Automation Systems

Thomas Wagner

Institute of Industrial Automation and Software Engineering
University of Stuttgart
Pfaffenwaldring 47, D-70550 Stuttgart – Germany
Fax: +49 711 685 7302
Phone: +49 711 685 7295
wagner@ias.uni-stuttgart.de

Abstract. Multi-agent systems have been successfully used in a number of industrial applications. However, in the domain of industrial automation systems, little practical experience about the use of agents in such a demanding environment exists. Nowadays automation systems face new challenges and therefore new concepts are needed to meet them. This paper presents a vision of using agents within industrial automation systems from the point of view of automation engineering. To this end, the characteristic structures of automation systems are analyzed and opportunities, as well as advantages, of applying agent-oriented concepts are investigated. A possible approach for the integration of a multi-agent system into existing automation systems is introduced and illustrated by an application example.

Keywords: agent applications, automation systems, process control software, plant design, plant engineering.

1 Introduction

Automation systems such as those deployed in industrial plants, are complex, large and persistent hardware software systems being stamped by the characteristics of the technical processes they are designed to control. The software applied today in industrial automation systems has grown incrementally and therefore distinguishes itself by static hierarchical structures with precisely separated tasks, specialized functions and information structures. The applied concepts are based on different functional software levels according to the hardware structure, and hardware devices are modeled as predefined software components that have to be configured and connected into complete applications. These concepts have been successful during the last 20 years.

However, the development of automation systems software faces new challenges. The rapid development of high-capacity hardware components such as microcontrollers leads to more decentralized systems structures. Besides that, a strong need for integration of new functionality from enterprise management and business levels into

R. Unland et al. (Eds.): Agent Technology Workshops 2002, LNAI 2592, pp. 314-328, 2003.

the automation system arises. This trend leads to an increasing complexity[1] in automation systems, appearing in a more extensive functionality of the used hardware devices in the plant (the so called field devices) and consequently in more complex interactions between hardware and software entities. This evolution effects an expanded effort for development, plant design[2], plant engineering[3] and maintenance. In order to decrease this effort and reduce the tasks of engineers, future software for automation systems has to be able to handle this complexity. Comparing these challenges to static structures and functional characteristics of existing software in automation systems, it is doubtful if actual software concepts are qualified to match the increasing complexity.

An interesting approach to handle complexity in automation systems could be to map the increased "intelligence"[4] of hardware components into the software system. However, therefore new concepts and methods for automation software engineering are needed. The paradigm of autonomous agents is a promising concept for engineering complex, distributed systems. This paper deals with the application of agent-oriented approaches to enhance the software of automation systems for simplification of both plant design, engineering and maintenance as well as the integration of functionality from external systems. Towards this objective, a clear understanding of the target area is necessary and two central questions have to be answered:

- *Is the agent-oriented approach well suited to design and build software for automation systems? (Or conversely: Do the characteristics of automation systems indicate the suitability of an agent-oriented approach?)*
- *How can agent-oriented concepts be integrated into the existing and persistent software / hardware structures of automation systems?*

In seeking to answer these questions, this paper is structured as follows. In Sect. 2 a detailed view at the software structures in automation systems is introduced, together with a general rating of the existing concepts compared to the challenges on future software development. In Sect. 3 the relevance of the agent-oriented paradigm to meet these challenges is examined and the fundamental possibility of applying agent-oriented concepts to legacy automation systems is investigated. The basic ideas for the introduction of multi-agent systems are described in Sect. 4. In Sect. 5 an example of using agents in asset management systems shows the advantages of an agent-oriented approach in this area.

---

[1] In this context, the term "complexity" is used in a general manner; not in the specific technical sense of algorithmic or computational complexity [1].

[2] Plant design contains activities of technical realization of a plant with design phase, implementation phase and plant commissioning phase.

[3] Plant engineering contains activities during plant operation with the aim to maintain or improve the technical operativeness of a plant. This includes condition monitoring and estimation of plant components, decision on measures and implementation of measures in the plant.

[4] The term "intelligence" refers to the extended functionality of field devices, like self-diagnosis and failure prediction [12]; in this article it is not used in the sense of artificial intelligence.

2   Problems in Engineering and Operation of Automation Systems

In this section the target area of automation systems in industrial plants will be introduced in detail to make clear the characteristics of the problem and compare them to
the merits of an agent-based approach.

2.1   What Is an Automation System?

Industrial plants are technical facilities, in which technical (e.g. energy-related,
chemical, production or material-handling) processes are controlled automatically.
Examples are pharmaceutical plants, automotive production plants or power plants.
The automation system used within a plant consists of three different elements (see
Fig.1):

- The technical process running in a technical system
- The automation system (control and communication system) for information flow
  linkage of hardware and software entities
- The process personnel for operating & managing

More detailed information on the automation system's elements is provided in [2].
There are also other types of automation systems, e.g. in the domain of product automation, where the technical process takes place within a single device. Within the
scope of this paper only automation systems used in plant automation are regarded.

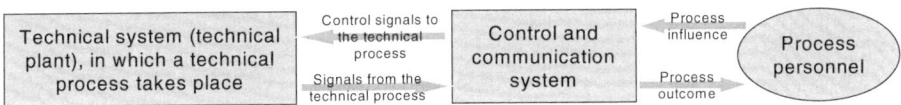

Fig. 1. Elements and information flows of an automation system

Nowadays deployed automation systems feature three outstanding characteristics:

1. High degree of automation, i.e. the operations to control the technical process are
   predominantly performed automatically. The tasks of process operators are limited
   to the superordinated process control and intervention in exceptional situations.
2. The automation system has to accomplish extensive and complex automation functions [2]. Automation functions are for instance: Control of the technical process,
   handling and observation of operating sequences, monitoring, diagnosis and maintenance. These automation functions are nearly solely fulfilled by software.
3. Plant development and design (including the configuration of hardware entities and
   communication systems) are often considered separately from the operation of
   automation systems. This leads to a lack of transmissibility between design phase
   and plant engineering phase (during operation) and therefore causes problems in
   data consistency together with a high manual engineering effort.

Additionally it should be considered that, from a technical point of view, an automation system often consists of hard and software entities of different manufactures and

their different technologies (e.g. bus protocols or field device interfaces). These entities have to be integrated into the complete system with manual engineering effort.

In order to determine the fundamental problems and challenges in plant design, in-
stallation, commissioning and engineering, a more detailed investigation of the inherent structure of automation systems will be given in the next section.

2.2   Characteristics of Modern Automation Systems

In the course of the development of automation technology more and more functions
which take place within technical systems were performed by automated equipment,
i.e. the degree of automation increased. At first automation functions were developed
at the operational level (like supervision of process values, control of sequences, etc.)
for the respective measurement and control systems. With the application of automation computer systems comprehensive functions from the "higher" levels were integrated into the automation system (like sequence planning and optimization) [2].

As a consequence of this evolution, modern automation systems have a hierarchical structure with different automation functions at each hierarchical level. This hierarchical structure is called the automation pyramid and represents a canonical abstract
model for the structure of complex multilayered plants [2,3]. Within this model the
several organizational levels of an automation system are shown (see Fig. 2). The
typical automation pyramid for automation systems in a plant consists of the following
levels (from the bottom up):

The *actuator-sensor level*, the *field level*, the *process control level* and the *management level*. These levels are networked between each other with different bus systems, which are adapted for the specific requirements of data flows at each level.

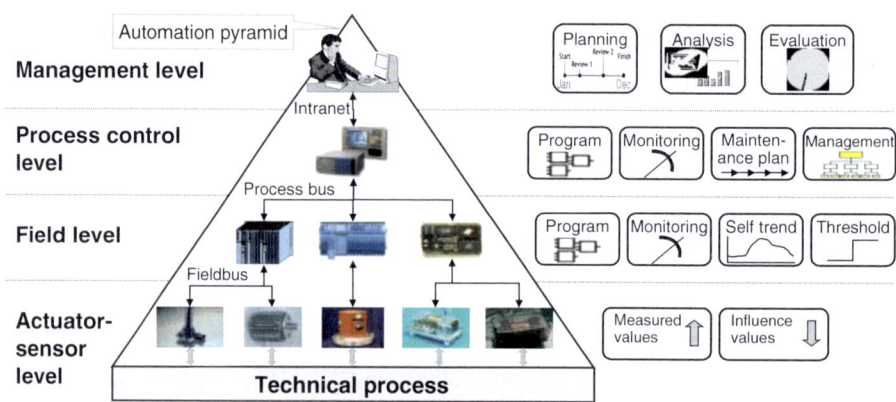

Fig. 2. Canonical structure of modern automation systems

At the *actuator-sensor level* (also called process level) measured values, i.e. physical
process variables like temperature, pressure and distance, are captured via sensors and

influence values for controlling the technical process are output via actuators (controlling elements) like valves, pumps, heatings and servo-motors.

The *field level* accommodates all tasks needed for processing measured values from the process and directly influencing of the process [3]. At the field level, typical automation computers like programmable logic controllers and microcontrollers are used. In addition the process can be influenced using monitoring and operating consoles.

At the *process control level* monitoring and process management of the overall process is performed, as well as production control, start up and shut down processes and superordinate malfunction handling. Due to the high number of plant hardware components, the management and engineering of the plant (e.g. setting of field devices and communication parameters) as well as maintenance planning has to be done.

Within the *management level* all higher-ranking enterprise levels are summarized. At this level tasks such as production planning, evaluation and analysis of process results and optimization of operational sequences are executed.

Regarding the software in the automation system, a clear separation of information processing can be identified: at each level different data is processed and different views on the technical process exist.

During the last 20 years the implementation of software in automation systems has been performed in the same way [2,6]: The overall functionality is decomposed into single functions, of which a lot of are available in the form of prefabricated universal program building blocks (i.e. standard blocks for control, monitoring, operating, supervision and recording tasks). These single functions are connected together with regard to data flows and temporal dependencies. This process is called configuration.

The quite different goals and specific conditions at the several levels and the resulting requirements to the software (e.g. amount of data, temporal restrictions) have caused a strong orientation of the means of description[5] used for implementation towards the automation functions and views that are realized at each level.

The prevalently applied means of description are some kind of visual net-based languages. Examples are Function Building Blocks (FBB), Continuous Function Charts (CFC) and Sequential Function Charts (SFC) [4,6]. The abstraction level of these means of description is very close to implementation and focused on the hardware- and machine-related properties of the system [5].

The same characteristics apply accordingly to application models[6] that are the basis for the means of description. All application models have in common to describe a system in the form of a composite (net) of passive entities and provide a function or sequence oriented view on the automation system.

Today the software systems in automation systems with their specific characteristics face new requirements as explained in the next section.

---

[5] Means of description picture certain issues in graphical or textual manner for visual perception and storage. The several description elements, their combinations or assignments are correlated to specific conditions or concepts of a certain technical context [5].

[6] An application model is a representation of a certain part of the real world which is subject to modeling and realization. It normally consists of a data model and a sequence model [7].

2.3   Challenges of Future Automation Systems

Two main factors have a strong influence on the evolution of automation systems:

1. Due to the rapid development of information technology, high-capacity hardware components like fast processors, large memories and wide-band networks are available today. This leads to an increasing decentralization within automation systems and a distribution of functionality into field devices. Resulting from this is an increasing amount of communication and increasing effort for the configuration of system entities as well as of the complete system. The dynamic development of hardware components in automation systems is in contradiction to the long life-cycle of a plant (up to 30 years). Hence, automation systems are subject to an ongoing partial modification. This affects particularly the automation system software which is intensely stamped by the structural aspects of the plant: several software entities are optimized or adapted to changes in the hardware structure or in processes. This results in strong requirements on flexibility and adaptability of the software during the whole life-cycle [8].

2. The requirements of the process control technology market are a driving force for future development directions and affect automation systems in several aspects:

- For further increase of the degree of automation, autonomous and adaptive automation systems are required. This includes support of human operators in complex tasks like condition monitoring of plant components and decision on measures.
- The vertical integration of additional functionality form higher enterprise levels (e.g. enterprise resource planning) requires stronger interconnection of the machine level with all business levels. Therefore, the interoperability of the different systems has to be accomplished. That means a higher coupling and transmissibility of information flows between the strictly separated levels of the automation system [17].
- A better support of users in plant design and engineering, especially configuration and parameterization[7] of system entities is needed (because of the increasing functionality in field devices, the efforts for configuration are even higher). The integration of new system components has to be supported and automated. The vision is a "plug'n'play" scenario comparable to the MS Windows world [10].

The two main factors mentioned above, that drive the evolution of automation systems, can be abstracted to the following consequence: Due to the extension of the former solely hierarchical communication within an automation system by additional services and the demand for a flexible communication to more "intelligent" hardware components (up to n:n communication correlations), more and complex interactions within the automation system become necessary. As automation systems cannot be built form scratch but are subject to ongoing change, these influences have to be compared to the static and hierarchical structure of contemporary systems. It seems obvious that as a result of these influences, both the scale and the complexity of automation software will strongly increase to fulfill all necessary automation functions. This complexity can be characterized and classified in two aspects:

---

[7]   Choice of variable values to characterize the system's behavior within a given structure [4].

- **Functional Aspects.** Organization and control of a distributed system at runtime (i.e. including the dynamic allocation of functionality).
- **Design and Engineering Aspects.** Development and design of a decentralized plant structure as well as configuration and management of single components.

The capabilities of existing software to manage the evolving complexity in comparison with the merits of an agent-oriented approach are presented in the next section.

3   Why Are Agent Concepts Relevant to Automation Systems?

In Sects. 2.2 and 2.3 the characteristics of modern automation systems and requirements for further evolution were shown. They can be summarized in properties that are intrinsic of all automation systems and their software used in industrial plants:

- Automation systems are complex and distributed systems.
- Automation systems require different views. Engineers, process personnel and managers have a individual view on structure, data and functionality.
- Automation systems require flexible and adaptive software. Changes affecting the software can refer to data, structure or operating sequences of the system.

These properties are comparable to the general properties of complex decentralized systems as described by Jennings in [11]. Figure 3 shows, on the left side, an overview of the inherent properties of complex systems: A complex system has a large number of parts that have many interactions. Complexity frequently takes the form of an organizational structure of interrelated subsystems and their components, relations between subsystems vary over time. Generally viewing the development of software these days, a wide range of software models and techniques have been devised to make it easier to handle this complexity. However, they fall short in two main ways [11]:

1. The interactions between the various computational entities are too rigidly defined.
2. There are insufficient mechanisms available for representing the system's inherent organizational structure.

Picking up the analysis aspects in Sect. 2.2, this statement also applies to software of automation systems: This software is hardware- and implementation-oriented and has strict restrictions to structure and interactions within clearly separated functional levels. Although there are means to represent organizational structures, these structures have to be defined at design time and are not adaptive to dynamic changes later at runtime. Also these structures implement a determinate view on functional levels within an automation system and rigid interactions between system entities. They do not provide the means for a flexible structural reorganization and new flexible interactions as needed to match the challenges mentioned in Sect. 2.3.

According to [1] the agent-oriented approach is a natural way of system decomposition and a reasonable alternative to contemporary approaches in software engineering. Figure 3 illustrates how the inherent properties of a complex system are reproduced in an agent-based system: Subsystems and subsystem components are mapped to agents and agent organizations; interactions between subsystems and subsystem components are mapped to cooperation, coordination and negotiation mechanisms; and relations between them are mapped to explicit mechanisms for representing organizational relationships. For further details, see [1,11].

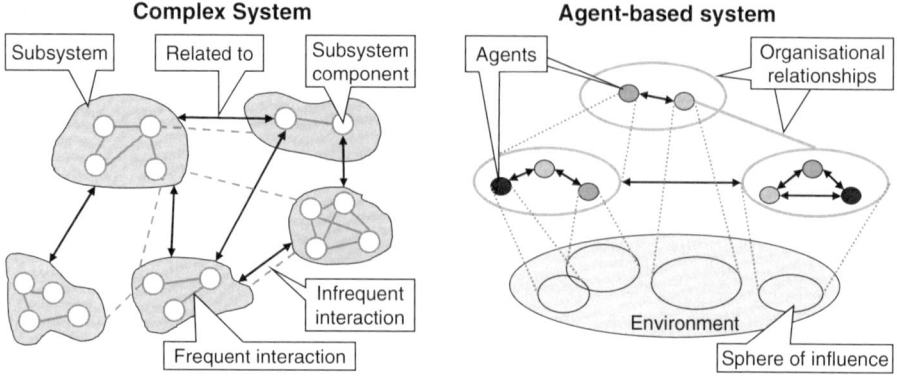

Fig. 3. Canonical view of a complex system compared to an agent-based system [11]

Two facets of the comparison between complex systems and agent oriented systems are of particular significance:

1. In an agent-oriented software system, multiple agents represent the decentralized nature of the problem, the multiple locations of control, the multiple perspectives or the competing interests. This abstraction reproduces the inherent properties of real-world systems and leads to a better handling of complexity within the software.
2. By high-level interactions, agent systems provide mechanisms for building flexible organizational structures. This effects dynamic bottom up coordination and enhances possibilities for dynamic adaptation of software to a changing environment.

For these reasons, it can be concluded that agent-oriented concepts are suitable to meet the requirements of modern automation systems. The agent-oriented approach seems applicable for the development of automation system software.

However, as automation systems and process control software have a long lifecycle, they can be subject to only slow and incremental changes. Hence, the question arises, can agent-oriented concepts be integrated in existing automation system structures and are they able to deal with legacy software or its underlying application models? As explained in Sect. 2.2, the means of description for the development of automation system software typically are specializations of some kind of visual net-based languages. With object oriented methods these languages can be described as networks of interconnected instances. In [5] the possibility is shown, that these models can always be mapped to an object-oriented reference model. In a reference model objects, encapsulate processes and functionality, but also can represent their dynamic input or output information [8].

There are certain similarities and conceptional correlations between object- and agent-oriented approaches. Both adhere the principles of encapsulation and information hiding and recognize the importance of interactions. Agents extend these concepts by autonomy (control over internal state and action choice), own thread of control and persuasion of objectives within themselves [1]. Hence, the ability of object-oriented concepts to reproduce application models of automation system software in principle can be assigned to agent-oriented concepts as well. In short: agent-oriented approaches basically can be integrated in existing automation systems.

In the next section the fundamental ideas of integrating agent-oriented approaches into existing automation systems are presented.

4   Application of the Agent-Oriented Approach to Existing Automation Systems

Within the domain of information technologies, new software concepts and technologies are deployed in real-world applications very rapidly. However, the innovation of automation system software structures is characterized by an explicitly slower progress than those in office automation or internet applications. This for several reasons:

- Software in automation systems evolves incrementally due to its long life-cylce.
- In contrast to the information technology domain, where mostly standard computers or servers are used, hardware components used in the automation domain are specific to the application area. For this reason software and hardware components have a much closer association (e.g. programmable logic controllers), changes in software often affect hardware and vice versa.
- Industrial plants are large heterogeneous hardware-software systems causing an enormous effort for development and operation and therefore they have a proportionate investment volume. Herein lies the reason for a significant skepticalness of plant operating managers against radical technical innovation.

Since the early nineties, cooperative multi-agent systems and intelligent agents are of increasing concern with respect to the software engineering of large scale distributed systems [13]. For an efficient introduction and a wide acceptance in the automation domain, new software concepts and software systems based on them have to support legacy systems to a large extent. Thus a number of constraints have to be considered:

- The new approach has to be adopted in a pragmatic manner. This means that the design of the new software system introduced has to be oriented to the basic structures of existing application models in order to meet the requirements of process control technology performantly and to be understandable for the user [10].
- Application models already used in the application domain have to be integrated while keeping interoperability. Existing implementations should be expandable [4].
- Adding new performance features has to be possible without deep changes to existing software and hardware architectures.

Considering these conditions leads to the approach for the integration of an agent-based system into the automation system structure explained in the next two sections.

4.1   The Automation System Seen as a Role System

The fundamental approach is based on the principle of a comprehensive complete representation of real entities by agents. This idea follows the insight that every real entity including its functional range can be regarded as a role within a role system. The notion of a role focuses on the position and responsibilities of an entity within an overall structure or system [14]. The role results from the task a certain entity has to fulfill within the system, in order to provide the system's overall functionality. Further it is important that a certain entity can - depending on its functional range - possibly obtain different roles, depending on the individual view (partial functional aspect) the system is regarded with. The analogy of roles within a system is very close to the human mental model[8] when decomposing, developing and realizing a system.

Due to the modeling concepts and implementation techniques applied so far, an automation system is viewed only as a composite of passive entities. That is, the activities necessary for the realization of several roles in a executable system have to be performed by humans during plant design and engineering phase. These activities include the customization of a system entity's functionality (both hardware and software entities) to the required role within the system and its environment (configuration). But further, the mentioned system entity fulfills the expected role in a passive manner. On changes within the system or its environment, the provided functionality of the entity has to be customized again by human activity according to the changes of the role.

If, in contrast, the role is realized by agents, substantial activities can be automated. The agent - aware of its own objectives - can dynamically adapt its behavior to changes and re-customize the represented entity. Beyond configuration, further activities can be automated such as condition monitoring of entities, overall management of system components and their configurations, or integration of external systems.

Below the basics for modeling existing automation systems with agents are explained.

4.2   Integration of an Agent System

Based on the abstraction of roles, the automation system is extended by an agent system in the form of an orthogonal communication level within the hierarchy (Fig. 4). The cooperating agents are designed to act very closely with the automation system and can be divided in two basic categories: *resource agents* and *service agents*.

---

[8] Mental models are representations of the real or virtual world in the human brain, built by the cognitive abilities of human [5].

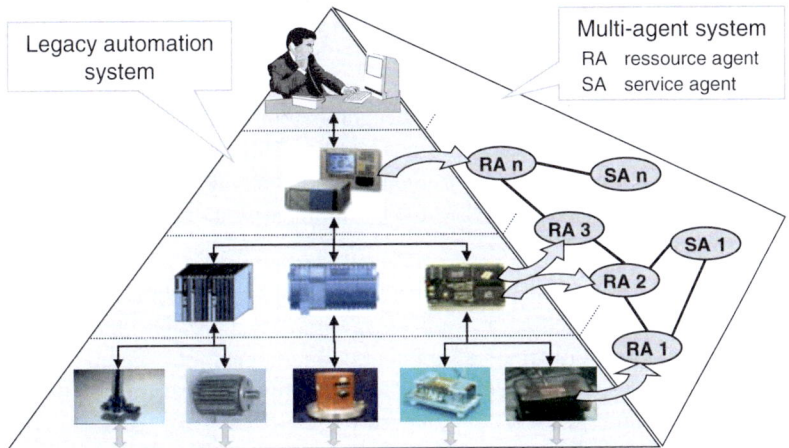

Fig. 4. Multi-agent system within the automation system hierarchy

- Within the agent system all of the real existing entities addressing the problem of the automation system are represented by resource agents. The entity represents the sphere of perception and influence of the agent. The granularity of this sphere depends from the goals to achieve (e.g. a single device, a device interface, a control program, a control element such as function building blocks or a set of variables).

- Resource agents encapsulate the information flow of the entity they represent. That is, they are able to communicate with the entity's (e.g. device's) specific interfaces and provide an abstract view on the entity within the agent system.

- Resource agents can be of different types in order to represent different views (roles) on the entity: *design agent, commissioning agent, condition monitoring agent, diagnosis agent*. These agents may have to cooperate in performing a role.

- Resource agents act autonomously in reference to the entity they represent. For example: agents representing a field device are capable of reacting to changing environmental conditions by self-adaptation of communication or physical parameters of the device. Device-specific status and diagnosis signals (as provided by modern field devices) can be recorded and interpreted by resource agents in order to generate diagnosis and maintenance information (e.g. life and failure prediction).

- Resource agents can autonomously build organizational structures reflecting the physical or logical plant structures, e.g. a compound of field devices within a certain machine. That way "plug'n'play" of devices can be realized. Within the compound of agents diagnosis and maintenance information can be used to determine the functional reserve[9] of single devices or of plant components. Pro-active acting can be applied for failure prediction or preventive maintenance planning.

- Service agents do not represent a certain real existing entity, but a certain task or a view on the automation system or part of it. They are designed to realize superordi-

---

[9] The functional reserve characterizes the instantaneous availability of an entity and allows conclusions on future plant or process behavior.

nate functionality that is not bound to a certain device or control element. Examples of such a functionality are management of devices, management of configurations or inspection of the overall system in order to generate maintenance information.

- The existing communication within the automation system (e.g. block-to-block communication via status signals) is maintained. This is necessary to accomplish the generally strict requirements for control of the technical process (e.g. real-time requirements) without restrictions on the flexibility within the agent system.

This way additional functionality and services like automation of configuration sequences or flexible management of devices, can be realized within the agent system. This functionality is enabled as a result of the high-level flexible interactions in an agent system compared to the signal oriented communication of the automation system. Operations and sequences that were restricted by the structure of the existing hardware and software system are reproduced in the orthogonal agent layer. By doing so, a closer coupling of information flows of the different levels in the automation system can be obtained without fundamental changes to existing hardware and software structures - an important aspect for the plant operating managers.

An interesting aspect for the manufacturers of plant components is the effect of implementation protection: The resource agents encapsulate all functionality and data specific to the device. The automation and engineering systems solely have access to the software interface of the agents, not to their implementation. This way in-house knowledge can be hidden, which is of great interest to plant manufacturers.

The next section illustrates, how to build applications for automation systems and which advantages can be achieved. As an application example the area of asset management is used, which is an upcoming trend in today's automation engineering world.

5   Application Example: Agent-Based Asset Management

In automation engineering, functions for administration and optimization of devices are gaining importance in comparison with pure process control functions. These functions are summarized in the term "asset management[10]" or "asset optimization". The ongoing developments in automation engineering concerning this area are targeted at supporting extensively plant commissioning and engineering (online asset management). That is, to utilize all plant equipment such as PLCs, machines, or field devices for adding value and provide the required availability with optimal effort.

An asset management system is a data processing system that collects, concentrates and rates heterogeneous information to conclude the condition of several plant components. Additional access exists to external systems like (electronic) plant documentation or business administration systems [15]. The NAMUR[11] [16] recommendation for asset management systems lists the functional requirements of these systems:

---

[10] In a broader sense the term "asset" is used in a more comprehensive way, describing all value added by entities of an enterprise; in this context "asset" refers to the plant equipment only.

[11] European association of users of process control technology - see http://www.namur.de

- Parameterization and configuration of field devices
- Documentation of field device history
- Condition monitoring of field devices and other plant components
- Access to components of the process control software
- Access to plant documentation, CAE-systems and business administration systems

This makes clear that asset management not only targets the maintenance of the existing plant, but starts with design and engineering of process control components and includes access to plant environment. To fulfill this functionality, the asset management system has to access process operating information. Sources of information are measured values from the technical process, status signals & diagnostic information of field devices and special monitoring facilities. Further sources of information that are not directly needed for process operation but for planning, design and engineering are:

- Engineering tools, CAE (computer-aided engineering) systems
- Parameterization tools for field devices
- Documentation of the plant and plant components
- Business administration systems

The data required for optimization of plant engineering must be completely extracted from the existing systems. Finally different user groups who have individual views on the automation systems have to be supported:

- Plant designers and plant commissioning experts
- Plant operating managers e.g. factory managers and plant operators
- Plant maintenance staff e.g. engineering and system experts

At present, enterprises have to bring up high costs for engineering their asset management applications to increase the availability and productivity of plants. The reasons are, on one side, the higher configuration effort due to the increased functionality of the field devices. On the other side, there is an increasing specialization of device data for the same reason. Thus, a divergent variety of control concepts and tools for field devices results. Another substantial problem is the access to existing engineering, CAE and business administration systems, as they are based on historically evolved "de facto standards" and have little in common regarding their data representation. In addition, there are problems of information protection that occur when information has to be accessible across the boundaries of different network infrastructure.

In short, two fundamental challenges exist in realizing an asset management system: 1. Support for varying views of different users groups, depending on their objectives and 2. Integration of heterogeneous distributed information sources.

In an agent-oriented approach, the different information sources, views, tasks and system abstractions are represented by resource and service agents. These agents encapsulate external data sources within the agent system and cooperate on a common semantic level. This way, the highly decentralized character of the asset management system can be abstracted to a flexible software structure. Figure 5 shows the basis structure of an agent-based asset management system: The blocks left and right show the system's environment, the different views of the system are illustrated on top.

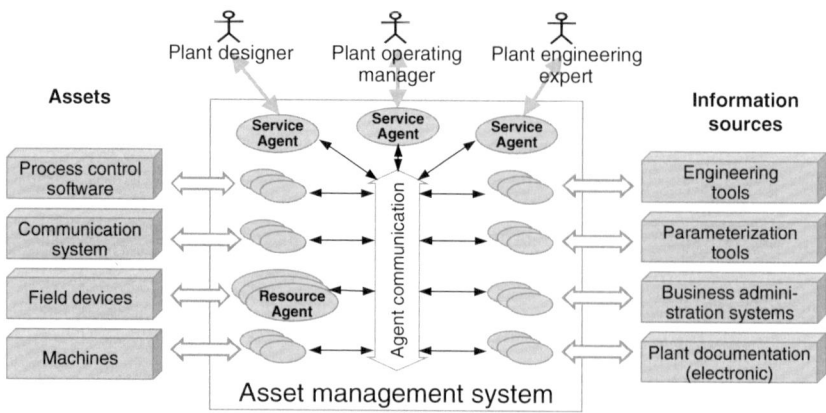

Fig. 5. Agent-based asset management system

The application of the agent-based approach, results in an integrated overall system that can also be introduced in an existing environment. Its properties are:

- The agent-based asset management system is a self-contained data processing system. The core functions of the automation system are not impaired.
- The heterogeneity of information sources has no impact on the entire asset management system, but is hidden by several agents.
- Changes in parts of the system have only local effects, new system parts or tasks can be added dynamically to the overall system.
- Every implementation of an asset management system shows a high degree of individualization, due to the particular environment. The use of agents facilitates the reuse of single system components or even partial systems.

6   Conclusion

In this paper the opportunities as well as the advantages, of introducing of agent-based concepts into automation systems have been investigated. The focus was on legacy, contemporary automation systems used in industrial plants. Due to the limits of current software concepts new approaches are needed to handle the increasing complexity in automation systems. It was argued that agents are well suited to solve the problems of automation systems and a basic approach for the introduction of a multi-agent system into existing systems was presented. Compared to current software concepts for automation systems, the merits of the presented approach are a transparent software structure, and a dynamic, adaptive application software.

For a successful application of the presented approach, the existing constraints regarding legacy automation system structures and applications have to be considered. For the implementation of the approach technical difficulties have to be overcome. These difficulties arise due to the limits of existing communication structures that restrict the access to field devices and process control data. Also challenges have to be met when introducing a new orthogonal agent-based communication layer. These challenges concern the data consistency between automation system and multi-agent system and the coordination of both without affecting existing automation functions.

References

1. Jennings, N.R.: On agent-based software engineering. Artificial Intelligence 117, 2000
2. Lauber, Rudolf; Göhner, Peter: Prozessautomatisierung I, 3. Vollst. Überarb. Aufl. - 1999, Springer-Verlag Berlin - Heidelberg - New York, 1999
3. Scherf, B.; Haese, E. und Wenzek, H.R.: Feldbussysteme in der Praxis: ein Leitfaden für den Anwender. Springer-Verlag Berlin - Heidelberg - New York, 1999
4. Meyer, Dirk: Innovation leittechnischer Softwarestrukturen. Fachkonferenz "Verteilte Automation 2000" (VA2000), Magdeburg, 2000
5. Mchouikha, M. et. al.: Klassifikation und Bewertung von Beschreibungsmitteln für die Automatisierungstechnik. at - Automatisierungstechnik, Nr. 46 (Heft12), 1998
6. Fay, A.: Methoden zur Unterstützung der Migration von Prozessleitsystem-Software. atp - Automatisierungstechnische Praxis, Nr.44 (Heft 6), 2002, S.39 - 44
7. Hesse, W. et. al.: Terminologie in der Softwaretechnik - Ein Begriffssystem für die Analyse und Modellierung von Anwendungssystemen. Teil 2: Tätigkeits- und ergebnisbezogene Elemente In: Informatik-Spektrum 17, pp. 96-105, Springer-Verlag 1994
8. Albrecht, H.; Meyer, D.: Ein Metamodell für den operativen Betrieb automatisierungs- und prozessleittechnischer Komponenten. at - Automatisierungstechnik, Heft 50, 3/2002
9. Tonshoff, H.; Woelk, P.-O.; Timm, I.; Herzog, O.: Flexible Process Planning and Production Control Using Co-operative Agent Systems. International Conference on Competitive Manufacturing (COMA '01), Stellenbosch, South Africa, 2001
10. Epple, U.: Agentensysteme in der Leittechnik. atp - Automatisierungstechnische Praxis, Nr.42 (Heft 8), 2000, S.45-51
11. Jennings, N.R.: An agent-based approach for building complex software systems. Communications of the ACM, 44 (4) 35-41, 2001
12. Döbel, U; Heidel, R: Strukturen künftiger verteilter leittechnischer Systeme am Beispiel der Feldtechnik. atp - Automatisierungs-technische Praxis, Heft 9 (42), 2001
13. Jennings, N.R. and Woolridge, M.J.: Agent Technology: Foundation, Applications, and Markets. Springer, New York, 1998
14. Kendall, E.: Role Modeling for Agent System Analysis, Design, and Implementation. First Internat. Symposium on Agent Systems and Applications (ASA'99), Palm Springs, 1999.
15. Nicklaus, E.; Zöller, M.: Anlagennahes Asset Management - Namur spricht Empfehlungen aus. atp - Automatisierungstechnische Praxis Messekompass, 2001
16. NAMUR-Recommendation: Requirements for Online Plant Asset Management Sytems, erstellt durch NAMUR-Arbeitskreis 2.7. "Asset Management", Leverkusen, 2001
17. Shen, W.; Norrie, D.: Agent-based systems for intelligent manufacturing: a state-of-the-art survey. Knowledge and Information Systems, 1(2), S. 129-156, 1999

Multi-agent Model to Control Production System: A Reactive and Emergent Approach by Cooperation and Competition between Agents

Mahmoud Tchikou and Eric Gouardères

Laboratoire d'Informatique Université de Pau et des Pays de l'Adour (LIUPPA)
U.F.R. Sciences et Techniques, Département Informatique, Avenue de l'Université
BP 1155 – 64013 Pau Cedex, France
{eric.gouarderes,tchikou}@univ-pau.fr

Abstract. Face to strong competitive market, current companies tend to new methods of production, switching from a logic of «projected planning» to a logic of "Just in time". In this context, the system that allows controlling the production has to be a modular, flexible and reactive system. The hierarchized and classical approaches don't permit any more to take into account the complexity linked to such a system. That's why, we propose an approach, which has reactive, distributive, and emergent properties to control the system of production, based on multi-agent system principles. After having introduced the context and reasoning work, we describe the different parts of our multi-agent model. Lastly, we illustrate this approach on a practical example of production cell.

1 Introduction

Flexibility, reactivity, and agility have become unavoidable qualities for many companies, which are confronted with ever more demanding constraints of quality and real time that are both varied and fluctuating. Indeed, the new manufacturing methods, in particular the production constrained by the demand (PULL), implies that at the level of production control, companies switch directly from a logic of «projected planning» to a logic of «just in time», directly led by the customer and the product in a process of development. This results in a new challenge for these companies, which must install modular and flexible production equipments with a control system able to manage them. The latter must, on the one hand, be able to adapt to the heterogeneity of available equipment (API, Computers, Automatically Programmed machine-tools, robots, etc.), equipment, which can be substituted, deleted, or reconfigured, according to needs. On the other hand, it must be sturdy when confronted with different malfunctions and disruptions, which can affect it. The control system is also inseparable from human beings, whose decisions and global vision lead to a well run system [7] [15], and from the company that placed the order, who nowadays is strongly and directly involved in the supply chain [14].

R. Kowalczyk et al. (Eds.): Agent Technology Workshops 2002, LNAI 2592, pp. 329–342, 2003.
© Springer-Verlag Berlin Heidelberg 2003

The problem of production systems control can be set out in the following way: how can we ensure that a group of elements from different origins are able to follow their goals, in agreement with the aims of the company? The main difficulty is to find a compromise between, on the one hand, maintaining the relevance outside the company through to the determination of clear performance criterions and on the other hand, to deal continually with the internal coherence of the collective actions.

The development of these systems of control remains very complex because of the great amount of data to process and the decisions to make, without forgetting the constraints of real time and the need to communicate with equipments in the shop and other functions within the company. Therefore, in order to answer simultaneous needs of reactivity, flexibility and robustness, a lot of researchers have neglected the prearranged, centralized and hierarchized structures to try to implement distributed structures. A control system is distributed between several decision-making centres, all of which have a degree of autonomy and cooperation and communicate with each other in order to well conclude the planned production. In this context, the approaches allowing self-configuration or configuration of a system are regarded nowadays as a major improvement. A recent study [11] shows that the multi-agent systems and the underlying emergent approach constitute actually one of the important research issues in the domain. See for example PABADIS European Project White Paper [20]. The second section introduces the approach with the main principles of reactivity, distribution and emergence for production activity control, and the third section presents models for the modelling of multi agent systems. Finally, we will illustrate our approach by using a concrete example to better explain the different mechanisms that were used at the time of the controlling production cell.

2   Presentation of the Approach

2.1   Principles: Reactivity, Distribution, and Emergence

The control of the production systems is synonymous with the action to run, to guide and to assure the pertinence and coherence of a system in a given environment. The production activity control can be considered as the art of adapting permanently the objectives of a company with the evolution of the environment through to the analysis of the of constraints and opportunities. This activity is delicate at the same time by combining aspects related to the organization of production, the multi-criterion aspect of decisions taken and finally, the management of uncertain data, inside as well as outside the production system. In order to determine these problems we propose to use a reactive operational approach, distributed and emergent. The reactive propriety is to be used for adapting the command to the different variations and disruptions of the system and its environment. The approach is distributive since it is made up of autonomous entities in order to give more flexibility. The approach is emergent in that the performance of the system is not globally planned, but the global plan will emerge from the dynamics of the interactions in real time between the entities (dynamic planning), using multi-agent technologies. In this way, it is not necessary for the system to alternate between planning and execution, but its behaviour is elaborated from competitive decision of the entities.

2.2 General Approach

A production system is a system (set of material or abstract elements in interaction) realizing production activity, which means transforming raw materials or components into a finished product. In order to conclude this operation of transformation, the production system uses a set of resources such as machines, operators, stocking area, and industrial tools [4]. From this definition, a distributed, reactive, and emergent model to control the production system will be proposed. This model is based on the multi agent approach. This approach was chosen because a lot of works and applications were done in the domain of distributed control. The reader can find a study and review of this work in [11,17,18,2].

Within this approach, a control process is associated with each resource and for each product present in the production system. Each process will be modelled by an entity. Each entity of the production system is represented by an autonomous agent which has individual behaviour and the capacity to make it's own local decisions. These agents gather the functions of action, decision and communication, as well as a local knowledge base. Each product is able to communicate and negotiate with the other agents to organize, plan and control the system of production. The product agents require services of the resources agents, which can accept or refuse these services. A population of resource agents and a population of product agents will be obtained.

The human operator is present in the production control loop by means of interface agent, which enables him to communicate with the other agents of the system. The system consists of three types of agents : product agents, resource agents and operator agents. In this approach the decision is distributed between all the agents, it is a team of agents in which there is no order relation. Only cooperation links exist. Each agent makes his decision cooperating with his neighbours in order to conserve the global coherence of the decisions and to respect the objectives fixed to the system. The advantage of this approach is its simplicity, flexibility, reactivity, its tolerance of faults, and its robustness. This system is able to adapt quickly to disruptions no matter if their origins are internal or external to the system.

3 Multi-agent Model

A multi-agent system is commonly characterized by: some agents, an environment, an organization and one or more interaction models. The Parunak model has been chosen as a basic model [16] to specify environment, agents and the coupling between them. But this model does not describe organizational aspect and agent behavioural. On this subject, much works use the concept of role, abstraction of a function, a service or a behaviour. The model of interaction being based then on the relations between these roles [3]. Among those, we have chosen the AALAADIN model [9] because it is a general model. There is no constraint or pre-requirement on the internal agent architecture and no particular model to describe agent behaviour. The agent behaviour can be produced by multiple ways: tasks made up of primitives then started by stimuli, sorters system, Petri nets [5], but to create agents which can adapt, will consist first and foremost to make their behaviours evolve in all their complexity. To tackle this problem, we should use an approach that allows the evolution of agent behaviour. The work of Picault and Landau [12] encouraged us to produce agent behaviour, by using a structure of oriented and stamped graph named ATN (Augmented Transition Network).

3.1 Basic Model of MAS

A multi-agents system according to Parunak can be defined like a triplet: a set of agents, an Environment, and a coupling, which defines the bond between them [16]. This model is used to specify relation between the control part (software agents) and the operative part (environment : resources, products, operators) which allows us to avoid a global controller of the production system.

3.1.1 Agents
Each agent is quadruplet (four-tuple): $<State_a, Input_a, Output_a, Process_a>$

- $State_a$: set of values that completely define the agent,
- $Input_a$ and $Output_a$: subsets of state of an agent's sensors and effectors,
- $Process_a$: an autonomously executed mapping that changes the agent's state. These Agents have complex behaviours, which will be defined afterwards with ATN graphs.

3.1.2 Environment
The environment is a duplet (two-tuple), $<State_e, Process_e>$, that is syntactically a subset of an agent. The environment has its own process that can change its state, independent of the actions of its embedded agents. Meaning that the environment itself is active.

3.1.3 Coupling
This system is Homodynamic Systems, meaning that both agents and environment are modelled the same way, (in our case discrete-event). The coupling of $Input_a$ and $Output_a$ to $state_e$ is simply a mapping of registers from environment to agent or vice-versa.

3.2 Method and Organisational Model

The dynamic aspect is very important in this approach because it ensures the emergence of the overall plan. AALAADIN model [9] was chosen to describe organisational model for interaction. This model provides a methodology based on the notions of agent-group-role. An agent is a communicating autonomous entity, which plays roles within different groups. An agent can have several distinct roles within several groups and the same role can be held by several agents, which makes possible the heterogeneity of the situations of interaction. A group is seen like a usual MAS and we identified four groups for production activity control:

- Product agents group,
- Resource agents group,
- Operator agents group,
- a group associating a product agent with the resources agents necessary to its transformation.

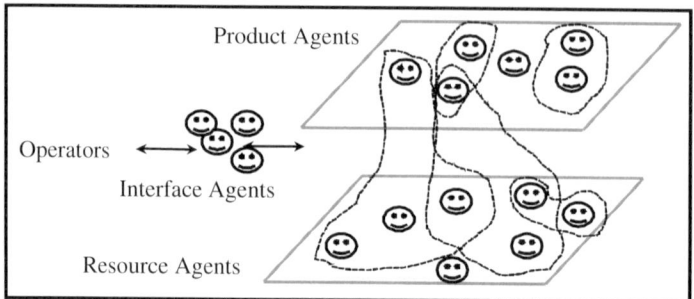

Fig. 1. Organisational structure of the multi-agent system to control the production system

3.2.1 Product Agents

Each product agent has its own procedure (range), these agents are created as soon as the product enters in the production system and destructed at the end of treatment. Several product agents can be found in the production system at the same time. The role of the product agents is to plan and to control the product in the production system in order to process all the treatments dictated by the procedure, respecting the time price and quality constraints.

3.2.2 Resource Agents

These agents control the resources of the production system, (for example: machines, robots, conveyor). In the production system resource agents can be found from the same type which has to cooperate with each other in order to avoid conflicts. The role of the resource agents is to process the treatment and task over part respecting the time constraints.

3.2.3 Operator Agents

The interface agent is designed to interpret the human operator's messages in order to configure the system and send back the interactions between the resource and product agents to the operator. The role of this agent is:

- to introduce new constraints on the resources or on the product,
- to favour the production of a product compared to another product,
- to add new data and external information,
- to release (the start and the stop of the system),

- to support simulation process, to introduce breakdowns on the resources, and to follow the behaviour of the system.

3.3 Interaction Model

The aim of the interaction model is to show and explain knowledge exchange, to resolve problems of conflict, to cooperate in order to reach their objective. This part presents the basic communicating system and negotiating process.

3.3.1 Communication System

The communication is a fundamental point of the multi-agent system. It is not reduced to data exchanged systems, but it's matter of an intentional act, which finds an expression in a modification of the agent knowledge. FIPA Agent Communication Language has been used; it is a high level of communication language and protocol, message oriented, independent of the syntax and semantic of the content (ontology). It is even independent of message transport mechanisms (ex: TCP/IP, SMTP, IIOP, HTTP...) and high-level protocol of negotiation (ex: Contract-Net). ACL is based on primitive of communication called "act of communication".

3.3.2 Negotiation System

Two large trends exist in the interaction between agents, an approach which tends to simulate a conversation between humans and a second approach resulting from work in the distributed systems. In the first case, the agents discuss freely, only guided by the objective which they laid down. There is no structure which comes to direct the conversation by imposing an answer to a particular question than another. The agents build their conversation during this one. The second approach employs protocols of interaction in order to direct the interaction. For a given state of the interaction, the agent can receive only one message belonging to the list of the awaited messages and the answer that it will make also belongs to a predetermined answer. Moreover, if the agent uses this protocol, it must accept his semantics. The contribution of the protocols of interaction is to allow a faster convergence towards the solution, because the agents are constrained in the choice of the messages to use. For our system of control we used the second approach. The essential goal is the modification of the local plans of the agents to reach a state of consensus on the execution of the tasks of the system. The Contract Net protocol was chosen as model of negotiation [19]. One agent takes the role of manager who wishes to have some task performed by one or more other agents and further wishes to optimise a function that characterizes the task. This characteristic is commonly expressed as the price, in some specific domain, but could also be the soonest time to completion, fair distribution of tasks.

3.3.2.1 Aadapted Contract-Net

We adapted the contract-net protocol to the multi-agent model for the production activity control. Being given one or more tasks (Range), a manager (product Agent), a group of contracting (resource agents):

1. The product agent solicits proposals from other agents (a group of resources agents) by issuing a *call for proposals* act (see [6]) which specifies the tasks (Range: describing the tasks to be treated and the description of the resources essential to its realization).
2. Each resource agent receiving the call for proposals evaluates if it is concerned by the Range depending on its abilities. Resource agents concerned by the Range (the resource agent can do one or more than one task of the Range) are viewed as potential contractors and are able to generate proposals to perform the task as *propose* acts (see [6]). A potential contractor proposes, according to its capacity, a bidding whose contents specify a date of beginning for the task (this date depends on the rules of production activity control see 3.3.2.2) and the list of the tasks already scheduled on the resource.
3. The product agent compares the biddings and chooses the resource which proposes the weakest date (respecting the time constraints and delay of production).
4. Once the deadline passes, the product agent evaluates all received proposals and selects resource agents to perform the tasks; one, several or no resource agents may be chosen. The resource agents of the selected proposal(s) will be received an *accept-proposal* act (see [6]) and the others will receive a *reject-proposal* act (see [6]).
5. The agent having obtained the contract, submits a report of execution when the task is supplemented.
6. At the end of the execution the product agent is destroyed. The global plan of production emerges from the plans planned by each product agent.
7. In the case of a breakdown or disruption the product agent cancels all the contracts and re-plans a new plan, launching a new *call for proposals* act to continue the treatments.

3.3.2.2 Scheduling Rules

The date of beginning for the task depends on the rules of production activity control based on several indicators listed in the tables below.
From these rules, the products are classified (tasks that must be realized) in one, or several queues by each resource :

- *SPT* : the tasks are classified from shortest to longest duration,
- *CEXSPT*: creation of 3 queues, the first queue for the tasks that are out of delay, the second for the urgent tasks, and the third for the normal tasks,
- *FIFO*: first in, first out.

Table 1. Rules for global indicators (can be announced by operator)

Indicators	Rules
Delay of current tasks are increased	SPT with the longest queue. (or CEXSPT)
The number of urgent tasks are increased	SPT with the longest queue

Table 2. Rules for local indicators (can be detected by the resource agents and product agents).

Indicators	Rules
There is only one urgent task	Priority task (head of queue)
A task is urgent	SPT (or CEXSPT)
There is a station neck (station saturated, rate of the use > 80%)	SPT (or CEXSPT)
A task waits since too a long time	FIFO

3.4 Process Model (Behavioural Model)

To describe the behaviour of the agents it has been decided that an oriented and stamped graph structure named ATN (Augmented Transition Network) will be used. At the beginning, these graphs were used within the context of automatic treatment of the language [21,22], and at a later stage, they were used to describe the process of the agents [1,8]. The choice of ATN is not by hazard but to produce the behavioural graph automatically using genetic algorithms [12]. As we have indicated previously, the behaviour of the agents are described by ATN graphs. The behaviour graph of an agent has a direct node of departure (called start), and a final node (end). The other nodes are linked by arcs, which can be associated by a whole range of conditions and by a sequence of actions. Our agents are initialised at the beginning. Also the agent activity consists for each node:

- to select between the arcs from the current node, the ones which are crossed, that is to say either without conditions or whose conditions are checked simultaneously,
- to chose one of its arcs randomly,
- to cross it (to take up on a node where it ends up) after having possibly realized actions linked at the arc in the order.

The agent searches continuously to get from one node to another, if no arc is surmountable, it stays in the first state, and will try again to the next step. It changes in a waking state when it is situated on the node called end and stops acting.

It can be noticed that two agents, which have a common ATN, can adopt very different behaviours (from the same abilities of reception and actions), in the situation when they can use different arcs if they are situated in the same conditions. We illustrate this part by giving actions, conditions and simplified ATN graph to represent the negotiating behaviour of a product agent.

4   Example: Control of a Production Cell

To test the approach, a simulator of a production cell was chosen [13]. The main objective is to control this cell in real time. In addition to reactivity, robustness, and efficiency, we test the coherence of the overall plan, which emerges from the local agent plan. Even the feasibility of the approach will be tested.

Table 3. Example of actions and conditions for product agent

Action!	Condition?
Ask a call for proposals! Choose proposals! Receipt proposals! Request resources list! Send Accept-proposal! Send Reject-proposal! Update resources list!	Breakdown? Blank pass all resources? Time out?

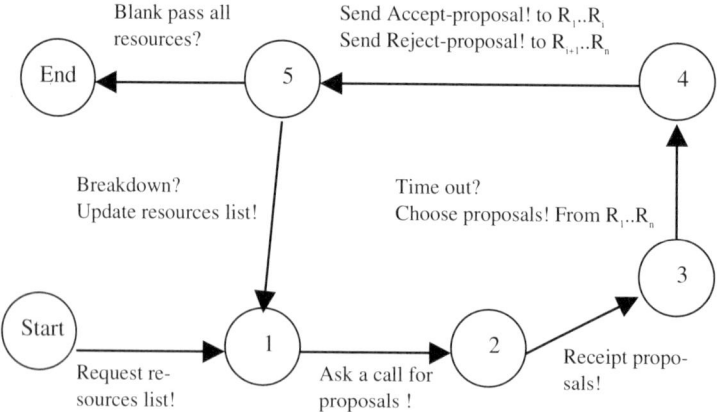

Fig. 2. Simplified ATN example for product agent

4.1 Presentation of Example

The production cell is equipped with two conveyors belts, four processing units and two portals with a travelling crane. The conveyor belts carries the blanks (products) in only one direction, from left to right. At the end of the feed belt, there is a light barrier and a code reader. The deposit belt contains a light barrier at it's beginning. The processing units are equipped with two sensors, one sensor which reports whether the unit is occupied or not and the second sensor which indicates whether the unit is working or not. There is two type of processing unit type 1 (drill or press) or type 2 (oven). Both travelling cranes can reach all four processing units and can be moved in three directions.

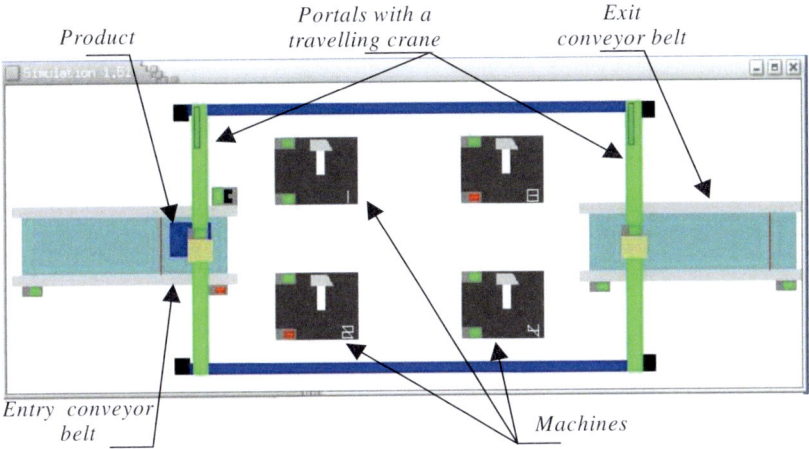

Portals with a travelling crane

Exit conveyor belt

Product

Entry conveyor belt

Machines

Fig. 3. Simulator of production cell

Blanks are introduced to the system via the feed belt, whenever a sensor reports a blank the belt must be stopped. Then, the blank is positioned directly in front of the bar code reader. Blanks have a bar code which contains information about the procedure for their processing. It tells which type of processing unit must be used and determines whether the processing order is correct or not. Additionally, there may be time constraints, which limit the time that can be used for processing the blanks. An additional time constraint gives a maximum limit on the total time a blank may spend in the whole system. The code reader transmits the information after having read it from each blank. The type of return value of the bar codes is tuplet

$$\left\langle n, \left\langle \min_i, \max_i \right\rangle_{i=1}^{n}, r, t_G \right\rangle$$ which tells how the blank must go through the system and

gives time constraints listed in Table 5.

4.2 Application of the Approach

Multi-agents system must control production cell in real time, solve the problem of conflict between the resources and avoid the collision between the two gantries (cranes). The agents must respect the procedure of treatment delivered by the bar code (order, and maximum duration that the blank should not exceed in the system, and each type of machine). To establish these agents, we use the last version of JADE (Java Agent DEvelopment framework)[10]. The main objective is to prove the coherence of the decision of each agent.

Agents model the blanks (products), the machines and one operator (Fig. 5). As soon as the blank arrives in the entry conveyor, the blank agent is created through an event caused by a light barrier sensor. The reader of bar code agent sends the procedure (range) to the blank agent, from this location the blank agent negotiates using the Contract-Net [6].

Table 5. The bar Code

Component	Type	Information
n	Integer	Number of type of processing units, which must be used (type 1 or type 2 or both).
min_i	Integer	Minimum processing time in seconds, which is necessary for type i
max_i	Integer	Maximum processing time in seconds, which is necessary for type i
r	Boolean	Indicates whether certain type of processing units must be used in an exact order during processing
t_G	Integer	Maximum time in secs, which may be spent in the whole system (measured between gripping from the feed belt and arrival on the deposit belt reported by the sensor

4.2.1 The Negotiation between Blank and Resources

Firstly the blank agent makes a bid to the machine agents in order to fulfil the procedure. According to its capability each machine agent proposes one bid, which specifies a starting date for the task and a list of tasks already scheduled on the machine. The blank agent analyses bids and chooses the machines, which propose the weakest date (the smallest date). All that is done, before the blank quits the entry conveyor and before that the blank suffers the treatment. As soon as the negotiation successes we pass to the execution of the task planed, at the end of the execution, the blank agent is destroyed by the exit conveyor agent. The global plan for the treatment in cell emerges from an organised plan by each blank agent.

In the case of a breakdown or perturbation, the blank agent cancels the contract and tries another new proposition with new bid in order to follow the treatment.

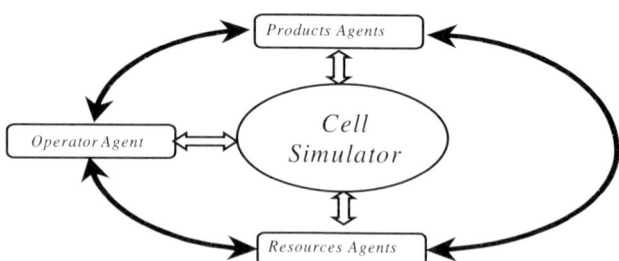

Fig. 4. The system with all actors

4.2.2 The Coordination between Resource Agents

The resource agents can coordinate their plan before giving the proposition to the blank agent in order to respect the constraints, for example collision between cranes.

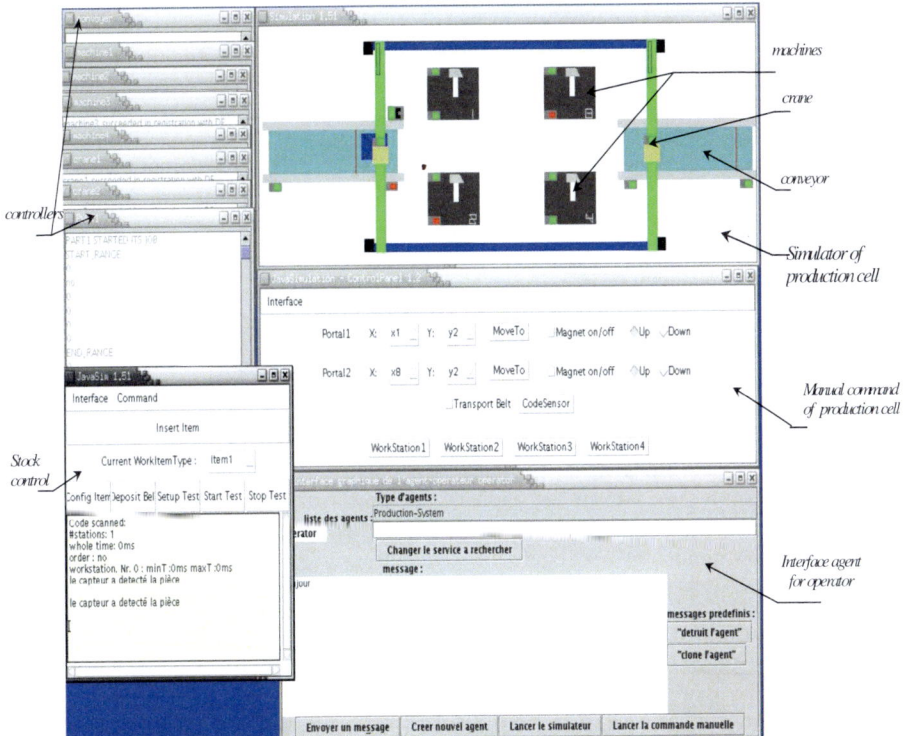

Fig. 5. The simulator with the resource agents, blank agents and interface agents

4.2.3 The Coordination between Blank Agents

To respect the production delays of some products, the system has to adapt and favour certain products; in that case, the blank agents must coordinate their plans and exchange them. The priority blank agent sends messages to the other blank agents, which are in the system, and negotiate, if possible, respecting the range constraints, of passing into resources before the others. In that case, the agents already in the system cancel contracts with some resources to leave a place for the priority blank agent.

5 Conclusion

In this paper, we have presented the modeling of multi-agent systems for controlling production systems. This approach is motivated by the evolution of production methods and increasing of complexity of products, which have consequently raised the complexity of the controller systems. The evolution towards the notion of (virtual company) does become worth this trend multiplying the number of parameters and the number of components, which are in interaction in the system. That leads the traditional methods of optimization to the combinatory explosion of counts, the number of interactions in-group of elements growing faster than the number of elements in this same group. Associating an entity at every element in interaction, the multi agent approach permits to substitute the explicit coding of the whole of interaction by its generation at the execution time. The combinatory aspect of behaviours is no longer planned at the design stage. That reduces drastically the quantity of coding at the production stage and so, the cost of developing the system. That is why this approach nowadays constitutes one of the most important types of research in the domain.

Considering the previous remarks, we have above all put the accent on complex activities as part of the aspect of interactions between agents instead of on the agent-thinking modes. Our main objective is to analyze; to realize a multi-agent system composed of autonomous agents, which are in interaction according to modes of co-operation more or less complex, conflict, challenge, etc, to lead to the achievement of global objective. This approach highlights the degrees of freedom available to the agents and makes it possible to manage the autonomy of decision of each agent while guaranteeing a total coherence of the system.

To reach this objective, the originality of our approach is to try to integrate different existing models in order to provide a complete model adapted to our problem. This allows us to describe all the aspects of a multi agent system: the agents, their behaviour, their organization, the environment in which they evolve, and the way in which they communicate to realize collective actions.

The realized work opens the way to several perspectives about:

- the evolution of models of behaviour: utilization of ATN allows us to plan the automatic generation and adaptation using the Picault Landau 's works based on an evolutionary approach,
- the resolution of the problem of the follow-up and possibly of recovery while a degraded functioning of the production system,
- the resolution of the agent autonomy problems compared to the global coherence of the system.

In this article, we have tried to give a global vision of the problems and the solution that we propose. However, as a work in process, we are conscious that a lot of work still lies ahead before we reach our objectives.

References

[1] Bouron, T. : Structures de communication et d'organisation pour la communication dans un univers multi-agent. PhD Thesis, University of Paris VI (1993)
[2] Bussmann, S., Schild, K.: An Agent-based Approach to the Control of Flexible Production Systems. In Proc. of the 8th IEEE Int. Conf. on Emergent Technologies and Factory Automation (ETFA 2001), Vol. 2 (2001) 481-488
[3] Casteran, J.C., Gleize, M.P., Glize, P. : Des méthodologies orientées multi-agent. In Proc. of JFIADSMA'00. Hermès Science Publications, ISBN 2-7462-0176-3, Paris (2000) 191-207
[4] Delattre, P.: Système, structure, fonction, évolution. Essai d'analyse épistémologique. PhD Thesis, Paris Maloine-Dion (1971)
[5] Ferber, J.: Les systèmes multi-agents, vers une intelligence collective. InterEditions, Paris (1995).
[6] FIPA Communicative Act Library Specification. Foundation for Intelligent Physical Agents, http://www.Fipa.org/specs/fipa00037/ (2000)

[7] Fox, M.S.: The tove project : Towards a common sense model of the enterprise. In Petrie C. (Ed.) : Enterprise Integration. AAAI Press (1992) 291-316

[8] Guessoum, Z. : Un environnement opérationnel de conception et de réalisation de systèmes multi-agents. PhD Thesis, University of Paris VI (1996)

[9] Gutknecht, 0., Ferber, J.: *MADKIT :* une expérience d'architecture de plateforme multi-agent générique. . In Proc. of JFIADSMA'00. Hermès Science Publications, ISBN 2-7462-0176-3, Paris (2000) 223-236

[10] Java Agent Development framework.
http://sharon.cselt.it/projects/jade/

[11] Kouiss, K., Gouardères, E., Massotte, P.: Organisations distribuées des systèmes de pilo-tage. In Pujo P., Kieffer J. P. (eds) : Fondements du pilotage des systèmes de production. Hermès Science Publications, ISBN 2-7462-0513-0, Paris (2002) 80-117

[12] Landau, S., Picault, S.: Modelling Adaptive Multi-Agent Systems Inspired by Develop-mental Biology. In Proc. of AEMAS'2001. Springer-Verlag, Berlin Heidelberg New York (2001) 238-246

[13] Lotzbeyer, A., Muhlfeld, R.: Task Description of Flexible Production Cell With Real Time Properties. Internal report, University of Karlsruhe (1996)

[14] Massotte, P.: Auto-organisation dans les structures et les systèmes In Proc. of MOSIM'99. SCS, Annecy, France (1999) 21-29

[15] Norrie, M., Wunieli, M., Montau, R., Leonhardt,U., Shaad, W., and Schek H.J.: Coordina-tion Approach for CIM. In Proc. of European Workshop on Integrated Manufacturing Systems Engineering (IMSE) (1994) 223-232

[16] Parunak, H.V.D.: Go to the Ant : Engineering Principles from Natural Multi-Agent Sys-tems. Annals of Operations Research, No. 75 (1997) 69-101

[17] Parunak, H.V.D.: Practical and Industrial Applications of Agent-Based Systems. Environ-mental Research Institute of Michigan (ERIM) (1998)

[18] Shen, W., Norrie, D.H.: Agent-based Systems for Intelligent Manufacturing : A State-of-the-Art Survey. Knowledge and Information Systems, an International Journal, Vol. 1 No. 2 (1999) 129-156

[19] Smith, R.G., Davis, R.: Framework for cooperation in distributed problem solving. IEEE Transactions on System, Man and Cybernetics, SMC, Vol. 11 No.1 (1981) 61-70

[20] Pabadis White Paper,
http://www.pabadis.org/downloads/pabadis_white_paper.pdf (2002)

[21] Winograd, T.: Language as a Cognitive Process. Addison-Wesley (1983)

[22] Woods, W.A.: Transition networks grammars for natural language analysis. Communica-tions of the ACM. Vol. 13 No. 10 (1970) 591-606

Author Index